Springer Handbook
of Crystal Growth

Govindhan Dhanaraj, Kullaiah Byrappa,
Vishwanath Prasad, Michael Dudley (Eds.)

Springer Handbook of Crystal Growth
Organization of the Handbook

Part A Fundamentals of Crystal Growth and Defect Formation
1 Crystal Growth Techniques and Characterization: An Overview
2 Nucleation at Surfaces
3 Morphology of Crystals Grown from Solutions
4 Generation and Propagation of Defects During Crystal Growth
5 Single Crystals Grown Under Unconstrained Conditions
6 Defect Formation During Crystal Growth from the Melt

Part B Crystal Growth from Melt Techniques
7 Indium Phosphide: Crystal Growth and Defect Control by Applying Steady Magnetic Fields
8 Czochralski Silicon Single Crystals for Semiconductor and Solar Cell Applications
9 Czochralski Growth of Oxide Photorefractive Crystals
10 Bulk Crystal Growth of Ternary III–V Semiconductors
11 Growth and Characterization of Antimony-Based Narrow-Bandgap III–V Semiconductor Crystals for Infrared Detector Applications
12 Crystal Growth of Oxides by Optical Floating Zone Technique
13 Laser-Heated Pedestal Growth of Oxide Fibers
14 Synthesis of Refractory Materials by Skull Melting Technique
15 Crystal Growth of Laser Host Fluorides and Oxides
16 Shaped Crystal Growth

Part C Solution Growth of Crystals
17 Bulk Single Crystals Grown from Solution on Earth and in Microgravity
18 Hydrothermal Growth of Polyscale Crystals
19 Hydrothermal and Ammonothermal Growth of ZnO and GaN
20 Stoichiometry and Domain Structure of KTP-Type Nonlinear Optical Crystals
21 High-Temperature Solution Growth: Application to Laser and Nonlinear Optical Crystals
22 Growth and Characterization of KDP and Its Analogs

Part D Crystal Growth from Vapor
23 Growth and Characterization of Silicon Carbide Crystals
24 AlN Bulk Crystal Growth by Physical Vapor Transport
25 Growth of Single-Crystal Organic Semiconductors
26 Growth of III–Nitrides with Halide Vapor Phase Epitaxy (HVPE)
27 Growth of Semiconductor Single Crystals from Vapor Phase

Part E Epitaxial Growth and Thin Films
28 Epitaxial Growth of Silicon Carbide by Chemical Vapor Deposition
29 Liquid-Phase Electroepitaxy of Semiconductors
30 Epitaxial Lateral Overgrowth of Semiconductors
31 Liquid-Phase Epitaxy of Advanced Materials
32 Molecular-Beam Epitaxial Growth of HgCdTe
33 Metalorganic Vapor-Phase Epitaxy of Diluted Nitrides and Arsenide Quantum Dots
34 Formation of SiGe Heterostructures and Their Properties
35 Plasma Energetics in Pulsed Laser and Pulsed Electron Deposition

Part F Modeling in Crystal Growth and Defects
36 Convection and Control in Melt Growth of Bulk Crystals
37 Vapor Growth of III Nitrides
38 Continuum-Scale Quantitative Defect Dynamics in Growing Czochralski Silicon Crystals
39 Models for Stress and Dislocation Generation in Melt Based Compound Crystal Growth
40 Mass and Heat Transport in BS and EFG Systems

Part G Defects Characterization and Techniques
41 Crystalline Layer Structures with X-Ray Diffractometry
42 X-Ray Topography Techniques for Defect Characterization of Crystals
43 Defect-Selective Etching of Semiconductors
44 Transmission Electron Microscopy Characterization of Crystals
45 Electron Paramagnetic Resonance Characterization of Point Defects
46 Defect Characterization in Semiconductors with Positron Annihilation Spectroscopy

Part H Special Topics in Crystal Growth
47 Protein Crystal Growth Methods
48 Crystallization from Gels
49 Crystal Growth and Ion Exchange in Titanium Silicates
50 Single-Crystal Scintillation Materials
51 Silicon Solar Cells: Materials, Devices, and Manufacturing
52 Wafer Manufacturing and Slicing Using Wiresaw

Subject Index

使 用 说 明

1.《晶体生长手册》原版为一册，分为A～H部分。考虑到使用方便以及内容一致，影印版分为6册：第1册—Part A，第2册—Part B，第3册—Part C，第4册—Part D、E，第5册—Part F、G，第6册—Part H。

2.各册在页脚重新编排页码，该页码对应中文目录。保留了原书页眉及页码，其页码对应原书目录及主题索引。

3.各册均给出完整6册书的章目录。

4.作者及其联系方式、缩略语表各册均完整呈现。

5.主题索引安排在第6册。

6.文前介绍基本采用中英文对照形式，方便读者快速浏览。

材料科学与工程图书工作室

联系电话　0451-86412421
　　　　　0451-86414559

邮　　箱　yh_bj@yahoo.com.cn
　　　　　xuyaying81823@gmail.com
　　　　　zhxh6414559@yahoo.com.cn

Springer手册精选系列

晶体生长手册

溶液法晶体生长技术

【第3册】

Springer
Handbook of
Crystal
Growth

〔美〕Govindhan Dhanaraj 等主编

（影印版）

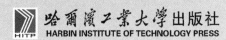
哈尔滨工业大学出版社
HARBIN INSTITUTE OF TECHNOLOGY PRESS

黑版贸审字08-2012-047号

Reprint from English language edition:
Springer Handbook of Crystal Growth
by Govindhan Dhanaraj, Kullaiah Byrappa, Vishwanath Prasad
and Michael Dudley
Copyright © 2010 Springer Berlin Heidelberg
Springer Berlin Heidelberg is a part of Springer Science+Business Media
All Rights Reserved

This reprint has been authorized by Springer Science & Business Media for distribution in China Mainland only and not for export there from.

图书在版编目（CIP）数据

　　晶体生长手册. 3, 溶液法晶体生长技术 = Handbook of Crystal Growth. 3, Solution Growth of Crystals: 英文 / (美) 德哈纳拉 (Dhanaraj,G.) 等主编. —影印本. —哈尔滨: 哈尔滨工业大学出版社, 2013.1
　　（Springer手册精选系列）
　　ISBN 978-7-5603-3868-2

　　Ⅰ.①晶… Ⅱ.①德… Ⅲ.①晶体生长－水溶液法－手册－英文 Ⅳ.①O78-62

　　中国版本图书馆CIP数据核字(2012)第292354号

责任编辑	杨　桦　许雅莹　张秀华
出版发行	哈尔滨工业大学出版社
社　　址	哈尔滨市南岗区复华四道街10号　邮编 150006
传　　真	0451-86414749
网　　址	http://hitpress.hit.edu.cn
印　　刷	哈尔滨市石桥印务有限公司
开　　本	787mm×960mm　1/16　印张 17.25
版　　次	2013年1月第1版　2013年1月第1次印刷
书　　号	ISBN 978-7-5603-3868-2
定　　价	52.00元

（如因印刷质量问题影响阅读，我社负责调换）

序言

多年以来，有很多探索研究已经成功地描述了晶体生长的生长工艺和科学，有许多文章、专著、会议文集和手册对这一领域的前沿成果做了综合评述。这些出版物反映了人们对体材料晶体和薄膜晶体的兴趣日益增长，这是由于它们的电子、光学、机械、微结构以及不同的科学和技术应用引起的。实际上，大部分半导体和光器件的现代成果，如果没有基本的、二元的、三元的及其他不同特性和大尺寸的化合物晶体的发展则是不可能的。这些文章致力于生长机制的基本理解、缺陷形成、生长工艺和生长系统的设计，因此数量是庞大的。

本手册针对目前备受关注的体材料晶体和薄膜晶体的生长技术水平进行阐述。我们的目的是使读者了解经常使用的生长工艺、材料生产和缺陷产生的基本知识。为完成这一任务，我们精选了50多位顶尖科学家、学者和工程师，他们的合作者来自于22个不同国家。这些作者根据他们的专业所长，编写了关于晶体生长和缺陷形成共计52章内容：从熔体、溶液到气相体材料生长；外延生长；生长工艺和缺陷的模型；缺陷特性的技术以及一些现代的特别课题。

本手册分为七部分。Part A介绍基础理论：生长和表征技术综述，表面成核工艺，溶液生长晶体的形态，生长过程中成核的层错，缺陷形成的形态。

Part B介绍体材料晶体的熔体生长，一种生长大尺寸晶体的关键方法。这一部分阐述了直拉单晶工艺、泡生法、布里兹曼法、浮区熔融等工艺，以及这些方法的最新进展，例如应用磁场的晶体生长、生长轴的取向、增加底基和形状控制。本部分涉及材料从硅和Ⅲ-Ⅴ族化合物到氧化物和氟化物的广泛内容。

第三部分，本书的Part C关注了溶液生长法。在前两章里讨论了水热生长法的不同方面，随后的三章介绍了非线性和激光晶体、KTP和KDP。通过在地球上和微重力环境下生长的比较给出了重力对溶液生长法的影响的知识。

Part D的主题是气相生长。这一部分提供了碳化硅、氮化镓、氮化铝和有机半导体的气相生长的内容。随后的Part E是关于外延生长和薄膜的，主要包括从液相的化学气相淀积到脉冲激光和脉冲电子淀积。

Part F介绍了生长工艺和缺陷形成的模型。这些章节验证了工艺参数和产生晶体质量问题包括缺陷形成的直接相互作用关系。随后的Part G展示了结晶材料特性和分析的发展。Part F和G说明了预测工具和分析技术在帮助高质量的大尺寸晶体生长工艺的设计和控制方面是非常好用的。

最后的Part H致力于精选这一领域的部分现代课题，例如蛋白质晶体生长、凝胶结晶、原位结构、单晶闪烁材料的生长、光电材料和线切割大晶体薄膜。

我们希望这本施普林格手册对那些学习晶体生长的研究生，那些从事或即将从事这一领域研究的来自学术界和工业领域的研究人员、科学家和工程师以及那些制备晶体的人是有帮助的。

我们对施普林格的Dr. Claus Acheron，Dr. Werner Skolaut和le-tex的Ms Anne Strobach的特别努力表示真诚的感谢，没有他们本书将无法呈现。

我们感谢我们的作者编写了详尽的章节内容和在本书出版期间对我们的耐心。一位编者（GD）感谢他的家庭成员和Dr. Kedar Gupta(ARC Energy 的CEO)，感谢他们在本书编写期间的大力支持和鼓励。还对Peter Rudolf, David Bliss, Ishwara Bhat和Partha Dutta在A、B、E部分的编写中所给予的帮助表示感谢。

Nashua, New Hampshire, April 2010	G. Dhanaraj
Mysore, India	K. Byrappa
Denton, Texas	V. Prasad
Stony Brook, New York	M. Dudley

Preface

Over the years, many successful attempts have been made to describe the art and science of crystal growth, and many review articles, monographs, symposium volumes, and handbooks have been published to present comprehensive reviews of the advances made in this field. These publications are testament to the growing interest in both bulk and thin-film crystals because of their electronic, optical, mechanical, microstructural, and other properties, and their diverse scientific and technological applications. Indeed, most modern advances in semiconductor and optical devices would not have been possible without the development of many elemental, binary, ternary, and other compound crystals of varying properties and large sizes. The literature devoted to basic understanding of growth mechanisms, defect formation, and growth processes as well as the design of growth systems is therefore vast.

The objective of this Springer Handbook is to present the state of the art of selected topical areas of both bulk and thin-film crystal growth. Our goal is to make readers understand the basics of the commonly employed growth processes, materials produced, and defects generated. To accomplish this, we have selected more than 50 leading scientists, researchers, and engineers, and their many collaborators from 22 different countries, to write chapters on the topics of their expertise. These authors have written 52 chapters on the fundamentals of crystal growth and defect formation; bulk growth from the melt, solution, and vapor; epitaxial growth; modeling of growth processes and defects; and techniques of defect characterization, as well as some contemporary special topics.

This Springer Handbook is divided into seven parts. Part A presents the fundamentals: an overview of the growth and characterization techniques, followed by the state of the art of nucleation at surfaces, morphology of crystals grown from solutions, nucleation of dislocation during growth, and defect formation and morphology.

Part B is devoted to bulk growth from the melt, a method critical to producing large-size crystals. The chapters in this part describe the well-known processes such as Czochralski, Kyropoulos, Bridgman, and floating zone, and focus specifically on recent advances in improving these methodologies such as application of magnetic fields, orientation of the growth axis, introduction of a pedestal, and shaped growth. They also cover a wide range of materials from silicon and III–V compounds to oxides and fluorides.

The third part, Part C of the book, focuses on solution growth. The various aspects of hydrothermal growth are discussed in two chapters, while three other chapters present an overview of the nonlinear and laser crystals, KTP and KDP. The knowledge on the effect of gravity on solution growth is presented through a comparison of growth on Earth versus in a microgravity environment.

The topic of Part D is vapor growth. In addition to presenting an overview of vapor growth, this part also provides details on vapor growth of silicon carbide, gallium nitride, aluminum nitride, and organic semiconductors. This is followed by chapters on epitaxial growth and thin films in Part E. The topics range from chemical vapor deposition to liquid-phase epitaxy to pulsed laser and pulsed electron deposition.

Modeling of both growth processes and defect formation is presented in Part F. These chapters demonstrate the direct correlation between the process parameters and quality of the crystal produced, including the formation of defects. The subsequent Part G presents the techniques that have been developed for crystalline material characterization and analysis. The chapters in Parts F and G demonstrate how well predictive tools and analytical techniques have helped the design and control of growth processes for better-quality crystals of large sizes.

The final Part H is devoted to some selected contemporary topics in this field, such as protein crystal growth, crystallization from gels, in situ structural studies, growth of single-crystal scintillation materials, photovoltaic materials, and wire-saw slicing of large crystals to produce wafers.

We hope this Springer Handbook will be useful to graduate students studying crystal growth and to researchers, scientists, and engineers from academia and industry who are conducting or intend to conduct research in this field as well as those who grow crystals.

We would like to express our sincere thanks to Dr. Claus Acheron and Dr. Werner Skolaut of Springer and Ms Anne Strohbach of le-tex for their extraordinary efforts without which this handbook would not have taken its final shape.

We thank our authors for writing comprehensive chapters and having patience with us during the publication of this Handbook. One of the editors (GD) would like to thank his family members and Dr. Kedar Gupta (CEO of ARC Energy) for their generous support and encouragement during the entire course of editing this handbook. Acknowledgements are also due to Peter Rudolf, David Bliss, Ishwara Bhat, and Partha Dutta for their help in editing Parts A, B, E, and H, respectively.

Nashua, New Hampshire, April 2010	G. Dhanaraj
Mysore, India	K. Byrappa
Denton, Texas	V. Prasad
Stony Brook, New York	M. Dudley

About the Editors

Govindhan Dhanaraj is the Manager of Crystal Growth Technologies at Advanced Renewable Energy Company (ARC Energy) at Nashua, New Hampshire (USA) focusing on the growth of large size sapphire crystals for LED lighting applications, characterization and related crystal growth furnace development. He received his PhD from the Indian Institute of Science, Bangalore and his Master of Science from Anna University (India). Immediately after his doctoral degree, Dr. Dhanaraj joined a National Laboratory, presently known as Rajaramanna Center for Advanced Technology in India, where he established an advanced Crystal Growth Laboratory for the growth of optical and laser crystals. Prior to joining ARC Energy, Dr. Dhanaraj served as a Research Professor at the Department of Materials Science and Engineering, Stony Brook University, NY, and also held a position of Research Assistant Professor at Hampton University, VA. During his 25 years of focused expertise in crystal growth research, he has developed optical, laser and semiconductor bulk crystals and SiC epitaxial films using solution, flux, Czochralski, Bridgeman, gel and vapor methods, and characterized them using x-ray topography, synchrotron topography, chemical etching and optical and atomic force microscopic techniques. He co-organized a symposium on Industrial Crystal Growth under the 17th American Conference on Crystal Growth and Epitaxy in conjunction with the 14th US Biennial Workshop on Organometallic Vapor Phase Epitaxy held at Lake Geneva, WI in 2009. Dr. Dhanaraj has delivered invited lectures and also served as session chairman in many crystal growth and materials science meetings. He has published over 100 papers and his research articles have attracted over 250 rich citations.

Kullaiah Byrappa received his Doctor's degree in Crystal Growth from the Moscow State University, Moscow in 1981. He is Professor of Materials Science, Head of the Crystal Growth Laboratory, and Director of the Internal Quality Assurance Cell of the University of Mysore, India. His current research is in crystal engineering of polyscale materials through novel solution processing routes, particularly covering hydrothermal, solvothermal and supercritical methods. Professor Byrappa has co-authored the Handbook of Hydrothermal Technology, and edited 4 books as well as two special editions of Journal of Materials Science, and published 180 research papers including 26 invited reviews and book chapters on various aspects of novel routes of solution processing. Professor Byrappa has delivered over 60 keynote and invited lectures at International Conferences, and several hundreds of colloquia and seminars at various institutions around the world. He has also served as chair and co-chair for numerous international conferences. He is a Fellow of the World Academy of Ceramics. Professor Byrappa is serving in several international committees and commissions related to crystallography, crystal growth, and materials science. He is the Founder Secretary of the International Solvothermal and Hydrothermal Association. Professor Byrappa is a recipient of several awards such as the Sir C.V. Raman Award, Materials Research Society of India Medal, and the Golden Jubilee Award of the University of Mysore.

Vishwanath "Vish" Prasad is the Vice President for Research and Economic Development and Professor of Mechanical and Energy Engineering at the University of North Texas (UNT), one of the largest university in the state of Texas. He received his PhD from the University of Delaware (USA), his Masters of Technology from the Indian Institute of Technology, Kanpur, and his bachelor's from Patna University in India all in Mechanical Engineering. Prior to joining UNT in 2007, Dr. Prasad served as the Dean at Florida International University (FIU) in Miami, where he also held the position of Distinguished Professor of Engineering. Previously, he has served as a Leading Professor of Mechanical Engineering at Stony Brook University, New York, as an Associate Professor and Assistant Professor at Columbia University. He has received many special recognitions for his contributions to engineering education. Dr. Prasad's research interests include thermo-fluid sciences, energy systems, electronic materials, and computational materials processing. He has published over 200 articles, edited/co-edited several books and organized numerous conferences, symposia, and workshops. He serves as the lead editor of the Annual Review of Heat Transfer. In the past, he has served as an Associate Editor of the ASME Journal of Heat. Dr. Prasad is an elected Fellow of the American Society of Mechanical Engineers (ASME), and has served as a member of the USRA Microgravity Research Council. Dr. Prasad's research has focused on bulk growth of silicon, III-V compounds, and silicon carbide; growth of large diameter Si tube; design of crystal growth systems; and sputtering and chemical vapor deposition of thin films. He is also credited to initiate research on wire saw cutting of large crystals to produce wafers with much reduced material loss. Dr. Prasad's research has been well funded by US National Science Foundation (NSF), US Department of Defense, US Department of Energy, and industry.

Michael Dudley received his Doctoral Degree in Engineering from Warwick University, UK, in 1982. He is Professor and Chair of the Materials Science and Engineering Department at Stony Brook University, New York, USA. He is director of the Stony Brook Synchrotron Topography Facility at the National Synchrotron Light Source at Brookhaven National Laboratory, Upton New York. His current research focuses on crystal growth and characterization of defect structures in single crystals with a view to determining their origins. The primary technique used is synchrotron topography which enables analysis of defects and generalized strain fields in single crystals in general, with particular emphasis on semiconductor, optoelectronic, and optical crystals. Establishing the relationship between crystal growth conditions and resulting defect distributions is a particular thrust area of interest to Dudley, as is the correlation between electronic/optoelectronic device performance and defect distribution. Other techniques routinely used in such analysis include transmission electron microscopy, high resolution triple-axis x-ray diffraction, atomic force microscopy, scanning electron microscopy, Nomarski optical microscopy, conventional optical microscopy, IR microscopy and fluorescent laser scanning confocal microscopy. Dudley's group has played a prominent role in the development of SiC and AlN growth, characterizing crystals grown by many of the academic and commercial entities involved enabling optimization of crystal quality. He has co-authored some 315 refereed articles and 12 book chapters, and has edited 5 books. He is currently a member of the Editorial Board of Journal of Applied Physics and Applied Physics Letters and has served as Chair or Co-Chair for numerous international conferences.

List of Authors

Francesco Abbona
Università degli Studi di Torino
Dipartimento di Scienze Mineralogiche
e Petrologiche
via Valperga Caluso 35
10125 Torino, Italy
e-mail: *francesco.abbona@unito.it*

Mohan D. Aggarwal
Alabama A&M University
Department of Physics
Normal, AL 35762, USA
e-mail: *mohan.aggarwal@aamu.edu*

Marcello R.B. Andreeta
University of São Paulo
Crystal Growth and Ceramic Materials Laboratory,
Institute of Physics of São Carlos
Av. Trabalhador Sãocarlense, 400
São Carlos, SP 13560-970, Brazil
e-mail: *marcello@if.sc.usp.br*

Dino Aquilano
Università degli Studi di Torino
Facoltà di Scienze Matematiche, Fisiche e Naturali
via P. Giuria, 15
Torino, 10126, Italy
e-mail: *dino.aquilano@unito.it*

Roberto Arreguín-Espinosa
Universidad Nacional Autónoma de México
Instituto de Química
Circuito Exterior, C.U. s/n
Mexico City, 04510, Mexico
e-mail: *arrespin@unam.mx*

Jie Bai
Intel Corporation
RA3-402, 5200 NE Elam Young Parkway
Hillsboro, OR 97124-6497, USA
e-mail: *jie.bai@intel.com*

Stefan Balint
West University of Timisoara
Department of Computer Science
Blvd. V. Parvan 4
Timisoara, 300223, Romania
e-mail: *balint@math.uvt.ro*

Ashok K. Batra
Alabama A&M University
Department of Physics
4900 Meridian Street
Normal, AL 35762, USA
e-mail: *ashok.batra@aamu.edu*

Handady L. Bhat
Indian Institute of Science
Department of Physics
CV Raman Avenue
Bangalore, 560012, India
e-mail: *hlbhat@physics.iisc.ernet.in*

Ishwara B. Bhat
Rensselaer Polytechnic Institute
Electrical Computer
and Systems Engineering Department
110 8th Street, JEC 6031
Troy, NY 12180, USA
e-mail: *bhati@rpi.edu*

David F. Bliss
US Air Force Research Laboratory
Sensors Directorate Optoelectronic Technology
Branch
80 Scott Drive
Hanscom AFB, MA 01731, USA
e-mail: *david.bliss@hanscom.af.mil*

Mikhail A. Borik
Russian Academy of Sciences
Laser Materials and Technology Research Center,
A.M. Prokhorov General Physics Institute
Vavilov 38
Moscow, 119991, Russia
e-mail: *borik@lst.gpi.ru*

Liliana Braescu
West University of Timisoara
Department of Computer Science
Blvd. V. Parvan 4
Timisoara, 300223, Romania
e-mail: lilianabraescu@balint1.math.uvt.ro

Kullaiah Byrappa
University of Mysore
Department of Geology
Manasagangotri
Mysore, 570 006, India
e-mail: kbyrappa@gmail.com

Dang Cai
CVD Equipment Corporation
1860 Smithtown Ave.
Ronkonkoma, NY 11779, USA
e-mail: dcai@cvdequipment.com

Michael J. Callahan
GreenTech Solutions
92 Old Pine Drive
Hanson, MA 02341, USA
e-mail: mjcal37@yahoo.com

Joan J. Carvajal
Universitat Rovira i Virgili (URV)
Department of Physics and Crystallography
of Materials and Nanomaterials (FiCMA–FiCNA)
Campus Sescelades, C/ Marcel·lí Domingo, s/n
Tarragona 43007, Spain
e-mail: joanjosep.carvajal@urv.cat

Aaron J. Celestian
Western Kentucky University
Department of Geography and Geology
1906 College Heights Blvd.
Bowling Green, KY 42101, USA
e-mail: aaron.celestian@wku.edu

Qi-Sheng Chen
Chinese Academy of Sciences
Institute of Mechanics
15 Bei Si Huan Xi Road
Beijing, 100190, China
e-mail: qschen@imech.ac.cn

Chunhui Chung
Stony Brook University
Department of Mechanical Engineering
Stony Brook, NY 11794-2300, USA
e-mail: chuchung@ic.sunysb.edu

Ted Ciszek
Geolite/Siliconsultant
31843 Miwok Trl.
Evergreen, CO 80437, USA
e-mail: ted_ciszek@siliconsultant.com

Abraham Clearfield
Texas A&M University
Distinguished Professor of Chemistry
College Station, TX 77843-3255, USA
e-mail: clearfield@chem.tamu.edu

Hanna A. Dabkowska
Brockhouse Institute for Materials Research
Department of Physics and Astronomy
1280 Main Str W.
Hamilton, Ontario L8S 4M1, Canada
e-mail: dabkoh@mcmaster.ca

Antoni B. Dabkowski
McMaster University, BIMR
Brockhouse Institute for Materials Research,
Department of Physics and Astronomy
1280 Main Str W.
Hamilton, Ontario L8S 4M1, Canada
e-mail: dabko@mcmaster.ca

Rafael Dalmau
HexaTech Inc.
991 Aviation Pkwy Ste 800
Morrisville, NC 27560, USA
e-mail: rdalmau@hexatechinc.com

Govindhan Dhanaraj
ARC Energy
18 Celina Avenue, Unit 77
Nashua, NH 03063, USA
e-mail: dhanaraj@arc-energy.com

Ramasamy Dhanasekaran
Anna University Chennai
Crystal Growth Centre
Chennai, 600 025, India
e-mail: rdhanasekaran@annauniv.edu;
rdcgc@yahoo.com

Ernesto Diéguez
Universidad Autónoma de Madrid
Department Física de Materiales
Madrid 28049, Spain
e-mail: *ernesto.dieguez@uam.es*

Vijay K. Dixit
Raja Ramanna Center for Advance Technology
Semiconductor Laser Section,
Solid State Laser Division
Rajendra Nagar, RRCAT.
Indore, 452013, India
e-mail: *dixit@rrcat.gov.in*

Sadik Dost
University of Victoria
Crystal Growth Laboratory
Victoria, BC V8W 3P6, Canada
e-mail: *sdost@me.uvic.ca*

Michael Dudley
Stony Brook University
Department of Materials Science and Engineering
Stony Brook, NY 11794-2275, USA
e-mail: *mdudley@notes.cc.sunysb.edu*

Partha S. Dutta
Rensselaer Polytechnic Institute
Department of Electrical, Computer
and Systems Engineering
110 Eighth Street
Troy, NY 12180, USA
e-mail: *duttap@rpi.edu*

Francesc Díaz
Universitat Rovira i Virgili (URV)
Department of Physics and Crystallography
of Materials and Nanomaterials (FiCMA-FiCNA)
Campus Sescelades, C/ Marcel·lí Domingo, s/n
Tarragona 43007, Spain
e-mail: *f.diaz@urv.cat*

Paul F. Fewster
PANalytical Research Centre,
The Sussex Innovation Centre
Research Department
Falmer
Brighton, BN1 9SB, UK
e-mail: *paul.fewster@panalytical.com*

Donald O. Frazier
NASA Marshall Space Flight Center
Engineering Technology Management Office
Huntsville, AL 35812, USA
e-mail: *donald.o.frazier@nasa.gov*

James W. Garland
EPIR Technologies, Inc.
509 Territorial Drive, Ste. B
Bolingbrook, IL 60440, USA
e-mail: *jgarland@epir.com*

Thomas F. George
University of Missouri-St. Louis
Center for Nanoscience,
Department of Chemistry and Biochemistry,
Department of Physics and Astronomy
One University Boulevard
St. Louis, MO 63121, USA
e-mail: *tfgeorge@umsl.edu*

Andrea E. Gutiérrez-Quezada
Universidad Nacional Autónoma de México
Instituto de Química
Circuito Exterior, C.U. s/n
Mexico City, 04510, Mexico
e-mail: *30111390@escolar.unam.mx*

Carl Hemmingsson
Linköping University
Department of Physics, Chemistry
and Biology (IFM)
581 83 Linköping, Sweden
e-mail: *cah@ifm.liu.se*

Antonio Carlos Hernandes
University of São Paulo
Crystal Growth and Ceramic Materials Laboratory,
Institute of Physics of São Carlos
Av. Trabalhador Sãocarlense
São Carlos, SP 13560-970, Brazil
e-mail: *hernandes@if.sc.usp.br*

Koichi Kakimoto
Kyushu University
Research Institute for Applied Mechanics
6-1 Kasuga-kouen, Kasuga
816-8580 Fukuoka, Japan
e-mail: *kakimoto@riam.kyushu-u.ac.jp*

Imin Kao
State University of New York at Stony Brook
Department of Mechanical Engineering
Stony Brook, NY 11794-2300, USA
e-mail: imin.kao@stonybrook.edu

John J. Kelly
Utrecht University,
Debye Institute for Nanomaterials Science
Department of Chemistry
Princetonplein 5
3584 CC, Utrecht, The Netherlands
e-mail: j.j.kelly@uu.nl

Jeonggoo Kim
Neocera, LLC
10000 Virginia Manor Road #300
Beltsville, MD, USA
e-mail: kim@neocera.com

Helmut Klapper
Institut für Kristallographie
RWTH Aachen University
Aachen, Germany
e-mail: klapper@xtal.rwth-aachen.de;
helmut-klapper@web.de

Christine F. Klemenz Rivenbark
Krystal Engineering LLC
General Manager and Technical Director
1429 Chaffee Drive
Titusville, FL 32780, USA
e-mail: ckr@krystalengineering.com

Christian Kloc
Nanyang Technological University
School of Materials Science and Engineering
50 Nanyang Avenue
639798 Singapore
e-mail: ckloc@ntu.edu.sg

Solomon H. Kolagani
Neocera LLC
10000 Virginia Manor Road
Beltsville, MD 20705, USA
e-mail: harsh@neocera.com

Akinori Koukitu
Tokyo University of Agriculture and Technology
(TUAT)
Department of Applied Chemistry
2-24-16 Naka-cho, Koganei
184-8588 Tokyo, Japan
e-mail: koukitu@cc.tuat.ac.jp

Milind S. Kulkarni
MEMC Electronic Materials
Polysilicon and Quantitative Silicon Research
501 Pearl Drive
St. Peters, MO 63376, USA
e-mail: mkulkarni@memc.com

Yoshinao Kumagai
Tokyo University of Agriculture and Technology
Department of Applied Chemistry
2-24-16 Naka-cho, Koganei
184-8588 Tokyo, Japan
e-mail: 4470kuma@cc.tuat.ac.jp

Valentin V. Laguta
Institute of Physics of the ASCR
Department of Optical Materials
Cukrovarnicka 10
Prague, 162 53, Czech Republic
e-mail: laguta@fzu.cz

Ravindra B. Lal
Alabama Agricultural and Mechanical University
Physics Department
4900 Meridian Street
Normal, AL 35763, USA
e-mail: rblal@comcast.net

Chung-Wen Lan
National Taiwan University
Department of Chemical Engineering
No. 1, Sec. 4, Roosevelt Rd.
Taipei, 106, Taiwan
e-mail: cwlan@ntu.edu.tw

Hongjun Li
Chinese Academy of Sciences
R & D Center of Synthetic Crystals,
Shanghai Institute of Ceramics
215 Chengbei Rd., Jiading District
Shanghai, 201800, China
e-mail: lh_li@mail.sic.ac.cn

Elena E. Lomonova
Russian Academy of Sciences
Laser Materials and Technology Research Center,
A.M. Prokhorov General Physics Institute
Vavilov 38
Moscow, 119991, Russia
e-mail: lomonova@lst.gpi.ru

Ivan V. Markov
Bulgarian Academy of Sciences
Institute of Physical Chemistry
Sofia, 1113, Bulgaria
e-mail: imarkov@ipc.bas.bg

Bo Monemar
Linköping University
Department of Physics, Chemistry and Biology
58183 Linköping, Sweden
e-mail: bom@ifm.liu.se

Abel Moreno
Universidad Nacional Autónoma de México
Instituto de Química
Circuito Exterior, C.U. s/n
Mexico City, 04510, Mexico
e-mail: carcamo@unam.mx

Roosevelt Moreno Rodriguez
State University of New York at Stony Brook
Department of Mechanical Engineering
Stony Brook, NY 11794-2300, USA
e-mail: roosevelt@dove.eng.sunysb.edu

S. Narayana Kalkura
Anna University Chennai
Crystal Growth Centre
Sardar Patel Road
Chennai, 600025, India
e-mail: kalkura@annauniv.edu

Mohan Narayanan
Reliance Industries Limited
1, Rich Branch court
Gaithersburg, MD 20878, USA
e-mail: mohan.narayanan@ril.com

Subramanian Natarajan
Madurai Kamaraj University
School of Physics
Palkalai Nagar
Madurai, India
e-mail: s_natarajan50@yahoo.com

Martin Nikl
Academy of Sciences of the Czech Republic (ASCR)
Department of Optical Crystals, Institute of Physics
Cukrovarnicka 10
Prague, 162 53, Czech Republic
e-mail: nikl@fzu.cz

Vyacheslav V. Osiko
Russian Academy of Sciences
Laser Materials and Technology Research Center,
A.M. Prokhorov General Physics Institute
Vavilov 38
Moscow, 119991, Russia
e-mail: osiko@lst.gpi.ru

John B. Parise
Stony Brook University
Chemistry Department
and Department of Geosciences
ESS Building
Stony Brook, NY 11794-2100, USA
e-mail: john.parise@stonybrook.edu

Srinivas Pendurti
ASE Technologies Inc.
11499, Chester Road
Cincinnati, OH 45246, USA
e-mail: spendurti@asetech.com

Benjamin G. Penn
NASA/George C. Marshall Space Flight Center
ISHM and Sensors Branch
Huntsville, AL 35812, USA
e-mail: benjamin.g.penndr@nasa.gov

Jens Pflaum
Julius-Maximilians Universität Würzburg
Institute of Experimental Physics VI
Am Hubland
97078 Würzburg, Germany
e-mail: jpflaum@physik.uni-wuerzburg.de

Jose Luis Plaza
Universidad Autónoma de Madrid
Facultad de Ciencias,
Departamento de Física de Materiales
Madrid 28049, Spain
e-mail: *joseluis.plaza@uam.es*

Udo W. Pohl
Technische Universität Berlin
Institut für Festkörperphysik EW5-1
Hardenbergstr. 36
10623 Berlin, Germany
e-mail: *pohl@physik.tu-berlin.de*

Vishwanath (Vish) Prasad
University of North Texas
1155 Union Circle
Denton, TX 76203-5017, USA
e-mail: *vish.prasad@unt.edu*

Maria Cinta Pujol
Universitat Rovira i Virgili
Department of Physics and Crystallography
of Materials and Nanomaterials (FiCMA-FiCNA)
Campus Sescelades, C/ Marcel·lí Domingo
Tarragona 43007, Spain
e-mail: *mariacinta.pujol@urv.cat*

Balaji Raghothamachar
Stony Brook University
Department of Materials Science and Engineering
310 Engineering Building
Stony Brook, NY 11794-2275, USA
e-mail: *braghoth@notes.cc.sunysb.edu*

Michael Roth
The Hebrew University of Jerusalem
Department of Applied Physics
Bergman Bld., Rm 206, Givat Ram Campus
Jerusalem 91904, Israel
e-mail: *mroth@vms.huji.ac.il*

Peter Rudolph
Leibniz Institute for Crystal Growth
Technology Development
Max-Born-Str. 2
Berlin, 12489, Germany
e-mail: *rudolph@ikz-berlin.de*

Akira Sakai
Osaka University
Department of Systems Innovation
1-3 Machikaneyama-cho, Toyonaka-shi
560-8531 Osaka, Japan
e-mail: *sakai@ee.es.osaka-u.ac.jp*

Yasuhiro Shiraki
Tokyo City University
Advanced Research Laboratories,
Musashi Institute of Technology
8-15-1 Todoroki, Setagaya-ku
158-0082 Tokyo, Japan
e-mail: *yshiraki@tcu.ac.jp*

Theo Siegrist
Florida State University
Department of Chemical
and Biomedical Engineering
2525 Pottsdamer Street
Tallahassee, FL 32310, USA
e-mail: *siegrist@eng.fsu.edu*

Zlatko Sitar
North Carolina State University
Materials Science and Engineering
1001 Capability Dr.
Raleigh, NC 27695, USA
e-mail: *sitar@ncsu.edu*

Sivalingam Sivananthan
University of Illinois at Chicago
Department of Physics
845 W. Taylor St. M/C 273
Chicago, IL 60607-7059, USA
e-mail: *siva@uic.edu; siva@epir.com*

Mikhail D. Strikovski
Neocera LLC
10000 Virginia Manor Road, suite 300
Beltsville, MD 20705, USA
e-mail: *strikovski@neocera.com*

Xun Sun
Shandong University
Institute of Crystal Materials
Shanda Road
Jinan, 250100, China
e-mail: *sunxun@icm.sdu.edu.cn*

Ichiro Sunagawa
University Tohoku University (Emeritus)
Kashiwa-cho 3-54-2, Tachikawa
Tokyo, 190-0004, Japan
e-mail: *i.sunagawa@nifty.com*

Xu-Tang Tao
Shandong University
State Key Laboratory of Crystal Materials
Shanda Nanlu 27, 250100
Jinan, China
e-mail: *txt@sdu.edu.cn*

Vitali A. Tatartchenko
Saint – Gobain, 23 Rue Louis Pouey
92800 Puteaux, France
e-mail: *vitali.tatartchenko@orange.fr*

Filip Tuomisto
Helsinki University of Technology
Department of Applied Physics
Otakaari 1 M
Espoo TKK 02015, Finland
e-mail: *filip.tuomisto@tkk.fi*

Anna Vedda
University of Milano-Bicocca
Department of Materials Science
Via Cozzi 53
20125 Milano, Italy
e-mail: *anna.vedda@unimib.it*

Lu-Min Wang
University of Michigan
Department of Nuclear Engineering
and Radiological Sciences
2355 Bonisteel Blvd.
Ann Arbor, MI 48109-2104, USA
e-mail: *lmwang@umich.edu*

Sheng-Lai Wang
Shandong University
Institute of Crystal Materials,
State Key Laboratory of Crystal Materials
Shanda Road No. 27
Jinan, Shandong, 250100, China
e-mail: *slwang@icm.sdu.edu.cn*

Shixin Wang
Micron Technology Inc.
TEM Laboratory
8000 S. Federal Way
Boise, ID 83707, USA
e-mail: *shixinwang@micron.com*

Jan L. Weyher
Polish Academy of Sciences Warsaw
Institute of High Pressure Physics
ul. Sokolowska 29/37
01/142 Warsaw, Poland
e-mail: *weyher@unipress.waw.pl*

Jun Xu
Chinese Academy of Sciences
Shanghai Institute of Ceramics
Shanghai, 201800, China
e-mail: *xujun@mail.shcnc.ac.cn*

Hui Zhang
Tsinghua University
Department of Engineering Physics
Beijing, 100084, China
e-mail: *zhhui@tsinghua.edu.cn*

Lili Zheng
Tsinghua University
School of Aerospace
Beijing, 100084, China
e-mail: *zhenglili@tsinghua.edu.cn*

Mary E. Zvanut
University of Alabama at Birmingham
Department of Physics
1530 3rd Ave S
Birmingham, AL 35294-1170, USA
e-mail: *mezvanut@uab.edu*

Zbigniew R. Zytkiewicz
Polish Academy of Sciences
Institute of Physics
Al. Lotnikow 32/46
02668 Warszawa, Poland
e-mail: *zytkie@ifpan.edu.pl*

Acknowledgements

C.17 Bulk Single Crystals Grown from Solution on Earth and in Microgravity
by Mohan D. Aggarwal, Ashok K. Batra, Ravindra B. Lal, Benjamin G. Penn, Donald O. Frazier

The authors are grateful for helpful discussions with a number of graduate students and other physics faculty in the Department of Physics at Alabama A&M University. Authors are thankful to Garland Sharp for his expert machining work and Jerry Johnson for his glass-blowing jobs in the design of various crystal growth systems described in this work. Spacelab-3 and IML-1 work was supported by NASA contracts. The optical holography work for these experiments was developed in collaboration of Dr. James Trolinger of MetroLaser, Inc. This work was partially supported under NSF-HBCU RISE program HRD-0531183 and US Army Space and Missile Defense Command, Contract W9113M-04-C-0005. Two of the authors (M.D.A. and R.B.L.) would like to acknowledge support from NASA Administrator's Fellowship Program (NAFP) through United Negro College Fund Special Programs (UNCFSP) Corporation under their Contract No. NNG06GC58A.

C.18 Hydrothermal Growth of Polyscale Crystals
by Kullaiah Byrappa

The author wishes to acknowledge Prof. M. Yoshimura (Tokyo Institute of Technology, Japan), Prof. Richard E. Riman (Rutgers University, USA), Prof. Yan Li (Tsinghua University, China), Prof. T. Adschiri, and Prof. Dirk Ehrentraut (Tohoku University, Japan) for providing photographs of crystals synthesized by them and also some fruitful discussion. Thanks are also due to my group members Prof. B. Basavalingu, Prof. K.M. Lokanatha Rai, and Prof. S. Ananda of Mysore University, India, for their assistance in preparing this chapter. The author acknowledges the help of Prof. Xu Haiyan, Anhui Institute of Architecture and Industry, Hefei, China, for providing some of the latest literature on solution processing of materials, and also for careful reading and constructive comments on this chapter. Also the authors acknowledges the help rendered by Dr. Jürgen Riegler, Tohoku University, Japan, in reading this manuscript and useful comments.

C.19 Hydrothermal and Ammonothermal Growth of ZnO and GaN
by Michael J. Callahan, Qi-Sheng Chen

The authors acknowledge collaborators whose work either influenced or was explicitly incorporated into this article. Special thanks are due to Govindhan Dhanaraj (SUNY-Stony Brook) who provided expertise on synchrotron white-beam x-ray topographs and to Dr. Buguo Wang (Solid State Sciences Corporation), Dr. Michael Alexander, Dr. David Bliss, and Michael Suscavage (Air Force Research Laboratory) for their expertise and many discussions with the authors on ammonothermal and hydrothermal research.

Finally, we thank the Air Force Office of Scientific Research (Drs. Jerry Witt, Dan Johnstone, Todd Steiner, and Don Silversmith) and the Naval Research Lab (Dr. Colin Wood) for their past and current support of research on wide-bandgap semiconductors.

C.20 Stoichiometry and Domain Structure of KTP-Type Nonlinear Optical Crystals
by Michael Roth

The author is grateful to Dr. N. Angert and Dr. M. Tseitlin for their long-term collaboration and numerous discussions on the science and technology of KTP-type crystals as well as for providing some crystal and domain photographs.

C.21 High-Temperature Solution Growth: Application to Laser and Nonlinear Optical Crystals
by Joan J. Carvajal, Maria Cinta Pujol, Francesc Díaz

The authors thank to our colleague Prof. Magdalena Aguiló for her relevant contribution specially in structural and crystallographic aspects. This work was supported by the Spanish Government under projects

MAT2008-06729-C02-02/NAN and the Catalan Authority under project 2009SGR235. J.J. Carvajal and M.C. Pujol are supported by the Education and Science Ministry of Spain and European Social Fund under the Ramon y Cajal program, RYC2006-858.

C.22 Growth and Characterization of KDP and Its Analogs
by Sheng-Lai Wang, Xun Sun, Xu-Tang Tao

We wish to thank the following for help in the preparation of this chapter and for helpful discussions: Bing Liu, Xiao-Min Mu, Bo Wang, Yong-Qiang Lu, Liang Li, Prof. Chang-Shui Fang, Prof. Xin-Guang Xu (Shandong Univ., Jinan, China), Dr. Natalia Zaitseva, and Dr. Jim De Yoreo (LLNL). The authors also wish to thank Dr. Govindhan Dhanaraj for critically reading the manuscript and helpful discussion.

目 录

缩略语

Part C 溶液法生长晶体

17 地球微重力下从溶液中生长体材料单晶 ... 3
17.1 结晶：成核和生长动力学 ... 5
17.2 低温溶液的晶体生长 ... 10
17.3 更低温度溶液的晶体生长 ... 11
17.4 硫酸三甘钛晶体生长：个案研究 ... 18
17.5 微重力下硫酸三甘钛晶体的溶液生长 ... 26
17.6 蛋白质晶体生长 ... 36
17.7 结 语 ... 38
参考文献 ... 38

18 水热法大尺寸晶体生长 ... 43
18.1 水热法晶体生长的历史 ... 47
18.2 水热法晶体生长的热力学基础 ... 50
18.3 水热法晶体生长的设备 ... 59
18.4 部分晶体的水热法生长 ... 64
18.5 精细晶体的水热法生长 ... 78
18.6 水热法生长纳米晶体 ... 81
18.7 结 语 ... 84
18.A 附 录 ... 85
参考文献 ... 90

19 水热法与氨热法生长ZnO和GaN ... 99
19.1 水热法与氨热法生长大晶体综述 ... 101
19.2 低缺陷大晶体的生长要求 ... 105
19.3 物理与数学模型 ... 110
19.4 过程模拟 ... 113
19.5 水热法生长ZnO晶体 ... 118
19.6 氨热法生长GaN ... 125
19.7 结 论 ... 129
参考文献 ... 129

20 KTP型非线性光学晶体的化学计量比和畴结构 ... 135
20.1 背 景 ... 135
20.2 化学计量比与铁电相转变 ... 141
20.3 生长引起的铁电畴 ... 147
20.4 人造畴结构 ... 152
20.5 非线性光学晶体 ... 157
参考文献 ... 160

21 高温溶液生长：用于激光和非线性光学的晶体 ... 169
21.1 基 础 ... 170
21.2 高温溶液生长 ... 175
21.3 用TSSG法生长激光体材料和NLO单晶 ... 180
21.4 液相外延：激光和NLO材料的外延膜的生长 ... 190
参考文献 ... 196

22 KDP及同类晶体的生长与表征 ... 203
22.1 背 景 ... 203
22.2 结晶机制和动力学 ... 205
22.3 单晶的生长技术 ... 213
22.4 生长条件对晶体缺陷的影响 ... 220
22.5 晶体质量检测 ... 227
参考文献 ... 233

Contents

List of Abbreviations

Part C Solution Growth of Crystals

17 Bulk Single Crystals Grown from Solution on Earth and in Microgravity
Mohan D. Aggarwal, Ashok K. Batra, Ravindra B. Lal, Benjamin G. Penn, Donald O. Frazier 559
- 17.1 Crystallization: Nucleation and Growth Kinetics 561
- 17.2 Low-Temperature Solution Growth 566
- 17.3 Solution Growth by Temperature Lowering 567
- 17.4 Triglycine Sulfate Crystal Growth: A Case Study 574
- 17.5 Solution Growth of Triglycine Sulfate Crystals in Microgravity 582
- 17.6 Protein Crystal Growth 592
- 17.7 Concluding Remarks 594
- **References** 594

18 Hydrothermal Growth of Polyscale Crystals
Kullaiah Byrappa 599
- 18.1 History of Hydrothermal Growth of Crystals 603
- 18.2 Thermodynamic Basis of the Hydrothermal Growth of Crystals 606
- 18.3 Apparatus Used in the Hydrothermal Growth of Crystals 615
- 18.4 Hydrothermal Growth of Some Selected Crystals 620
- 18.5 Hydrothermal Growth of Fine Crystals 634
- 18.6 Hydrothermal Growth of Nanocrystals 637
- 18.7 Concluding Remarks 640
- 18.A Appendix 641
- **References** 646

19 Hydrothermal and Ammonothermal Growth of ZnO and GaN
Michael J. Callahan, Qi-Sheng Chen 655
- 19.1 Overview of Hydrothermal and Ammonothermal Growth of Large Crystals 657
- 19.2 Requirements for Growth of Large, Low-Defect Crystals 661
- 19.3 Physical and Mathematical Models 666
- 19.4 Process Simulations 669
- 19.5 Hydrothermal Growth of ZnO Crystals 674
- 19.6 Ammonothermal GaN 681
- 19.7 Conclusion 685
- **References** 685

20 Stoichiometry and Domain Structure of KTP-Type Nonlinear Optical Crystals
Michael Roth 691
- 20.1 Background 691
- 20.2 Stoichiometry and Ferroelectric Phase Transitions 697
- 20.3 Growth-Induced Ferroelectric Domains 703
- 20.4 Artificial Domain Structures 708
- 20.5 Nonlinear Optical Crystals 713
- **References** 716

21 High-Temperature Solution Growth: Application to Laser and Nonlinear Optical Crystals
Joan J. Carvajal, Maria Cinta Pujol, Francesc Díaz 725
- 21.1 Basics 726
- 21.2 High-Temperature Solution Growth 731
- 21.3 Growth of Bulk Laser and NLO Single Crystals by the TSSG Method 736
- 21.4 Liquid-Phase Epitaxy: Growth of Epitaxial Films of Laser and NLO Materials 746
- **References** 752

22 Growth and Characterization of KDP and Its Analogs
Sheng-Lai Wang, Xun Sun, Xu-Tang Tao 759
- 22.1 Background 759
- 22.2 Mechanism and Kinetics of Crystallization 761
- 22.3 Growth Techniques for Single Crystals 769
- 22.4 Effect of Growth Conditions on Defects of Crystals 776
- 22.5 Investigations on Crystal Quality 783
- **References** 789

List of Abbreviations

μ-PD	micro-pulling-down
1S-ELO	one-step ELO structure
2-D	two-dimensional
2-DNG	two-dimensional nucleation growth
2S-ELO	double layer ELO
3-D	three-dimensional
4T	quaterthiophene
6T	sexithienyl
8MR	eight-membered ring
8T	hexathiophene

A

a-Si	amorphous silicon
A/D	analogue-to-digital
AA	additional absorption
AANP	2-adamantylamino-5-nitropyridine
AAS	atomic absorption spectroscopy
AB	Abrahams and Burocchi
ABES	absorption-edge spectroscopy
AC	alternate current
ACC	annular capillary channel
ACRT	accelerated crucible rotation technique
ADC	analog-to-digital converter
ADC	automatic diameter control
ADF	annular dark field
ADP	ammonium dihydrogen phosphate
AES	Auger electron spectroscopy
AFM	atomic force microscopy
ALE	arbitrary Lagrangian Eulerian
ALE	atomic layer epitaxy
ALUM	aluminum potassium sulfate
ANN	artificial neural network
AO	acoustooptic
AP	atmospheric pressure
APB	antiphase boundaries
APCF	advanced protein crystallization facility
APD	avalanche photodiode
APPLN	aperiodic poled LN
APS	Advanced Photon Source
AR	antireflection
AR	aspect ratio
ART	aspect ratio trapping
ATGSP	alanine doped triglycine sulfo-phosphate
AVT	angular vibration technique

B

BA	Born approximation
BAC	band anticrossing
BBO	BaB_2O_4
BCF	Burton–Cabrera–Frank
BCT	$Ba_{0.77}Ca_{0.23}TiO_3$
BCTi	$Ba_{1-x}Ca_xTiO_3$
BE	bound exciton
BF	bright field
BFDH	Bravais–Friedel–Donnay–Harker
BGO	$Bi_{12}GeO_{20}$
BIBO	BiB_3O_6
BLIP	background-limited performance
BMO	$Bi_{12}MO_{20}$
BN	boron nitride
BOE	buffered oxide etch
BPD	basal-plane dislocation
BPS	Burton–Prim–Slichter
BPT	bipolar transistor
BS	Bridgman–Stockbarger
BSCCO	Bi–Sr–Ca–Cu–O
BSF	bounding stacking fault
BSO	$Bi_{20}SiO_{20}$
BTO	$Bi_{12}TiO_{20}$
BU	building unit
BaREF	barium rare-earth fluoride
BiSCCO	$Bi_2Sr_2CaCu_2O_n$

C

C–V	capacitance–voltage
CALPHAD	calculation of phase diagram
CBED	convergent-beam electron diffraction
CC	cold crucible
CCC	central capillary channel
CCD	charge-coupled device
CCVT	contactless chemical vapor transport
CD	convection diffusion
CE	counterelectrode
CFD	computational fluid dynamics
CFD	cumulative failure distribution
CFMO	Ca_2FeMoO_6
CFS	continuous filtration system
CGG	calcium gallium germanate
CIS	copper indium diselenide
CL	cathode-ray luminescence
CL	cathodoluminescence
CMM	coordinate measuring machine
CMO	$CaMoO_4$
CMOS	complementary metal–oxide–semiconductor
CMP	chemical–mechanical polishing
CMP	chemomechanical polishing

COD	calcium oxalate dihydrate
COM	calcium oxalate-monohydrate
COP	crystal-originated particle
CP	critical point
CPU	central processing unit
CRSS	critical-resolved shear stress
CSMO	$Ca_{1-x}Sr_xMoO_3$
CST	capillary shaping technique
CST	crystalline silico titanate
CT	computer tomography
CTA	$CsTiOAsO_4$
CTE	coefficient of thermal expansion
CTF	contrast transfer function
CTR	crystal truncation rod
CV	Cabrera–Vermilyea
CVD	chemical vapor deposition
CVT	chemical vapor transport
CW	continuous wave
CZ	Czochralski
CZT	Czochralski technique

D

D/A	digital to analog
DBR	distributed Bragg reflector
DC	direct current
DCAM	diffusion-controlled crystallization apparatus for microgravity
DCCZ	double crucible CZ
DCPD	dicalcium-phosphate dihydrate
DCT	dichlorotetracene
DD	dislocation dynamics
DESY	Deutsches Elektronen Synchrotron
DF	dark field
DFT	density function theory
DFW	defect free width
DGS	diglycine sulfate
DI	deionized
DIA	diamond growth
DIC	differential interference contrast
DICM	differential interference contrast microscopy
DKDP	deuterated potassium dihydrogen phosphate
DLATGS	deuterated L-alanine-doped triglycine sulfate
DLTS	deep-level transient spectroscopy
DMS	discharge mass spectroscopy
DNA	deoxyribonucleic acid
DOE	Department of Energy
DOS	density of states
DPH-BDS	2,6-diphenylbenzo[1,2-*b*:4,5-*b*']diselenophene
DPPH	2,2-diphenyl-1-picrylhydrazyl
DRS	dynamic reflectance spectroscopy
DS	directional solidification
DSC	differential scanning calorimetry
DSE	defect-selective etching
DSL	diluted Sirtl with light
DTA	differential thermal analysis
DTGS	deuterated triglycine sulfate
DVD	digital versatile disk
DWBA	distorted-wave Born approximation
DWELL	dot-in-a-well

E

EADM	extended atomic distance mismatch
EALFZ	electrical-assisted laser floating zone
EB	electron beam
EBIC	electron-beam-induced current
ECE	end chain energy
ECR	electron cyclotron resonance
EDAX	energy-dispersive x-ray analysis
EDMR	electrically detected magnetic resonance
EDS	energy-dispersive x-ray spectroscopy
EDT	ethylene dithiotetrathiafulvalene
EDTA	ethylene diamine tetraacetic acid
EELS	electron energy-loss spectroscopy
EFG	edge-defined film-fed growth
EFTEM	energy-filtered transmission electron microscopy
ELNES	energy-loss near-edge structure
ELO	epitaxial lateral overgrowth
EM	electromagnetic
EMA	effective medium theory
EMC	electromagnetic casting
EMCZ	electromagnetic Czochralski
EMF	electromotive force
ENDOR	electron nuclear double resonance
EO	electrooptic
EP	EaglePicher
EPD	etch pit density
EPMA	electron microprobe analysis
EPR	electron paramagnetic resonance
erfc	error function
ES	equilibrium shape
ESP	edge-supported pulling
ESR	electron spin resonance
EVA	ethyl vinyl acetate

F

F	flat
FAM	free abrasive machining
FAP	$Ca_5(PO_4)_3F$
FCA	free carrier absorption
fcc	face-centered cubic
FEC	full encapsulation Czochralski

FEM	finite element method	HIV-AIDS	human immunodeficiency virus–acquired immunodeficiency syndrome
FES	fluid experiment system		
FET	field-effect transistor	HK	high potassium content
FFT	fast Fourier transform	HLA	half-loop array
FIB	focused ion beam	HLW	high-level waste
FOM	figure of merit	HMDS	hexamethyldisilane
FPA	focal-plane array	HMT	hexamethylene tetramine
FPE	Fokker–Planck equation	HNP	high nitrogen pressure
FSLI	femtosecond laser irradiation	HOE	holographic optical element
FT	flux technique	HOLZ	higher-order Laue zone
FTIR	Fourier-transform infrared	HOMO	highest occupied molecular orbital
FWHM	full width at half-maximum	HOPG	highly oriented pyrolytic graphite
FZ	floating zone	HOT	high operating temperature
FZT	floating zone technique	HP	Hartman–Perdok
		HPAT	high-pressure ammonothermal technique
		HPHT	high-pressure high-temperature
		HRTEM	high-resolution transmission electron microscopy

G

GAME	gel acupuncture method
GDMS	glow-discharge mass spectrometry
GE	General Electric
GGG	gadolinium gallium garnet
GNB	geometrically necessary boundary
GPIB	general purpose interface bus
GPMD	geometric partial misfit dislocation
GRI	growth interruption
GRIIRA	green-radiation-induced infrared absorption
GS	growth sector
GSAS	general structure analysis software
GSGG	$Gd_3Sc_2Ga_3O_{12}$
GSMBE	gas-source molecular-beam epitaxy
GSO	Gd_2SiO_5
GU	growth unit

HRXRD	high-resolution x-ray diffraction
HSXPD	hemispherically scanned x-ray photoelectron diffraction
HT	hydrothermal
HTS	high-temperature solution
HTSC	high-temperature superconductor
HVPE	halide vapor-phase epitaxy
HVPE	hydride vapor-phase epitaxy
HWC	hot-wall Czochralski
HZM	horizontal ZM

I

IBAD	ion-beam-assisted deposition
IBE	ion beam etching
IC	integrated circuit
IC	ion chamber
ICF	inertial confinement fusion
ID	inner diameter
ID	inversion domain
IDB	incidental dislocation boundary
IDB	inversion domain boundary
IF	identification flat
IG	inert gas
IK	intermediate potassium content
ILHPG	indirect laser-heated pedestal growth
IML-1	International Microgravity Laboratory
IMPATT	impact ionization avalanche transit-time
IP	image plate
IPA	isopropyl alcohol
IR	infrared
IRFPA	infrared focal plane array
IS	interfacial structure
ISS	ion-scattering spectroscopy
ITO	indium-tin oxide
ITTFA	iterative target transform factor analysis
IVPE	iodine vapor-phase epitaxy

H

HA	hydroxyapatite
HAADF	high-angle annular dark field
HAADF-STEM	high-angle annular dark field in scanning transmission electron microscope
HAP	hydroxyapatite
HB	horizontal Bridgman
HBM	Hottinger Baldwin Messtechnik GmbH
HBT	heterostructure bipolar transistor
HBT	horizontal Bridgman technique
HDPCG	high-density protein crystal growth
HE	high energy
HEM	heat-exchanger method
HEMT	high-electron-mobility transistor
HF	hydrofluoric acid
HGF	horizontal gradient freezing
HH	heavy-hole
HH-PCAM	handheld protein crystallization apparatus for microgravity
HIV	human immunodeficiency virus

J

JDS	joint density of states
JFET	junction FET

K

K	kinked
KAP	potassium hydrogen phthalate
KDP	potassium dihydrogen phosphate
KGW	$KY(WO_4)_2$
KGdP	$KGd(PO_3)_4$
KLYF	$KLiYF_5$
KM	Kubota–Mullin
KMC	kinetic Monte Carlo
KN	$KNbO_3$
KNP	$KNd(PO_3)_4$
KPZ	Kardar–Parisi–Zhang
KREW	$KRE(WO_4)_2$
KTA	potassium titanyl arsenate
KTN	potassium niobium tantalate
KTP	potassium titanyl phosphate
KTa	$KTaO_3$
KTaN	$KTa_{1-x}Nb_xO_3$
KYF	KYF_4
KYW	$KY(WO_4)_2$

L

LACBED	large-angle convergent-beam diffraction
LAFB	L-arginine tetrafluoroborate
LAGB	low-angle grain boundary
LAO	$LiAlO_2$
LAP	L-arginine phosphate
LBIC	light-beam induced current
LBIV	light-beam induced voltage
LBO	LiB_3O_5
LBO	$LiBO_3$
LBS	laser-beam scanning
LBSM	laser-beam scanning microscope
LBT	laser-beam tomography
LCD	liquid-crystal display
LD	laser diode
LDT	laser-induced damage threshold
LEC	liquid encapsulation Czochralski
LED	light-emitting diode
LEEBI	low-energy electron-beam irradiation
LEM	laser emission microanalysis
LEO	lateral epitaxial overgrowth
LES	large-eddy simulation
LG	$LiGaO_2$
LGN	$La_3Ga_{5.5}Nb_{0.5}O_{14}$
LGO	$LaGaO_3$
LGS	$La_3Ga_5SiO_{14}$
LGT	$La_3Ga_{5.5}Ta_{0.5}O_{14}$
LH	light hole
LHFB	L-histidine tetrafluoroborate
LHPG	laser-heated pedestal growth
LID	laser-induced damage
LK	low potassium content
LLNL	Lawrence Livermore National Laboratory
LLO	laser lift-off
LLW	low-level waste
LN	$LiNbO_3$
LP	low pressure
LPD	liquid-phase diffusion
LPE	liquid-phase epitaxy
LPEE	liquid-phase electroepitaxy
LPS	$Lu_2Si_2O_7$
LSO	Lu_2SiO_5
LST	laser scattering tomography
LST	local shaping technique
LT	low-temperature
LTa	$LiTaO_3$
LUMO	lowest unoccupied molecular orbital
LVM	local vibrational mode
LWIR	long-wavelength IR
LY	light yield
LiCAF	$LiCaAlF_6$
LiSAF	lithium strontium aluminum fluoride

M

M–S	melt–solid
MAP	magnesium ammonium phosphate
MASTRAPP	multizone adaptive scheme for transport and phase change processes
MBE	molecular-beam epitaxy
MBI	multiple-beam interferometry
MC	multicrystalline
MCD	magnetic circular dichroism
MCT	HgCdTe
MCZ	magnetic Czochralski
MD	misfit dislocation
MD	molecular dynamics
ME	melt epitaxy
ME	microelectronics
MEMS	microelectromechanical system
MESFET	metal-semiconductor field effect transistor
MHP	magnesium hydrogen phosphate-trihydrate
MI	morphological importance
MIT	Massachusetts Institute of Technology
ML	monolayer
MLEC	magnetic liquid-encapsulated Czochralski

MLEK	magnetically stabilized liquid-encapsulated Kyropoulos	NTRS	National Technology Roadmap for Semiconductors
MMIC	monolithic microwave integrated circuit	NdBCO	NdBa$_2$Cu$_3$O$_{7-x}$

O

OCP	octacalcium phosphate
ODE	ordinary differential equation
ODLN	opposite domain LN
ODMR	optically detected magnetic resonance
OEIC	optoelectronic integrated circuit
OF	orientation flat
OFZ	optical floating zone
OLED	organic light-emitting diode
OMVPE	organometallic vapor-phase epitaxy
OPO	optical parametric oscillation
OSF	oxidation-induced stacking fault

MNA	2-methyl-4-nitroaniline
MNSM	modified nonstationary model
MOCVD	metalorganic chemical vapor deposition
MOCVD	molecular chemical vapor deposition
MODFET	modulation-doped field-effect transistor
MOMBE	metalorganic MBE
MOS	metal–oxide–semiconductor
MOSFET	metal–oxide–semiconductor field-effect transistor
MOVPE	metalorganic vapor-phase epitaxy
mp	melting point
MPMS	mold-pushing melt-supplying
MQSSM	modified quasi-steady-state model
MQW	multiple quantum well
MR	melt replenishment
MRAM	magnetoresistive random-access memory
MRM	melt replenishment model
MSUM	monosodium urate monohydrate
MTDATA	metallurgical thermochemistry database
MTS	methyltrichlorosilane
MUX	multiplexor
MWIR	mid-wavelength infrared
MWRM	melt without replenishment model
MXRF	micro-area x-ray fluorescence

P

PAMBE	photo-assisted MBE
PB	proportional band
PBC	periodic bond chain
pBN	pyrolytic boron nitride
PC	photoconductivity
PCAM	protein crystallization apparatus for microgravity
PCF	primary crystallization field
PCF	protein crystal growth facility
PCM	phase-contrast microscopy
PD	Peltier interface demarcation
PD	photodiode
PDE	partial differential equation
PDP	programmed data processor
PDS	periodic domain structure
PE	pendeo-epitaxy
PEBS	pulsed electron beam source
PEC	polyimide environmental cell
PECVD	plasma-enhanced chemical vapor deposition
PED	pulsed electron deposition
PEO	polyethylene oxide
PET	positron emission tomography
PID	proportional–integral–differential
PIN	positive intrinsic negative diode
PL	photoluminescence
PLD	pulsed laser deposition
PMNT	Pb(Mg, Nb)$_{1-x}$Ti$_x$O$_3$
PPKTP	periodically poled KTP
PPLN	periodic poled LN
PPLN	periodic poling lithium niobate
ppy	polypyrrole
PR	photorefractive
PSD	position-sensitive detector
PSF	prismatic stacking fault

N

N	nucleus
N	nutrient
NASA	National Aeronautics and Space Administration
NBE	near-band-edge
NBE	near-bandgap emission
NCPM	noncritically phase matched
NCS	neighboring confinement structure
NGO	NdGaO$_3$
NIF	National Ignition Facility
NIR	near-infrared
NIST	National Institute of Standards and Technology
NLO	nonlinear optic
NMR	nuclear magnetic resonance
NP	no-phonon
NPL	National Physical Laboratory
NREL	National Renewable Energy Laboratory
NS	Navier–Stokes
NSF	National Science Foundation
nSLN	nearly stoichiometric lithium niobate
NSLS	National Synchrotron Light Source
NSM	nonstationary model

PSI	phase-shifting interferometry	RTV	room temperature vulcanizing
PSM	phase-shifting microscopy	R&D	research and development
PSP	pancreatic stone protein		
PSSM	pseudo-steady-state model		
PSZ	partly stabilized zirconium dioxide		
PT	pressure–temperature		
PV	photovoltaic		
PVA	polyvinyl alcohol		
PVD	physical vapor deposition		
PVE	photovoltaic efficiency		
PVT	physical vapor transport		
PWO	$PbWO_4$		
PZNT	$Pb(Zn, Nb)_{1-x}Ti_xO_3$		
PZT	lead zirconium titanate		

S

S	stepped
SAD	selected area diffraction
SAM	scanning Auger microprobe
SAW	surface acoustical wave
SBN	strontium barium niobate
SC	slow cooling
SCBG	slow-cooling bottom growth
SCC	source-current-controlled
SCF	single-crystal fiber
SCF	supercritical fluid technology
SCN	succinonitrile
SCW	supercritical water
SD	screw dislocation
SE	spectroscopic ellipsometry
SECeRTS	small environmental cell for real-time studies
SEG	selective epitaxial growth
SEM	scanning electron microscope
SEM	scanning electron microscopy
SEMATECH	Semiconductor Manufacturing Technology
SF	stacking fault
SFM	scanning force microscopy
SGOI	SiGe-on-insulator
SH	second harmonic
SHG	second-harmonic generation
SHM	submerged heater method
SI	semi-insulating
SIA	Semiconductor Industry Association
SIMS	secondary-ion mass spectrometry
SIOM	Shanghai Institute of Optics and Fine Mechanics
SL	superlattice
SL-3	Spacelab-3
SLI	solid–liquid interface
SLN	stoichiometric LN
SM	skull melting
SMB	stacking mismatch boundary
SMG	surfactant-mediated growth
SMT	surface-mount technology
SNR	signal-to-noise ratio
SNT	sodium nonatitanate
SOI	silicon-on-insulator
SP	sputtering
sPC	scanning photocurrent
SPC	Scientific Production Company
SPC	statistical process control
SR	spreading resistance
SRH	Shockley–Read–Hall
SRL	strain-reducing layer
SRS	stimulated Raman scattering

Q

QD	quantum dot
QDT	quantum dielectric theory
QE	quantum efficiency
QPM	quasi-phase-matched
QPMSHG	quasi-phase-matched second-harmonic generation
QSSM	quasi-steady-state model
QW	quantum well
QWIP	quantum-well infrared photodetector

R

RAE	rotating analyzer ellipsometer
RBM	rotatory Bridgman method
RC	reverse current
RCE	rotating compensator ellipsometer
RE	rare earth
RE	reference electrode
REDG	recombination enhanced dislocation glide
RELF	rare-earth lithium fluoride
RF	radiofrequency
RGS	ribbon growth on substrate
RHEED	reflection high-energy electron diffraction
RI	refractive index
RIE	reactive ion etching
RMS	root-mean-square
RNA	ribonucleic acid
ROIC	readout integrated circuit
RP	reduced pressure
RPI	Rensselaer Polytechnic Institute
RSM	reciprocal space map
RSS	resolved shear stress
RT	room temperature
RTA	$RbTiOAsO_4$
RTA	rapid thermal annealing
RTCVD	rapid-thermal chemical vapor deposition
RTP	$RbTiOPO_4$
RTPL	room-temperature photoluminescence
RTR	ribbon-to-ribbon

List of Abbreviations

SRXRD	spatially resolved XRD		TTV	total thickness variation
SS	solution-stirring		TV	television
SSL	solid-state laser		TVM	three-vessel solution circulating method
SSM	sublimation sandwich method		TVTP	time-varying temperature profile
ST	synchrotron topography		TWF	transmitted wavefront
STC	standard testing condition		TZM	titanium zirconium molybdenum
STE	self-trapped exciton		TZP	tetragonal phase
STEM	scanning transmission electron microscopy			

U

STM	scanning tunneling microscopy
STOS	sodium titanium oxide silicate
STP	stationary temperature profile
STS	space transportation system
SWBXT	synchrotron white beam x-ray topography
SWIR	short-wavelength IR
SXRT	synchrotron x-ray topography

UC	universal compliant
UDLM	uniform-diffusion-layer model
UHPHT	ultrahigh-pressure high-temperature
UHV	ultrahigh-vacuum
ULSI	ultralarge-scale integrated circuit
UV	ultraviolet
UV-vis	ultraviolet–visible
UVB	ultraviolet B

T

TCE	trichloroethylene
TCNQ	tetracyanoquinodimethane
TCO	thin-film conducting oxide
TCP	tricalcium phosphate
TD	Tokyo Denpa
TD	threading dislocation
TDD	threading dislocation density
TDH	temperature-dependent Hall
TDMA	tridiagonal matrix algorithm
TED	threading edge dislocation
TEM	transmission electron microscopy
TFT-LCD	thin-film transistor liquid-crystal display
TGS	triglycine sulfate
TGT	temperature gradient technique
TGW	Thomson–Gibbs–Wulff
TGZM	temperature gradient zone melting
THM	traveling heater method
TMCZ	transverse magnetic-field-applied Czochralski
TMOS	tetramethoxysilane
TO	transverse optic
TPB	three-phase boundary
TPRE	twin-plane reentrant-edge effect
TPS	technique of pulling from shaper
TQM	total quality management
TRAPATT	trapped plasma avalanche-triggered transit
TRM	temperature-reduction method
TS	titanium silicate
TSC	thermally stimulated conductivity
TSD	threading screw dislocation
TSET	two shaping elements technique
TSFZ	traveling solvent floating zone
TSL	thermally stimulated luminescence
TSSG	top-seeded solution growth
TSSM	Tatarchenko steady-state model
TSZ	traveling solvent zone

V

VAS	void-assisted separation
VB	valence band
VB	vertical Bridgman
VBT	vertical Bridgman technique
VCA	virtual-crystal approximation
VCSEL	vertical-cavity surface-emitting laser
VCZ	vapor pressure controlled Czochralski
VDA	vapor diffusion apparatus
VGF	vertical gradient freeze
VLS	vapor–liquid–solid
VLSI	very large-scale integrated circuit
VLWIR	very long-wavelength infrared
VMCZ	vertical magnetic-field-applied Czochralski
VP	vapor phase
VPE	vapor-phase epitaxy
VST	variable shaping technique
VT	Verneuil technique
VTGT	vertical temperature gradient technique
VUV	vacuum ultraviolet

W

WBDF	weak-beam dark-field
WE	working electrode

X

XP	x-ray photoemission
XPS	x-ray photoelectron spectroscopy
XPS	x-ray photoemission spectroscopy
XRD	x-ray diffraction
XRPD	x-ray powder diffraction
XRT	x-ray topography

Y

YAB	YAl$_3$(BO$_3$)$_4$
YAG	yttrium aluminum garnet
YAP	yttrium aluminum perovskite
YBCO	YBa$_2$Cu$_3$O$_{7-x}$
YIG	yttrium iron garnet
YL	yellow luminescence
YLF	LiYF$_4$
YOF	yttrium oxyfluoride
YPS	(Y$_2$)Si$_2$O$_7$
YSO	Y$_2$SiO$_5$

Z

ZA	Al$_2$O$_3$-ZrO$_2$(Y$_2$O$_3$)
ZLP	zero-loss peak
ZM	zone-melting
ZNT	ZN-Technologies
ZOLZ	zero-order Laue zone

Part C Solution Growth of Crystals

17 Bulk Single Crystals Grown from Solution on Earth and in Microgravity
Mohan D. Aggarwal, Normal, USA
Ashok K. Batra, Normal, USA
Ravindra B. Lal, Normal, USA
Benjamin G. Penn, Huntsville, USA
Donald O. Frazier, Huntsville, USA

18 Hydrothermal Growth of Polyscale Crystals
Kullaiah Byrappa, Mysore, India

19 Hydrothermal and Ammonothermal Growth of ZnO and GaN
Michael J. Callahan, Hanson, USA
Qi-Sheng Chen, Beijing, China

20 Stoichiometry and Domain Structure of KTP-Type Nonlinear Optical Crystals
Michael Roth, Jerusalem, Israel

21 High-Temperature Solution Growth: Application to Laser and Nonlinear Optical Crystals
Joan J. Carvajal, Tarragona, Spain
Maria Cinta Pujol, Tarragona, Spain
Francesc Díaz, Tarragona, Spain

22 Growth and Characterization of KDP and Its Analogs
Sheng-Lai Wang, Jinan, Shandong, China
Xun Sun, Jinan, China
Xu-Tang Tao, Jinan, China

17. Bulk Single Crystals Grown from Solution on Earth and in Microgravity

Mohan D. Aggarwal, Ashok K. Batra, Ravindra B. Lal, Benjamin G. Penn, Donald O. Frazier

The growth of crystals has been of interest to physicists and engineers for a long time because of their unique properties. Single crystals are utilized in such diverse applications as pharmaceuticals, computers, infrared detectors, frequency measurements, piezoelectric devices, a variety of high-technology devices, and sensors. Solution crystal growth is one of the important techniques for the growth of a variety of crystals when the material decomposes at the melting point and a suitable solvent is available to make a saturated solution at a desired temperature. In this chapter an attempt is made to provide some fundamentals of growing crystals from solution, including improved designs of various crystallizers.

Since the same solution crystal growth techniques could not be used in microgravity, authors had proposed a new cooled sting technique to grow crystals in space. Authors' experiences of conducting two Space Shuttle experiments relating to solution crystal growth are also detailed in this work. The complexity of these solution growth experiments to grow crystals in space are discussed. These were some of the earliest experiments performed in space, and various lessons learnt are described.

A brief discussion of protein crystal growth, which also shares the basic principles of the solution growth technique, is given along with some flight hardware information for such growth in microgravity.

17.1	**Crystallization: Nucleation and Growth Kinetics**	561
	17.1.1 Expression for Supersaturation	561
	17.1.2 Effects of Convection in Solution Growth	563
	17.1.3 Effect of Impurities	564
17.2	**Low-Temperature Solution Growth**	566
	17.2.1 Solution Growth Methods	566
17.3	**Solution Growth by Temperature Lowering**	567
	17.3.1 Solvent Selection and Solubility	567
	17.3.2 Design of a Crystallizer	569
	17.3.3 Solution Preparation and Starting a Growth Run	573
17.4	**Triglycine Sulfate Crystal Growth: A Case Study**	574
	17.4.1 Growth of Single Crystals of Triglycine Sulfate	574
	17.4.2 Growth Kinetics and Habit Modification	576
17.5	**Solution Growth of Triglycine Sulfate Crystals in Microgravity**	582
	17.5.1 Rationale for Solution Crystal Growth in Space	583
	17.5.2 Solution Crystal Growth Method in Space	583
	17.5.3 Results and Discussion	588
17.6	**Protein Crystal Growth**	592
	17.6.1 Protein Crystal Growth Methods	592
	17.6.2 Protein Crystal Growth Mechanisms	593
	17.6.3 Protein Crystal Growth in Microgravity	593
17.7	**Concluding Remarks**	594
	References	594

The growth of crystals with tailored physical and chemical properties, the characterization of crystals with advanced instrumentation, and their eventual conversion into devices play vital roles in science and technology. Crystal growth is an important field of materials science, involving controlled phase transformation. Growth of crystals from solution at low temperature is one of the important techniques in the fields of pharmaceutical, agriculture, and materials science. Crystal growth acts as a bridge between science and technology for practical applications. In the past few decades, there has been growing interest in the crystal growth process, particularly in view of the increasing demand for materials for technological applications. The strong influence of single crystals in present-day technology is evident from recent advances in the fields of semiconductors, transducers, infrared detectors, ultrasonic amplifiers, ferrites, magnetic garnets, solid-state lasers, nonlinear optics, piezoelectrics, acousto-optics, photosensitive materials, and crystalline thin films for microelectronics and computer applications. All these developments could only be achieved due to the availability of single crystals such as silicon, germanium, and gallium arsenide, and also through the discovery of nonlinear optical properties in some inorganic, semiorganic, and organic crystals. Researchers have always sought new materials for growth of single crystals for new applications and the modification of present crystals for various applications. Any crystal growth process is complex; it depends on many parameters, which can interact. A complete description of a process may well be impossible, since this would require the specification of too many variables. This is why crystal growth is sometimes called an art as well as a science, but like other crafts, it can provide great satisfaction after successful crystal growth of a desired material.

Solid-state materials can be classified into single crystals, polycrystalline, and amorphous materials depending upon the arrangement of their constituent molecules, atoms or ions. An ideal crystal is one in which the surroundings of any atom would be exactly the same as the surroundings of every similar atom in three dimensions. However, real crystals are finite and contain defects. The consistency of the characteristics of devices fabricated from a crystal depends on the homogeneity and defect contents of the crystal. Hence, the process of producing single crystals, which offer homogeneous media at the atomic level with directional properties, attracts more attention than any other process. The methods of growing crystals are mainly dictated by the characteristics of the material and the desired size of the crystal. The method of growing crystals at low and high temperature can be broadly divided into the following six categories:

1. Growth from aqueous solution (low-temperature growth)
2. Growth by gel method (low-temperature growth)
3. Growth from flux or top-seeded solution growth method (high-temperature growth)
4. Hydrothermal growth (high-temperature growth)
5. High-pressure growth (high-temperature growth)
6. Growth by electrodeposition.

Growth of bulk crystals from aqueous solution is technically very important. Besides bulk crystal growth, this method is also used for purification of materials and separation of impurities. Growth of large single crystals from aqueous solution is of interest for essentially two reasons. First, there is a growing need for solution-grown crystals in the area of high-power laser technology, such as potassium dihydrogen phosphate (KDP)-type crystals. Second, research in this area of crystal growth and the corresponding in-depth examination of several key parameters provides fundamental case studies generating theory and technology that are applicable to all solution crystal growth processes, including new aqueous growth systems and high-temperature solution growth.

In this chapter, the fundamental aspects of solution growth and the various methods of bulk crystal growth from solution are described, along with solution crystal growth in the microgravity environment of space. Based on extensive experience of the authors in growing inorganic and organic crystals on Earth and in space, the authors have tried to provide a lucid explanation of the fundamentals of solution crystal growth and crystal growth systems. However, enough details are given on fabrication of crystallizers, associated instruments, and techniques that new researchers may be able to design and set up his/her own solution crystal growth system after review of this chapter. Furthermore, growth and perfection of a technologically important crystal from aqueous solution based on the case study of triglycine sulfate is presented. The effects of various parameters such as the design of the seed holder, seed morphology, the characteristics of the solution such as pH, the growth temperature, dopants, impurities, and microgravity on the physical properties are presented in detail.

17.1 Crystallization: Nucleation and Growth Kinetics

The study and investigation of crystal growth implies the determination of growth laws and growth mechanisms, and the explanation of the final result, i.e., the crystal habit. These aspects are interconnected. Since the growth rate of a face depends on its growth mechanisms and contributes to define the crystal habit, detailed knowledge of these aspects is essential for production of crystals of specific physical or morphological properties. Crystal growth is due to deposition of solute particles on crystal faces, which can grow layer by layer at different rates. The growth rate of a face, i.e., the advancement of its surface in the normal direction per unit time, depends upon both internal and external factors. Internal factors include the surface structure of faces, which in turn are related to the bulk crystal structure, and their degree of perfection. Defects usually occur in the crystals and can emerge at the surface, affecting the growth kinetics. The external factors include the supersaturation, the solute concentration, which is related to solubility, the temperature of the solution, the composition of the solution, mechanical conditions such as the use of stirring, and the presence of impurities or a magnetic or gravitational field. The crystal growth of a face is a succession of complex processes that take place at the interface between the liquid and solid phase. This therefore implies transport of matter and energy across the interface, which is the site of major importance in crystal growth.

In the following section, the fundamentals of nucleation and crystal growth in low-temperature solution are described.

17.1.1 Expression for Supersaturation

The supersaturation of a system can be expressed in a number of ways. A basic unit of concentration as well as temperature must be specified. The concentration driving force (ΔC), the supersaturation ratio (S), and the relative supersaturation (σ) are related to each other as follows.

The concentration driving force is

$$\Delta C = C - C^*, \qquad (17.1)$$

where C is the actual concentration of the solution and C^* is the equilibrium concentration at a given temperature.

The supersaturation ratio is

$$S = \frac{C}{C^*}. \qquad (17.2)$$

The relative supersaturation is

$$\sigma = \frac{C - C^*}{C^*} \quad \text{or} \quad \sigma = S - 1. \qquad (17.3)$$

If the concentration of a solution can be measured at a given temperature and the corresponding equilibrium saturation concentration is known, then the supersaturation can be estimated.

The required supersaturation can be achieved either by cooling/evaporation or by the addition of a precipitant. *Meirs* and *Isaac* reported a detailed investigation on the relationship between supersaturation and spontaneous crystallization [17.1]. The results of their analysis are shown in Fig. 17.1, which shows three zones, termed regions I, II, and III. The lower continuous line indicates the normal solubility of the salt concerned. The temperature and concentration at which spontaneous crystallization occurs are represented by the upper broken curve, generally referred to as the supersolubility curve. This curve is not well defined, as the solubility curve and its position in the diagram depend on the degree of agitation of the solution. The three zones are defined as:

I. The stable (undersaturated) zone, where crystallization is not possible.
II. The metastable zone, where spontaneous crystallization is improbable. However, if a seed crystal is placed in such a metastable solution, growth will occur.

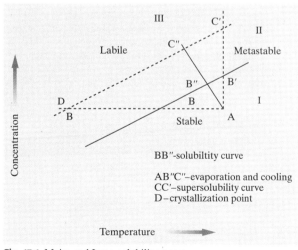

Fig. 17.1 Meirs and Issac solubility curve

BB''–solubiltity curve
$AB''C''$–evaporation and cooling
CC'–supersolubility curve
D–crystallization point

III. The unstable or labile (supersaturation) zone, where spontaneous crystallization is more probable.

The achievement of supersaturation is not sufficient to initiate crystallization. The formation of embryos or nuclei with a number of minute solid particles present in the solution, often termed centers of crystallization, is a prerequisite. Nucleation may occur spontaneously or can be induced artificially. Broadly speaking, nucleation can be classified into primary and secondary. All types of nucleation, whether homogeneous or heterogeneous, in systems that do not contain crystalline matter are described as primary. On the other hand, nucleation generated in the vicinity of crystals present in a supersaturated system is termed secondary nucleation.

The formation of stable nuclei occurs only by the addition of a molecule (A_1) until a critical cluster is formed

$$A_{n-1} + A_1 \rightarrow A_n \quad \text{(critical cluster)}. \tag{17.4}$$

Subsequent additions to the critical cluster result in nucleation followed by growth. The growth units (ions or molecules) in a solution can interact with one another, resulting in a short-lived cluster. Short chains or flat monolayers may be formed initially, and eventually the lattice structure is built up. This process occurs very rapidly and continues in regions of very high supersaturation. Many nuclei fail to achieve maturity and simply dissolve due to their unstable nature. If the nuclei grow beyond a certain critical size, they become stable under the average conditions of supersaturation in the bulk of the solution.

The formation of a solid particle within a homogeneous solution results from the expenditure of a certain quantity of energy. The total quantity of work W required for the formation of a stable nucleus is equal to the sum of the work required to form the surface W_S (a positive quantity) and the work required to form the bulk of the particle W_V (a negative quantity).

$$W = W_S + W_V. \tag{17.5}$$

The change in Gibbs free energy (ΔG) between the crystalline phase and the surrounding mother liquor results in a driving force, which stimulates crystallization. This ΔG is the sum of the surface free energy and the volume free energy

$$\Delta G = \Delta G_S + \Delta G_V. \tag{17.6}$$

For a spherical nucleus

$$\Delta G = 4\pi r^2 \gamma + \frac{4}{3}\pi r^2 \Delta G_V, \tag{17.7}$$

where r is the radius of the nucleus, γ is the interfacial tension, and ΔG_V is the free energy change per unit volume.

For rapid crystallization, $\Delta G < 0$; the first term in the above equation expresses the formation of new surface, and the second term expresses the difference in chemical potential between the crystalline phase (μ) and the surrounding mother liquor (μ_0). At the critical condition, the free energy formation obeys the condition $d\Delta G/dr = 0$. Hence the radius of the critical nucleus is expressed as

$$r^* = \frac{2\gamma}{\Delta G_V}. \tag{17.8}$$

The critical free energy barrier is

$$\Delta G^* = \frac{16\pi\gamma^3 v^2}{3(\Delta\mu)^2}. \tag{17.9}$$

The number of molecules in the critical nucleus is

$$I^* = \frac{4}{3}\pi\gamma(r^*)^3. \tag{17.10}$$

The crucial parameter between a growing crystal and the surrounding mother liquor is the interfacial tension γ. This complex parameter can be determined by conducting nucleation experiments.

Growth of crystals from the vapor, melt or solution occurs only when the medium is supersaturated. The process involves at least two stages [17.2]:

1. Formation of stable three-dimensional (3-D) nuclei
2. Development of the stable 3-D nuclei into crystals with well-developed faces.

The formation of 3-D nuclei is usually discussed in terms of reduction in the Gibbs free energy of the system. At a given supersaturation and temperature, there is a critical value of the free energy at which 3-D nuclei of a critical radius are formed. Only those nuclei which are greater than the critically sized nucleus are capable of growing into crystals of visible size by the attachment of growth species (i.e., molecules, atoms or ions) at energetically favorable growth sites such as kinks (K) in the ledges (L) of a surface. The surfaces of growing crystals may be flat (F), stepped (S) or kinked (K). However, crystals of visible size are usually bounded by slowly growing F-faces which grow by the attachment of growth units at energetically favorable sites. Figure 17.2 shows different positions for the attachment of growth units at a flat crystal–medium interface of a simple cubic lattice. A growth unit attached at a surface terrace (T), smooth ledge (L) or kink (K) site has one, two or three out of the six potential nearest neighbors, respectively. Therefore, a growth unit arriving at a surface terrace, ledge or kink simply loses one, two or three degrees of freedom. If ϕ is the binding energy per pair, the corresponding binding energy of a growth unit attached at these sites is ϕ, 2ϕ, and 3ϕ, respectively. Since the probability of capture of a growth unit at a given site depends on the term $\exp(n\phi/k_\mathrm{B}T)$ (where n is the number of bonds formed, k_B is the Boltzmann constant, and T is the temperature in Kelvin), the growth unit has a much higher probability of becoming a part of the crystal at a kink site than at a ledge or surface terrace. Consequently, in contrast to ledges, the contribution of kinks is overwhelmingly high in the rate v of displacement of a step along the surface and in the rate R of displacement of the surface normal to it. Similarly, the contribution to the face growth rate R by the direct attachment of growth units at the terrace is negligible.

From the above discussion, it may be concluded that the kinetics of crystal growth may, in general, be considered to occur in the following stages:

1. Transport of growth units to the growing surface by bulk diffusion and their capture onto the surface terrace
2. Migration of growth units adsorbed onto the terrace to the step by surface diffusion and their capture at the step
3. Migration of growth units adsorbed onto the step to the kink site and their integration into the kink
4. Transport of the released heat of the reaction and solvent molecules from the solvated atoms/molecules.

One or more of the above stages may control the growth rate but the slowest one is always rate limiting. However, it should be noted that growth kinetics, characterized by the rates v' and R, depends on the crystal structure, the structure of the crystal–medium interface (i.e., rough or smooth), the presence of dislocations emerging on the growing face, the supersaturation of the growth medium, the growth temperature, stirring, and impurities present in the growth medium. It is also these factors that ultimately determine the surface morphology of the crystal.

To explain the crystal growth processes, various theories, models, and role of impurities have been proposed in the past. Some of these are listed below. For details, one can refer to various excellent references, and references therein [17.2–36]. The important growth models are: (i) the two-dimensional nucleation model, (ii) the

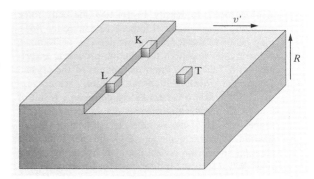

Fig. 17.2 Different positions for the attachment of growth units at a flat crystal–medium interface of a simple cubic lattice

spiral growth model, (iii) the bulk diffusion model, and (iv) growth by a group of cooperating screw dislocations.

17.1.2 Effects of Convection in Solution Growth

Convection is comprised of two mechanisms: energy transfer due to random molecular motion (diffusion) and energy transferred by bulk or macroscopic motion of the fluid. This fluid motion is associated with the fact that, at any instant, large numbers of molecules are moving collectively or as aggregates. Such motion, in the presence of a temperature gradient, contributes to heat transfer. Because the molecules in the aggregate retain their random motion, the total heat transfer is then due to a superposition of energy transport by the random motion of the molecules and by the bulk motion of the fluid. It is customary to use the term *convection* when referring to this cumulative transport and the term *advection* when referring to transport due to bulk fluid motion.

Natural Convection
Convectional heat flow can be classified as natural (or free) convection or forced convection according to the nature of the fluid flow. Natural convection is due to the density difference of a solution near a crystal and far from it. This density difference is due primarily to the change in concentration of a solution during growth or the dissolution of a crystal; and, secondly, due to the absorption or evolution of heat in the fluid. In natural convection, fluid motion is due to buoyancy forces within the fluid. Buoyancy is due to the combined presence of a fluid density gradient and a body force that

is proportional to density. In practice, the body force is usually the gravitational force. Free convection flows may occur in the form of a plume. The well-known convective flow pattern for solution growth is associated with fluid rising from the bottom of the crystal. During the growth of a crystal, solution rises because the solution near a crystal is less dense as a result of a reduction in its concentration, and the temperature is higher because of the evolution of the heat of crystallization. With this depletion of the heavier solute, the solution around the crystal becomes lighter and, thus, rises. When a crystal is dissolved, the direction of motion is opposite (downward). Under these conditions, the diffusion of molecules is supplemented by the more energetic convective transport of matter.

Diffusion is the distribution of a substance by random motion of individual particles. It is due to the presence of a gradient of the chemical potential in the system. A gradient is defined as the increment of a function in an infinitely short distance, along the direction of the most rapid variation of the function. Diffusion always reduces this gradient. Molecular diffusion is observed in viscous media and at low supersaturations, as well as in the growth of crystals, in thin films of liquids, and in capillaries. In molecular diffusion the transport of matter to a crystal is slower than under other diffusion conditions. The thickness of the boundary layer increases with time and the concentration gradient gradually decreases. Therefore, the rate of growth decreases with time. The time interval during the formation of a boundary diffusion layer represents the non-steady-state condition. During this initial period, the rate of growth varies considerably. The thickness of the boundary layer depends on the difference between the densities of different parts of the solution (i.e., on the rate of growth of a crystal), the viscosity of the solution, and the dimensions of the crystal. The presence of the boundary near the crystal and the orientation of the crystal itself affect the nature of the convection currents and the thickness of the boundary layer at different crystal faces.

Forced Convection

Forced convection is produced by the action of external forces such as the forced motion of a crystal in the solution. There is no basic difference between forced and natural convection. When the velocity of motion of a solution with respect to a crystal is increased, the thickness of the boundary layer increases and the supply of matter to a face of the crystal increases. Therefore, by increasing the rate of motion of a solution, we can increase the growth rate of the crystal faces. However, we cannot continue this process indefinitely. A temperature gradient constitutes the driving potential for heat transfer. Similarly, a concentration gradient of a species in a mixture (or solution) provides the driving potential for mass transfer. Both conduction heat transfer and mass diffusion are transport processes that originate from molecular activities. Crystal growers are actually concerned with two aspects of nutrient-to-crystal transport:

1. The mass flux across an interface, which we will call the interfacial flux and which determines the crystal growth rate
2. The concentration profile of growth species in the nutrient adjacent to the crystal, which is an essential parameter in morphological stability discussions.

Let us now introduce the dimensionless numbers that govern forced and free convection. The Grashof number Gr is

$$\mathrm{Gr} = \frac{g\beta \Delta T L^3}{v^2}, \qquad (17.11)$$

where g is gravitational acceleration (m/s²), $\beta = 1/\rho(\delta\rho/\delta T)$ is the thermal expansion coefficient (where ρ is density), ΔT is the temperature difference between the horizontal surfaces that are separated by L, and v is kinematic viscosity (m²/s). The Grashof number Gr plays the same role in free convection that the Reynolds number plays in forced convection. The Reynolds number Re is

$$\mathrm{Re} = \frac{VL}{v} = \frac{\rho V L}{\mu}, \qquad (17.12)$$

where V is velocity (m/s), L is the characteristic length (m), v is kinematic viscosity (m²/s), ρ is density, and μ is viscosity (kg/(s m)). The Reynolds number Re provides a measure of the ratio of the inertial to viscous forces acting on a fluid element. In contrast, the Grashof number Gr, indicates the ratio of the buoyancy force to the viscous force acting on the fluid.

17.1.3 Effect of Impurities

We will now define impurities, which are inherently present, and additives or dopants, which are deliberately added. The former are naturally present in the growth environment and are unwanted; the latter are deliberately added in order to control nucleation, improve crystal quality, increase the size, and change the crystal habit or other physical properties. This topic has received great attention since it is of relevant theoretical and practical interest in the growth of crystals of industrial importance. The ability of impurities to change the growth behavior has been studied by many authors [17.23–36]. It is well known that the influence of impurities on the crystal form and the growth rate is based on the adsorption of ions, atoms or molecules of foreign species at kinks, ledges, and terraces of a growing crystal. The change of the crystal form is based on a difference in adsorption energies on different faces. Impurity molecules will be preferentially adsorbed on surfaces where the free adsorption energy is maximum. It has been possible to predict this preferred surface using computational approaches [17.37]. Recently the mechanisms and models of adsorption of impurities during the growth of bulk crystals have been surveyed by *Sangwal*, including kinetic effects of impurities on the growth of single crystals from solution [17.36].

The solvent itself is an impurity. High temperatures and high supersaturations increase growth rate, but in the presence of a solvent the effect of temperature is stronger, since it promotes water desorption and growth kinetics much more than supersaturation, as found for sucrose [17.38, 39]. Anomalies found by *Chernov* and *Sipyagin* [17.40] at 10 and 40 °C in growth rates disappeared when ethanol, which is known to disrupt the bulk structure of water, was added to the solution. Indeed, water adsorbed on crystal surfaces has properties differing from those of free water. This is attributed to the different structures of the adsorbed layer which undergo phase-like transformations at these temperatures.

Impurity adsorption can be studied indirectly through the adsorption isotherm, i.e., the fraction θ of adsorbed sites which are occupied as the impurity concentrations C_i increase. The simplest model of localized adsorption, i.e., situated at lattice sites, is the Langmuir isotherm

$$K_a C_i = \frac{\theta}{1+\theta}, \quad (17.13)$$

where K_a is the temperature-dependent adsorption constant, which is different for each crystal face. Other models, which take into account the interactions between adsorbed impurities or the occupation probability, have been proposed.

Impurities can act in different ways. When they interact with the solute or solvent, they can have strong influences on the solubility and consequently on the supersaturation and kinetic processes. When impurities are adsorbed on crystals, they can have thermodynamic and kinetic effects. The dominant effect is on the rates of exchange in which the adsorbed molecule or ions and growth units are involved. If the former are exchanged more rapidly than the latter, adsorption mainly affects surface and edge free energy. For a face, a decrease of γ_i (interfacial energy of face i) results in, according to the Gibbs equation,

$$\Delta \gamma_i = k_B T \ln(1-\theta)/S, \quad (17.14)$$

where S is the area of the adsorption site. Similarly, the edge free energy is decreased. These effects should cause an increase in the nucleation and growth rate. If the exchange rate of the adsorbed molecules is slower, impurities can strongly decrease the kinetic coefficients ($R_F = K\sigma^2$ at low supersaturation and $R_F = K'\sigma$ at high supersaturation, where K and K' are kinetic coefficients that depend on the temperature and growth mechanisms, R_F is the growth rate of face F, and σ is the relative supersaturation). So that as a final result the kinetic effects dominate the thermodynamic ones and a decrease in growth rate and impingement flux occur. The interpretation of impurity effects can be done on a structural and kinetic basis:

1. Low concentrations of impurity can form an adsorbed monolayer on the surface even in undersaturated solutions, due to the structural relationship between the two-dimensional (2-D) crystal face and the adsorbed layer (as in the case of NaCl grown in the presence of $CdCl_2$, where a monolayer of $Na_2CdCl_2 \cdot 3H_2O$ is formed). The main influence is on the crystal habit.
2. The kinetic interpretation considers the possibility of adsorption on the different surface sites. If impurities are adsorbed at the kinks, the advancement rate of the edge is hindered even at very low impurity concentrations and the growth rate is strongly decreased and even blocked. Adsorption can also occur on the surface with so strong bonds that impurity molecules cannot move and form a barrier through which the steps have to filter. The spreading of steps beyond this barrier demands supersaturation higher than a critical value for each impurity concentration. In this case impurities are incorporated. Such additives are tailor-made to modify the crystal habit for industrial needs. The molecules of these impurities are similar to those of crystals, but contain some structural differences, so that when they are incorporated into the crystal they disrupt some bonds and change the growth rate of the faces.

17.2 Low-Temperature Solution Growth

Among the various methods of growing single crystals [17.41–43], solution growth at low temperature occupies a prominent place owing to its versatility and simplicity. Growth from solution occurs close to equilibrium conditions and hence crystals of high perfection can be grown.

Solution growth is the most widely used method for the growth of crystals when the starting materials are unstable or decompose at high temperatures. This method demands that the materials must crystallize from solution with prismatic morphology. In general, this method involves seeded growth from a saturated solution. The driving force, i.e., the supersaturation, is achieved either by lowering the temperature or solvent evaporation. This method is widely used to grow bulk crystals that have high solubility and solubility variation with temperature. After many modifications and refinements, the process of solution growth now yields good-quality crystals for a variety of applications. Growth of crystals from solution at room temperature has many advantages over other growth methods, though the rate of crystallization is slow. Since growth is carried out close to room temperature, the density of structural imperfections in solution-grown crystals is relatively low.

17.2.1 Solution Growth Methods

Low-temperature solution growth can be subdivided into the following categories: (i) the slow cooling method, (ii) the slow evaporation method, (iii) the temperature gradient method, and (iv) the chemical/gel method.

Slow Cooling Method

Slow cooling is the best way to grow crystals by solution technique. The main disadvantage of the slow cooling method is the need to use a narrow temperature range. The possible range of temperature is usually narrow and hence much of the solute remains in the solution at the end of the growth run. To compensate for this effect, a large volume of solution is required. A wide range of temperature may not be desirable because the properties of the grown crystal may vary with temperature. Even though this method has the technical difficulty of requiring a programmable temperature control, it is widely used with great success. In this method, growth occurs without any secondary nucleation in the solution, if the supersaturation is fixed within the metastable zone limit. A large cooling rate changes the solubility beyond the metastable zone width and multinucleation occurs at the expense of the seed crystal. A balance between temperature lowering and the growth rate has to be maintained. Growth at low supersaturation prevents strain and dislocation formation at the interface. Supersaturation can be increased after initial growth to achieve a reasonable growth rate.

Slow Evaporation Method

This method is similar to the slow cooling method in terms of the apparatus requirements. The tem-

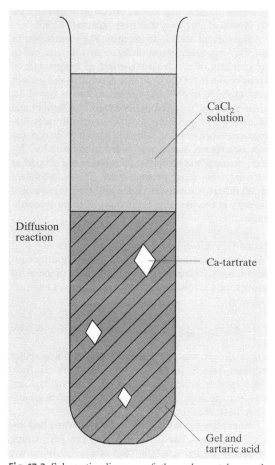

Fig. 17.3 Schematic diagram of the gel crystal growth process

perature is fixed and provision is made for evaporation. With nontoxic solvents such as water, it is permissible to allow evaporation into the atmosphere. Typical growth conditions involve a temperature stabilization of about $\pm 0.05\,°\text{C}$ and rates of evaporation of a few mm^3/h. The evaporation technique has the advantage that the crystals grow at fixed temperature, but inadequacies of the temperature control system still have a major effect on the growth rate. This method can be used effectively for materials with very low temperature coefficient of solubility.

Temperature Gradient Method

This method involves the transport of materials from a hot region containing the solute material to be grown in a cooler region, where the solution is supersaturated and the crystal grows. The main advantages of this method are

1. Crystal growth at fixed temperature
2. Insensitivity to changes in temperature, provided that both the source and growing crystal undergo the same change
3. Economy of solvent and solute.

On the other hand, a small temperature difference between the source and the crystal zones has a large effect on the growth rate.

Chemical/Gel Method

The gel method is exceedingly simple. One procedure is to prepare a gel using commercial waterglass, adjusted to a specific gravity of $1.06\,\text{g}/\text{cm}^3$. Gel is then mixed with 1 M tartaric acid and allowed to gel in a test tube. Once the gel is formed, some other solution can be placed on top (1 M $CaCl_2$ solution), as shown in Fig. 17.3. In the course of time, crystals of calcium tartrate tetrahydrate are formed in the gel. In a nutshell, one solution diffuses through the gel and reacts with the other solution to form crystals of appropriate chemicals [17.44].

17.3 Solution Growth by Temperature Lowering

The growth of crystals from low-temperature solutions occupies a prominent place, especially when materials are not stable at elevated temperatures. A number of concepts for solution crystal growth systems can be found in the literature. One of the best concepts for growth of both inorganic and organic crystals from solution is by temperature lowering of a solution, provided that the material has a positive temperature coefficient of solubility. In this method, a saturated solution of the material to be grown is prepared at a chosen temperature and kept at this temperature for 24 h. Then a seed holding rod is inserted into the growth chamber and its rotation is initiated. The growth process is initiated by lowering the temperature slowly. The temperature of the solution is lowered at a preprogrammed rate, typically $0.05-2.0\,°\text{C/day}$, depending on the solubility of the chosen material. The complete crystallization process may take from one to several weeks. To terminate the growth process the grown crystals are taken out of the solution without thermal shock.

Solution crystal growth is a highly complex process and depends on various growth parameters such as the quality of the seed, the growth temperature, the temperature lowering rate, the character of the solution, the seed rotation speed, and stirring of the solution, besides other conditions. To grow good-quality crystals, these parameters have to be optimized for each crystal.

17.3.1 Solvent Selection and Solubility

A solution is a homogeneous mixture of a solute in a solvent. Generally, the solute is the component present in a smaller quantity. For a given solute, there may be different solvents. Apart from high-purity starting materials, solution growth requires a good solvent. The solvent must be chosen by taking into account the factors:

1. High solubility for the given solute
2. Good solubility gradient
3. Low viscosity
4. Low volatility
5. Low corrosion.

If the solubility is too high, it is difficult to grow bulk single crystals and, if it is too small, solubility restricts the size and growth rate of the crystals. Solubility data at various temperatures is essential to determine the level of supersaturation. Hence, the solubility of the solute in the chosen solvent must be determined before starting the growth process. If the solubility gradient is very

Table 17.1 Solubility parameters δ of water and some organic solvents at 25 °C

Solvent	δ (MPa$^{1/2}$)	Solvent	δ (MPa$^{1/2}$)
Water	47.9	Acetic acid	20.7
Methanol	29.6	1,4-Dioxane	20.5
Ethanol	26.0	Carbondisulfide	20.4
Formamide	39.3	Cyclohexanone	20.3
N-Methyl-formamide	32.9	Acetone	20.2
1,2-Ethanediol	29.9	1,2-Dichloroethane	20.0
Tetrahydrothiophene-1,1-dioxide	27.4	Chlorobenzene	19.4
N,N-Dimethylformamide	24.8	Chloroform	19.0
Dimethylsulfoxide	24.5	Benzene	18.8
Acetonitrile	24.3	Ethylacetate	18.6
1-Butanol	23.3	Tetrahydrofuran	18.6
Cyclohexanol	23.3	Tetrachloromethane	17.6
Pyridine	21.9	Cyclohexane	16.8
tert-Butanol	21.7	n-Hexane	14.9
Aniline	21.1	Perfluoro-n-heptane	11.9

small, slow evaporation of the solvent is the other option for crystal growth to maintain supersaturation in the solution. Growth of crystal from solution is mainly a diffusion-controlled process; the medium must be less viscous to enable faster transference of the growth units from the bulk solution to the growth site by diffusion. Hence a solvent with lower viscosity is preferable. Most important single crystals, such as potassium dihydrogen phosphate (KH$_2$PO$_4$) and L-arginine phosphate monohydrate (LAP), are grown in aqueous solutions or in solvents that are mixtures of water and miscible organic solvents. Of all known substances, water was the first to be considered for use as a solvent because it is nontoxic, most abundant, and low cost. A proper choice of solvent based on a knowledge of its chemical reactivity helps one to avoid undesired reactions between solute and solvent. Except that, in general, the solubility of the growth materials in solvents is required to be sufficiently large, the solubility parameter δ can often be used in estimating the solubility of nonelectrolytes in organic solvents

$$\delta = \left(\frac{\Delta U}{V_m}\right)^{1/2} = \left(\frac{\Delta H - RT}{V_m}\right)^{1/2}, \quad (17.15)$$

where V_m is the molar volume of the solvent, ΔU is the molar energy, and ΔH is the molar enthalpy; δ is a solvent property that measures the work necessary to separate the solvent molecules. Often a mixture of two solvents, one having a δ value higher than that of a solute and the other lower, is a better solvent than either of the two solvents separately [17.45]. A selection of δ-values is given in Table 17.1.

Another property, that is, the dipole moments between the solute and solvent, may also be considered for selecting solvent for crystal growth. Most typical organic solvents have a dipole moment less than about 3 debye. Therefore, in the case of a solute having a similar value of dipole moment, a much wider choice of solvents is possible.

Solubility Determination

Solubility is an important parameter for crystal growth from solution at low temperature. Before any solution growth technique can be applied, determination of congruent or incongruent solubility and the establishment of absence of compound formation with pure or mixed solvents must be achieved. In the latter cases, a special compositional and thermal regime will be necessary to crystallize the desired phase. A simple apparatus for solubility studies is shown in Fig. 17.4. Visual inspection allows the determination of the solubility. Upon cooling, crystallized material is obtained for solid-phase analysis. This apparatus is easily fabricated and is very convenient for measuring solubility. The following is a description of how this has been achieved. The solute and solvent are weighed into a glass ampoule. The ampoule is seated and rotated in a bath controlled by a thermostat, the temperature of which is increased in steps of 0.5 °C every 1–2 h. The final disappearance of the solute yields the saturation temperature. The accuracy of this measurement is within ±0.5 °C.

However, the time needed to reach equilibrium for most covalent organic materials is usually shorter than that of sparingly soluble salts, although the settling times before analyses may be longer. In many soluble salts, such as potassium dihydrogen phosphate, KH_2PO_4 (KDP), triglycine sulfate $(NH_2CH_2COOH)_3H_2SO_4$ (TGS), and ethylene dithiotetrathiafulvalene $(CH_2NH_2)_2C_2H_4O_6$ (EDT), the solubility is strongly temperature dependent. On the other hand, for some soluble salts, such as $LiIO_3$ and $Li_2SO_4 \cdot H_2O$, the solubility is not dependent on temperature and even has a negative slope.

Various techniques for measuring solubility, such as methods based on the vortex flow caused by concentration and optical effects, can be found in the literature. However, accurate measurement of supersaturation is usually difficult. Some new methods such as holographic phase-contrast interferometric microphotography and trace fluorescent probe have been developed. Using these techniques, the concentration distributions and thickness of the boundary layers under different convection conditions can be measured with greater accuracy. Although these methods still need more development and refinement to become more generally applicable, they are promising alternatives for the determination of supersaturation of easily soluble compounds. Of course, if the solubility is known, supersaturation can be calculated by measuring the temperature of the solution and its equilibrium temperature. The problem is that equilibrium temperature measurements are not always easy.

17.3.2 Design of a Crystallizer

When designing a crystallizer for growing crystals from solution by the temperature lowering method, the following conditions should be met [17.43, 46, 47]:

1. A range of operating temperature from room temperature to $80\,°C$, depending on the solvent
2. The choice of hydrodynamic conditions in the solution
3. Measurements of growth parameters such a growth rate
4. Arrangement for taking grown crystals out of the crystallizer without any thermal shock
5. Arrangement for changing the saturation/temperature decrease rate
6. The possibility of interchanging different kinds of seed holders
7. Long-term operating reliability.

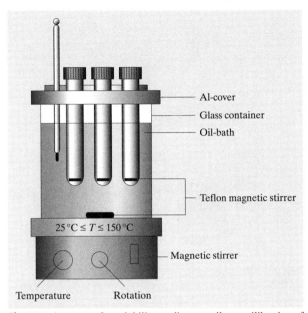

Fig. 17.4 Apparatus for solubility studies as well as equilibration of feed material and growth solution

Since these types of solution crystallizers are not available on the market, one has to design and fabricate one's own system based on one's requirements. A description of a modified crystallizer for growing large crystals from solutions along with the design of a versatile electronic reciprocating control system to change and reciprocate the motor speed containing the seed holding rod for solution growth crystallizer is given below.

In this system, the rotation rate and number of revolutions in the clockwise and counterclockwise direction can be adjusted as desired. This electronic system alleviates the problem of jerky motion of the seed holder [17.48] during reciprocation as in earlier electromechanical systems designed by the authors. Good-quality crystals of important nonlinear optical materials such as 2-methyl-4-nitroaniline:methyl-(2,4-dinitrophenyl)-aminopropanoate (MNA:MAP), L-arginine phosphate (LAP), L-histidine tetrafluoroborate (LHFB), L-arginine tetrafluoroborate (LAFB), and others such as triglycine sulfate and potassium dihydrogen sulfate have been successfully grown in authors' laboratory using this system [17.47, 49–54]. The complete crystallization apparatus along with the electronic circuit can be easily fabricated in the laboratory with readily available components.

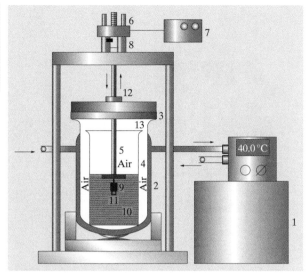

Fig. 17.5 Schematic diagram of a new type of crystallizer for growing organic crystals. (1) Circulating bath, (2) jacked reaction kettle, (3) RTV/Teflon seal, (4) crystallizer jar, (5) Teflon seed holder, (6) reversible motor, (7) circuit for reciprocating and controlling the stirring rate of seed holder, (8) arrangement for pulling the crystal during growth, (9) Teflon tape cover, (10) solution, (11) seed crystal, (12) Teflon seal, and (13) glass lid

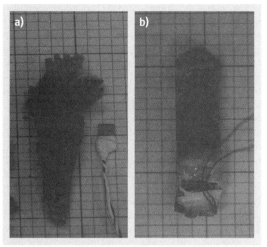

Fig. 17.6a,b MNA:MAP seed: (**a**) with aloe-vera-tree-type growth and (**b**) without aloe-vera-tree-type growth

A Typical Solution Crystal Growth Crystallizer

A schematic diagram of a modified solution crystal growth system that the authors designed and fabricated after designing a number of crystallizers [17.48, 55–57] in our laboratory is shown in Fig. 17.5. It consists of a 250 ml crystallizer jar (4), which holds the growth solution that is placed inside a 2.5 l glass-jacketed kettle (2). The linear and reciprocating motion of the Teflon seed holder (5) is controlled by a rack–pinion arrangement (8) and electronic circuit (7), respectively. A reversible motor (6) is used for rotating the seed holder. The temperature of the growth solution is controlled and programmed by circulating water using a NesLab bath (1). To prevent evaporation of the solvent, a specially designed oil–Teflon seal (3) and/or room temperature vulcanizing silicone RTV/Teflon seal (3) are used. The main advantages of our crystal growth system are: (i) better temperature stability even with sudden fluctuations in room temperature, (ii) better control over evaporation of organic solvents, (iii) a mechanical screw-type arrangement for pulling the seed crystal at a controlled rate, (iv) the possibility of varying the seed orientation and type, and (v) a versatile electronic reciprocating control system to change and reciprocate the motor speed containing the seed holding rod. Better temperature stability was accomplished by loading the growth solution into a beaker kept inside the jacketed vessel.

An air gap provides extra insulation. Moreover, spontaneous nucleation at the bottom of the growth vessel, which hampers growth and impacts on the crystal yield, is completely eliminated. By providing an extra lid on the inside beaker and a Teflon seal over the jacketed vessel, the evaporation of the solvent was dramatically reduced. The inner beaker is filled halfway with solution rather than three-quarters as is usually done, and the growing crystal is pulled in a controlled fashion. Since filling of the inner beaker to three-quarters is not required, not only is the crystal annealed in situ but also spurious aloe-vera-tree-like growth near the seed in some crystals such as MNA:MAP is greatly reduced or completely eliminated.

Figure 17.6a shows the seed crystal along with a MNA:MAP crystal grown using the usual technique, i.e., without pulling the growing crystal. Figure 17.6b shows the same crystal grown with pulling where aloe-vera-tree-type growth is avoided. Furthermore, large crystals can be grown from a smaller amount of expensive mother liquor when the crystal is pulled while growing.

Another modified solution growth crystallizer was also designed in our laboratory, whose three-dimensional cutout view with reciprocating seed arrangement and other components is illustrated in Fig. 17.7. It uses a magnetic stirrer to keep the temperature of the water bath uniform at a particular temperature. A layer of silicon oil on the surface of water was found to reduce the evaporation of water to a minimum, which is a big improvement over earlier designs.

Besides temperature control, the uniform rotation of seeds is required so that stagnant regions or recirculating flows are not produced, otherwise inclusions in the crystals will be formed. To study and achieve uniform and optimum transport of solute to the growing crystals, various seed rotation mechanisms have been used in the past. Unidirectional rotation of the seed leads to the formation of cavities in central regions of a crystal face because of lesser solute transport to this region than to edges and corner of the growing crystal. Furthermore, nonuniform solute supply favors the formation of thick layers which subsequently lead to the trapping of inclusions and the generation of dislocations. Periodic rotation of the growing crystal in opposite directions suppresses edge formation but does not eliminate the formation of the central cavity. To avoid these defects and stagnant regions in the solution, eccentric or clockwise and counterclockwise motion of the seed holder is used when growing crystals from solutions. Several mechanisms [17.48, 55, 58] have been used in the past to generate reciprocating motion of the seed holder, such as electromechanical [17.55] and rack–pinion methods [17.48]. In the electromechanical system, a connection of the motor polarity is reversed mechanically by using a microswitch. In this mechanical system, there is a jerky motion on reversal, which sometimes causes seeds to fall down. The jerky motion also creates a turbulent flow in the fluid and hence nonuniform transfer of solute to the growing faces, which may lead to the formation of defective crystals. Also, the microswitch has to be changed frequently due to mechanical failure. Furthermore, the effect of seed rotation rates on the growth rate and the quality of the crystals cannot be systematically studied because the rotation rate cannot be varied. In the rack–pinion arrangement, there is no jerking motion but one has to change gears to change rotation and reversal rate, which is quite an involved process.

To address these drawbacks, in our Crystal Growth Laboratory at Alabama A&M University, a versatile solid-state electronic circuit for reciprocating the direction of the seed holder was designed along with added features in such a way as to vary the rotation rate and stopping time on reversal, and control the timing for

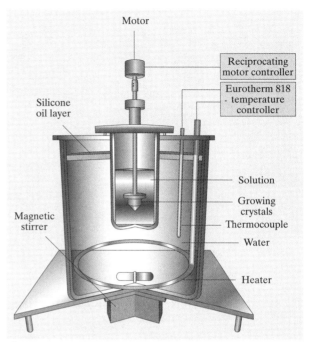

Fig. 17.7 A modified crystallizer with arrangement to stop water evaporation

clockwise and counterclockwise motion of the crystal/seed holder [17.56]. These design features will allow crystal growers to study more clearly the effect of seed rotation rate on the growth and quality of the grown crystals, thereby optimizing this important parameter for growing better-quality crystals.

A schematic diagram of the basic electronic circuit for reciprocating motion control is shown in Fig. 17.8. In Fig. 17.8, the timer (chip LM 555, U3) produces a square-wave timing pulse. It may be set to a particular frequency (POT1) and duty cycle (POT2) in combination with a timing capacitor (C3) and reset if necessary by a switch (S2). The timing waveform is divided by the J–K flip-flop chip 74LS112 (U1) to one-half of the timer frequency. Parasitic oscillations are suppressed by three capacitors (C1, C2, and C4). The two waveforms are combined by a NAND gates chip SN7400 (U2) to turn on two transistors (Q1 and Q2) alternately to control the solid-state relays (1 and 2) which connect alternate sides of the motor capacitor to the 110 V_{AC} return line. Similarly, the transistors (Q3 or Q4) alternately turn on the indicator lamps (LED1 and LED2). Current limiting and bias is provided by resistors (R1 through R8). Motor rotation speed is controlled by a potentiome-

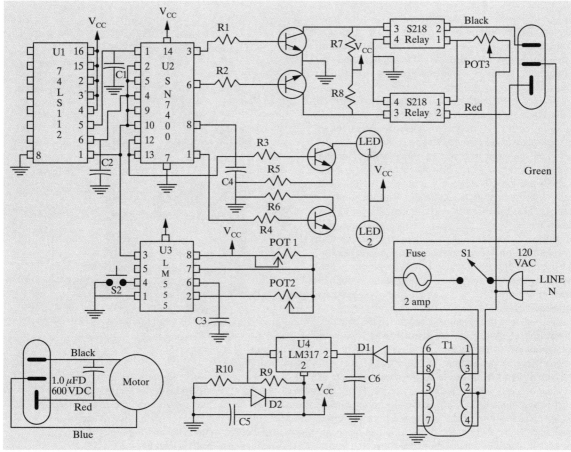

Fig. 17.8 Diagram of electronic circuit for control of the reciprocating motion of the seed holder for solution crystallizer

ter (POT3). A power supply consisting of a step-down transformer (T1), a voltage regulator (U4), and an associated filtering circuit (D1, D2, C5, and C5), and voltage setting divider (R9 and R10) provides 5 V_{DC} to the circuit. The operation of the circuit causes the following sequence of states in the system: during first interval, the seed holder motor runs counterclockwise; during the second interval, the motor comes to a stop; during the third interval, the motor runs clockwise; and during the fourth interval, the motor again comes to stop. Then the entire cycle of operation in repeated, and the intervals can be varied as needed for a particular crystal growth experiment.

Crystal Seed Holder

In order to ensure the best growth conditions, it is necessary to use a special crystal holder because the success of an experiment may depend upon its suitability. The selection of the crystal holder and the method for attach-

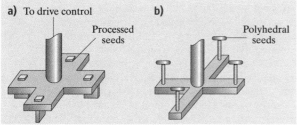

Fig. 17.9a,b Plexiglas seed holders for solution-growth crystallizers: (**a**) with processed seeds and (**b**) with polyhedral seeds

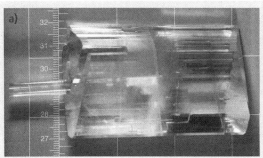
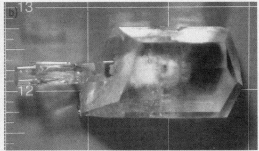

Fig. 17.10a,b Photographs of crystals grown at Alabama A&M University: (**a**) L-histidine tetrafluoroborate and (**b**) L-pyroglutamic acid crystals

ing a seed to it are no less important than the selection of the growth method. A crystal holder should ensure that a seed is held securely in a desired orientation and that the seed and therefore the growing crystal can be moved in any required manner. Also, the crystal holder should not become deformed at the selected speed and direction of the motion or by the weight of the final crystal grown on it. The crystal holder material should be chemically inert in the solution of the substance being crystallized.

A schematic diagram of two plexiglas seed holders that were specially designed, fabricated, and successfully used by the authors for aqueous solution crystal growth are shown in Fig. 17.9.

Preparation of the Seed Crystal and Mounting
A seed is a small fragment of a crystal or a whole crystal which is used to start the growth of a larger crystal in a solution. This seed must meet the following requirements:

1. It should be a single crystal free of cracks or boundaries.
2. It should be free of inclusions.
3. Its surface should be free of any sharp cleaved edges.
4. It must be grown under the same conditions as those to be used in growing the desired single crystals.

Following the above requirements in preparing the seed crystals will result in the growth of high-quality crystals, if other criteria such as solution preparation are performed carefully as well. Prior to crystal growth, seed crystals are mounted on plexiglas rods using 100% silicon rubber Dow Corning Silastic 732 RTV adhesive.

17.3.3 Solution Preparation and Starting a Growth Run

For solution preparation it is essential to have the solubility data of the growth material at different temperatures. Sintered glass filters of different pore sizes are used for solution filtration. The clear solution, saturated at the desired temperature, is poured into the growth vessel. For growth by slow cooling, the vessel is sealed to prevent solvent evaporation. Before starting the crystal growth process, a small crystal suspended in the solution is used to test the saturation. By varying the temperature, a situation is obtained where neither growth nor dissolution occurs. The test seed is then replaced with a good-quality seed. All unwanted nuclei and the surface damage on the seed are removed by dissolving at a temperature slightly above the saturation point. Growth is initiated after lowering the temperature to the equilibrium saturation. Controlled solvent evaporation can also be used in initiating the growth. The quality of the grown crystal depends on (a) the nature of the seed, (b) the cooling rate employed, and (c) the agitation of the solution.

Various new nonlinear optical crystals which hold promise for use in optical processing devices such as L-arginine phosphate, L-histidine tetrafluoroborate, L-arginine tetrafluoroborate, 2-methyl-4-nitroaniline:methyl-(2,4-dinitrophenyl)-aminopropanoate (MNA:MAP) and L-pyroglutamic acid have been successfully grown using the above-mentioned reciprocating system in combination with the temperature lowering technique, as described by the authors [17.47, 49–51]. Some of these crystals are shown in Fig. 17.10. In the investigators' observation and experience, there is significant improvement in the quality of grown crystals and success rates of the growth runs, as evident from the transparency and lower scattering observed using laser illumination. This electronic system for reciprocal motion control of solution growth crystallizers has been in use in our laboratory for several years, and continues to work satisfactorily.

It is worth mentioning that this simple and versatile crystallization apparatus can be fabricated in any college, university or scientific laboratory from readily available components. Besides its use in physics or chemistry laboratory experiments, it can also be used for doing extensive research on the effect of important parameters such as seed rotation rate, stopping time of reversal, and number of rotations in the clockwise or counterclockwise direction on the quality and growth rate of technologically important crystals.

17.4 Triglycine Sulfate Crystal Growth: A Case Study

Triglycine sulfate (TGS) is one of the most important ferroelectric materials. The ferroelectric nature of triglycine sulfate, $(NH_2CH_2COOH)_3 \cdot H_2SO_4$, usually abbreviated as TGS, was discovered by Matthias et al. and discussed by *Jona* and *Shirane* [17.59]. The crystal structure of TGS was reported by Hoshino, Okaya, and Pepinsky and discussed in the above reference. In the ferroelectric phase below the Curie temperature ($T_C \approx 49\,°C$), the symmetry is monoclinic with space group $P2_1$. Above the Curie temperature, the structure gains an additional set of mirror planes in the space group $P2_{1/m}$. It has been reported that the lattice parameters are $a = 9.42\,Å$, $b = 12.64\,Å$, $c = 5.73\,Å$, and $\beta = 110°23'$ and that the structure contains three independent glycine molecules. One of the structures, designated as glycine II, has a zwitterion configuration $(NH_3)^+CH_2OO^-$, and the other two, $(NH_3)^+CH_2COOH$. TGS may be called glycine-diglycinium sulfate with chemical formula $((NH_3)^+CH_2COO^-) \cdot ((NH_3^+)CH_2COOH)_2 \cdot SO_4^{2-}$. The projection of the structure along the c-direction is illustrated in Fig. 17.11. Glycine I deviates only slightly from the plane m' at $y = \frac{1}{4}$ on which the $[SO_4]^{2-}$ tetrahedra also lie, whereas glycine II and III are approximately related by inversion through $(\frac{1}{2}, \frac{1}{2}, \frac{1}{2})$ [17.60].

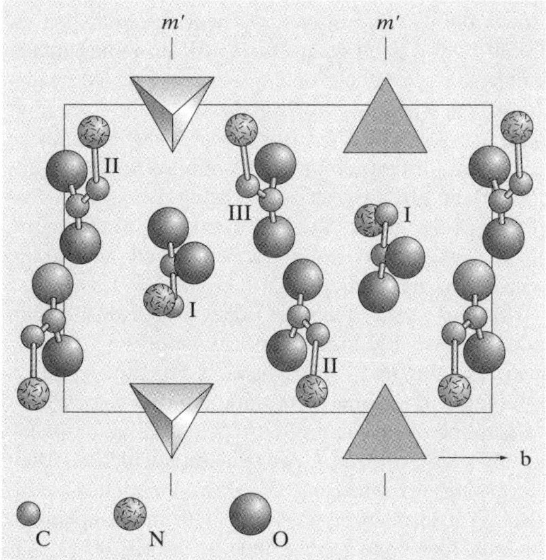

Fig. 17.11 Projection of the structure of TGS crystal along the c-direction: m' represents the set of pseudomirror planes in which glycine I molecules are inverted on ferroelectric switching

17.4.1 Growth of Single Crystals of Triglycine Sulfate

Single crystals of TGS have usually been grown from aqueous solution by the temperature lowering or solvent evaporation method. The authors have successfully grown TGS crystals using the crystallizer whose schematic diagram is shown in Fig. 17.12 [17.57]. The outside water bath, with a capacity of about 12 l, and the inside smaller cubical growth cell with 1 l capacity were made out of Plexiglas. Temperature control of the crystallizer was achieved using 250 W immersion heaters controlled by YSI 72 proportional temperature controllers to an accuracy of $\pm 0.1\,°C$. Uniformity of the temperature throughout the bath was achieved with the help of a fluid circulation pump. The bath temperature is monitored at two points during the crystal growth using NIST (National Institute of Standards and Technology) calibrated thermometers. The crystals were grown by slow cooling of the solution at any desired rate.

TGS crystals are doped with L-alanine to enhance its performance and check depoling for their use in infrared sensor element. A rotating disc technique [17.61] has been applied to grow uniformly L-alanine-doped TGS crystals using a large-area seed crystal having large

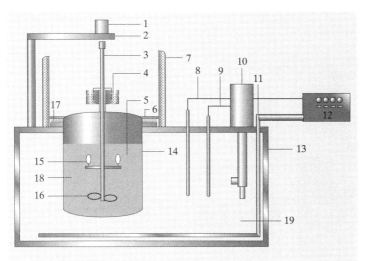

Fig. 17.12 Schematic diagram of reciprocating motion crystallizer

1. Reciprocating motor
2. Motor mounting stand
3. Seed mounting stand
4. Seal to limit evaporation
5. Glass thermometer
6. Air tight glass cover
7. Post growth anealing heater
8. Control thermister
9. Recording thermister
10. Fluid circulation pump
11. Water heater
12. Temperature controller/programmer
13. Plexiglass water bath
14. Pyrex crystal growig cell
15. Seed/crystal
16. Stirrer blades
17. Teflon gasket
18. Solution
19. Water

(010) face. A conventional crystallizer was modified to allow growth under suitable hydrodynamic conditions in order to stabilize growth on the (010) face. Such a crystallizer is shown in Fig. 17.13. In this crystallizer, a seed crystal in the form of a disc was held in a circular holder with the (010) face exposed to the solution. The disc was attached to the end of a spindle that was rotated at 340 rpm. This creates a uniform boundary layer of solution over the crystal's exposed face. The container with 30 l capacity was heated by a hot plate spaced from its bottom surface and regulated to hold the temperature within $\pm 0.01\,°C$. The solution rises from the bottom of the vessel but hotter liquid is prevented from reaching the crystal directly by a plexiglas baffle. A growth rate of 1 mm/day was maintained by lowering the temperature uniformity at $0.05\,°C$/day. The resulting crystals were found to be visibly of good quality, without defects propagating from the seed. In addition to uniform doping and growth of high-quality crystal, the method has several other useful features such as short growth time, with decreased cost and reuse of seeds, and that growth occurs within a narrow temperature range. *Brezina* et al. [17.62] designed a crystallizer for growing L-alanine-doped deuterated triglycine sulfate (DTGS) crystals by isothermal evaporation of D_2O. *Satapathy* et al. [17.63] have described a novel technique for mounting the TGS seeds and a crystallizer. *Banan* [17.64] has also described a crystallizer and a seed holder for growing pure and doped TGS crystals.

TGS crystals weighing more than 100 g have been grown from solution with ethyl alcohol additions [17.65]. When alcohol is mixed in an aqueous solution of TGS, part of the water in the solution associates with alcohol, which concentrates the solution. Thus, the supersaturation can be controlled to a certain degree, making it easier to grow TGS crystals.

To achieve success in growing crystals from aqueous solutions, it is important to prepare a solution with a well-determined saturation temperature, solubility profile, and absence of any foreign particles. For our investigation, TGS solution was prepared using high-purity crystalline triglycine sulfate from BDH, UK. The solubility of TGS at various temperature were determined and compared with information available from various sources. TGS solution was prepared at 40 °C saturation temperature. To prepare saturated solution,

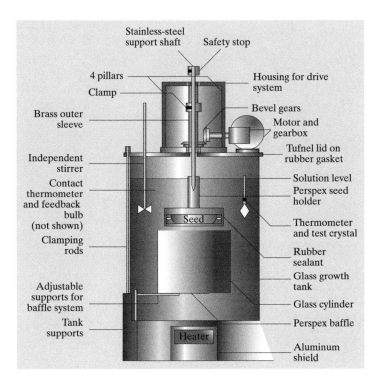

Fig. 17.13 Apparatus for spinning disc growth

464 g TGS was weighed and dissolved in 1000 cm^3 distilled water. The mixture was heated to 50 °C and mixed thoroughly using a Teflon-coated magnetic stirrer. The solution was then filtered through a 5 μm filter funnel using a vacuum unit. After filtration, this solution was transferred to the growth chamber. To start the growth run, the bath temperature was kept at 45 °C. The solution was poured into the growth cell. Then the temperature was reduced to 41.5 °C, 1.5 °C above the saturation temperature and allowed to stabilize over night. The saturation temperature was again checked by the technique of crystal insertion into the solution as well as refractive index measurements. For each saturation point the refractive index was measured at different temperatures beforehand using an Abbe refractometer. The starting growth temperature was adjusted based on the result of this procedure. After this the seed crystal holder was placed in an oven and heated to 45 °C. Prior to transfer to the growth cell, all precautions were taken to keep the seeds as well as the holder surface free of dust and foreign particles. The preheated seed crystals holder were then inserted into the growth cell and holder attached to the reciprocating apparatus. The seed crystals were slightly dissolved and the growth run started.

The bath temperature was reduced by 0.1 °C/day initially and at the final stage of growth by 0.2 °C/day. The removal of the grown crystals from the mother liquor requires some care. Mishandling may induce defects, thus destroying the scientific value of the crystal or even fracturing it altogether. To avoid cracking the crystals due to thermal shock, the crystals were wrapped in a lint-free paper towel maintained at final growth temperature. The crystals were then transferred to an oven kept at an appropriate temperature. The temperature of the oven was slowly lowered to room temperature. Grown crystals can be easily removed from the seed holder by slight finger force, as RTV 732 adhesive was used for mounting the seed crystal.

17.4.2 Growth Kinetics and Habit Modification

Triglycine sulfate crystal normally grows with the habit shown schematically in Fig. 17.14a. A photograph of the TGS crystal grown at authors' laboratory is shown in Fig. 17.14b. It is observed that the $V_{(010)}$ growth rate is much faster than $V_{(001)}$. So the (010) face, as seen in Fig. 17.14, is very small or not present. Both

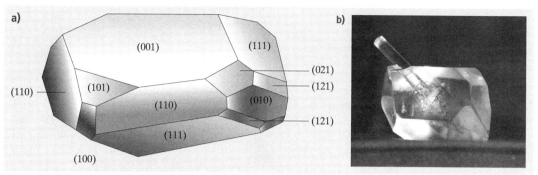

Fig. 17.14a,b Normal growth habit of TGS crystal. (**a**) Normal growth habit of TGS crystal, (**b**) photograph of as grown crystal of TGS at Alabama A&M University

growth kinetics and habit modifications of TGS have been extensively studied over the past few decades. The work published so far has resulted in a description sufficient for reliable growth of this crystal as described above. A number of studies of the growth kinetics of TGS grown from aqueous solution have been reported in the literature [17.66–71]. *Novotny* et al. [17.69] studied the growth of the (110) face of TGS crystals grown isothermally above the phase transition, at higher supersaturation ($\sigma > 10^{-3}$), and under constant hydrodynamically controlled conditions. The researchers observed that the ratio of growth rates along the individual axes is $V_a : V_b : V_c = 0.67 : 1 : 0.25$. On the basis of the measured dependence of the linear growth rate on the supersaturation (σ), it was found that the growth of the (110) face is probably controlled by volume diffusion of TGS molecules towards the surface of the growing crystal. Increasing the supersaturation caused a reduction of the number of faces in the prismatic zone of the crystal and an increase of the dislocation density in the (110) faces. Measurements of the growth rates [17.69] of (110) and (001) faces as a function of supersaturation of the solution were also analyzed on the basis of the surface diffusion model of *Burton, Cabrera,* and *Frank* (BCF) [17.67]. It was shown that surface diffusion is responsible for the low growth rates of (001) faces; in the case of (110) faces, the mechanism is less important at higher values of supersaturation than volume diffusion. *Rashkovich* [17.68] investigated the growth of (001) faces below the transition temperature. The results were qualitatively consistent with the dislocation model of crystal growth, but the growth at low supersaturation did not agree with the BCF model [17.67]. *Reiss* et al. [17.71] studied the growth of crystals at $33.55\,°\mathrm{C}$ at relative supersaturations of 0.004 and 0.045. In their study, the BCF law is fitted to the growth rate data. They also found that qualitative aspects of the growth are consistent with the BCF model.

The role of pH, impurities, degree of supersaturation, growth temperature and technical parameters, including seed preparation and attachment etc. on growth kinetics has also been quantitatively investigated by various investigators [17.72–90]. The results are described below.

Effect of Seed Crystal

It has been observed that morphology does not change much for seed crystals obtained at different temperatures [17.80]. However, at higher temperatures (35–45 °C) seeds tend to be elongated in the (001) direction, while seeds grown at lower temperatures are nearly isometric. Morphological study of the crystals grown using the above-cited seeds showed dependency of the crystal habit on the characteristics of the seed. The grown crystals tended to be elongated when elongated seeds were used. Crystals with large (010) faces grew when cleaved platelets were used for seeding. Crystals with high transparency and lower dislocation densities were obtained when the crystal growth tem-

Table 17.2 Crystal growth data for TGS crystals grown on poled and unpoled seeds [17.72]

TGS crystals	Crystal yield weight (g/(day °C))	Growth velocity $V_{(010)}$ (mm/(day °C))
(010) poled seed	0.618	1.05
(010) unpoled seed	0.621	1.16
($0\bar{1}0$) poled seed	0.624	1.20
($0\bar{1}0$) unpoled seed	0.637	1.25

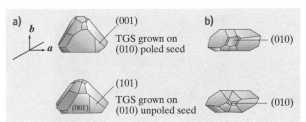

Fig. 17.15a,b Growth habits of TGS crystals grown on (**a**) poled and (**b**) unpoled seeds

Table 17.3 Growth rates of various faces of TGS versus solution pH [17.76]

pH	$V_{(001)}$ (10^{-3} mm/h)	$V_{(010)}$ (10^{-3} mm/h)	$V_{(100)}$ (10^{-3} mm/h)
2.70	71.6	291.5	260.9
2.14	49.6	118.2	117.6
1.25	207.0	262.0	109.0
1.00	144.6	156.2	43.9
0.30	109.9	120.9	40.5

perature was kept the same as that used to grow the seed. Crystal growth was seriously impaired when cleaved platelets were used as seeds, because of unwanted nucleation that started growing during the growth process. *Banan* et al. [17.82] studied the effect of using poled seed on the morphology and growth rate of TGS crystals. Table 17.2 summarizes the normalized growth data for two crystal growth runs using poled and unpoled seeds, and Fig. 17.15 gives the morphology of resulting TGS crystals. A number of interesting effects on the growth rate and morphology of these crystals were observed. Generally, the growth rate along the ($-b$-axis) (010) was faster than along the (0$\bar{1}$0) ($+b$-axis). The well-developed (010)/(0$\bar{1}$0) faces, which are generally not present or less developed in pure TGS crystals, were prominent and large in crystals grown on poled seed. In this way, the identification of the ferroelectric axis in the TGS crystal becomes easier, and cleaving normal to the ferroelectric axis for preparation of pyroelectric infrared (IR) element can be economically accomplished. It can be inferred from Table 17.2 that growth velocity along the [010] axis of TGS crystal is affected by using a poled seed crystal. The decrease in growth rate along the [010] direction in the case of poled seed helps in the emergence of larger (010) face.

Effect of Growth Temperature and Supersaturation

The effect of crystal morphology and quality on growth temperature using the seed also grown at the same growth temperature, and from the same solution, has also been studied [17.80]. The change in morphology was not substantial, but the rate of growth in different directions changed with temperature, and relative change in the size of the faces was observed. Extra nuclei hindered growth at higher temperature (40 °C), and the crystals were of poor quality with low transparency. The change in habit of TGS crystals [17.76] as a function of temperature and supersaturation is shown in Fig. 17.16.

Effect of Solution pH

The influence of solution pH on the growth, morphology, and quality of TGS crystals has been studied by a number of workers. It was observed that crystal quality is not greatly affected by pH variation [17.80]. The influence of growth solution pH on growth rates of various faces: {(001), (010), (100)} and habit of TGS was studied by *Tsedrik* et al. [17.76].

At pH < 1, diglycine sulfate (DGS) was formed. Table 17.3 gives average values of the growth rate V of (001), (010), and (100) faces of TGS crystals versus pH of the solution as well as DGS grown at pH 0.3. Values of $V_{(100)}$ decreased monotonously with lowering pH, and $V_{(001)}$ and $V_{(100)}$ had a local minima near the pH value corresponding to the stoichiometric (pH = 2.14) value and a local maxima around pH = 1.25. The crystal habit is defined by the growth rates of the faces. Figure 17.17 shows the dependence of crystal habit on pH [17.76]. The most isometric crystals were obtained at pH = 1.55, when $V_{(100/001)} \approx 1$ (Fig. 17.17c). Almost all crystals at low pH had gaps on the (111) and (11$\bar{1}$) faces (Fig. 17.17d,e). The above observed changes in morphology of TGS single crystals with the pH of the solution were apparently affected by different capture of incidental impurities, which are always present in the solutions. Chemical (structural) impurities captured by the crystal faces reduced the growth rates of the corresponding faces, and mechanical impurities (defects) increased these rates. At low pH values,

Fig. 17.16a–c Change of growth habits of TGS with growth temperature and supersaturation: (**a**) 32 °C, 0.7×10^{-3} (**b**) 32 °C, 3.0×10^{-3}, and (**c**) 52 °C, 3.0×10^{-3}

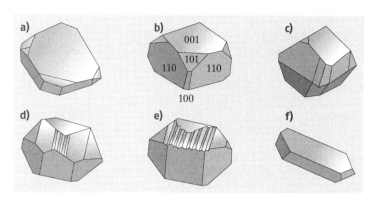

Fig. 17.17a–f Change of habit of TGS crystals with solution pH: (**a**) 2.75, (**b**) 2.1, (**c**) 1.55, (**d**) 1.23, (**e**) 1.0, and (**f**) 0.3 (DTGS)

chemical impurities played the predominant role. Their entry into the growing crystal was increased with reducing pH. Table 17.3 clearly shows that the growth rate of all faces decreased with reducing pH, starting with pH = 1.25. The gaps on the (111) and (11$\bar{1}$) faces were connected with strong hindering of the growth layers by the adsorbed impurities (Fig. 17.17d,e). At high pH (> 2) another kind of impurity (mechanical defects) has a predominant influence on crystal morphology. Their entry increased with rising pH, so the growth rates of all faces increased (Table 17.3). At pH = 1.55 the action of impurities of both kinds was comparable, and mostly isometric crystals were formed (Fig. 17.17c). Recently, it has been shown [17.91] that the growth rate of (010) face of TGS and L-alanine doped triglycine sulfo-phosphate (ATGSP) crystals varies with the pH of the solution.

With the same supersaturation, the growth rate of TGS crystals was slowest in the neutral solution (pH = 2.25). It grew faster in both acidic (pH = 1.73–2.25) and alkaline solution (pH = 2.25–2.52). In alkaline solution, the growth rate of TGS varied faster with changing pH value. However, if the pH was too high, then the (010) face capped quickly. The variation of growth rate of L-alanine-doped triglycine sulfo-phosphate (ATGSP) with pH was not similar to that of TGS. The growth rate of ATGSP crystals in a neutral solution (pH = 2.5) was the fastest, and it was slower in both acidic (pH = 2.20–2.50) and alkaline (pH = 2.5–2.85) solutions. The above results demonstrate that, on the basis of the pH of a solution, one can grow crystals at higher growth rates.

Effect of Impurities on TGS Crystal Growth

The presence of impurities in the process of crystal growth results in modification of the crystal shape and growth rates. Various workers have studied the effects of inorganic and organic impurities on the kinetics of growth of doping TGS crystals. It was observed that Ni-doped crystals were very similar in habit to pure TGS crystals, while in the case of Cu- and Fe-doped crystals, the numbers of developed faces were strongly reduced [17.70]. In the presence of Ni, Co, and Cu ions, the rate of crystallization decreased [17.81]. An odd behavior was found while growing Cr-doped crystals. The addition of Cr with a concentration of 1% changed the regime of crystallization owing to the high chemical activity of these ions. At a concentration of about 3%, the rate of crystallization became very fast even without lowering the temperature [17.81]. In Pd-doped crystals, the ratios of the growth rate along the c-axis to the growth rate along the a- and b-axes slightly decreased as the crystal grew larger [17.85]. For medium-sized crystals (≈ 30 g), the average relative growth rate along the c-axis was larger by more than an order of magnitude in Pd-doped crystals than in pure TGS. Pd-doped crystals also developed other faces which have not been observed before. *Banan* et al. [17.82] studied the effect of Ce-, Cs-, L-alanine, and L-alanine + Cs on the growth and morphology of TGS crystals. Table 17.4 shows the crystal growth data, and Fig. 17.17 shows their habit. The well-developed (010)/(0$\bar{1}$0) faces, which were generally not present or less developed in pure TGS crystals are obtained with L-alanine- or Cs-doped crystals. Moreover, (101) faces obtained in crystals doped with L-alanine and in crystals doped with Cs and L-alanine were more dominant than pure TGS crystals. Also, the axial velocities, $V_{(001)}$ and $V_{(100)}$, were affected by doping (Table 17.4). Lower growth rates were especially obvious in Cs-doped crystals. The crystals became plate-like for $V_{(010)}/V_{(001)} \approx 28.0$; and the habits was strongly disturbed (Fig. 17.18). L-Alanine-doped crystals developed a habit which was asymmetric about the (010) plane. The growth rate in

Table 17.4 Growth data of doped TGS crystals [17.82]

	Crystal yield (g/(day °C))	$V_{(010)}$ (mm/(day °C))	$V_{(001)}$ (mm/(day °C))	$V_{(010)}$ (mm/(day °C))
TGS	0.171	0.88	0.34	2.58
TGS + Ce	0.021	0.079	0.047	1.68
TGS + Cs	0.009	0.198	0.007	28.20
TGS + L-alanine	0.192	0.89	0.44	2.02
TGS + L-alanine + Cs	0.132	0.65	0.063	10.30

the positive b-direction was higher than in the negative b-direction [17.29]. In D-alanine and L-alanine-doped TGS crystals, (101) faces developed more prominently than (001) faces, so the deuterated L-alanine-doped triglycine sulfate (DLATGS) crystals seemed to be thinner than pure TGS crystals. The (010) faces were more developed in aniline-doped crystals [17.84]. Recently, *Seif* et al. [17.92] studied the dependence of growth rate of the faces of TGS and KDP crystals on concentration of Cr(III) [17.92]. They proposed the following hypothesis to explain the effect of impurities on TGS and KDP crystals. It has long been known that, when a solute crystallizes from its supersaturated solution, the presence of impurities can often have a spectacular effect on the crystal growth kinetics and the habit of the crystalline phase. The impurities exhibit a marked specificity in their action as they are absorbed onto growing crystal surfaces. Adsorption of impurities onto crystal faces changes the relative surface free energies of the face and may block sites essential to the incorporation of new solute molecules into the crystal lattice and hence slow down the growth. The habit is thus determined by slow-growing faces. Furthermore, in the TGS:Cr(III) system dope with metal ions, metal–glycine complexes are formed in solution and enter the crystal lattice in the process of growth. The structure and type of metal ion complexes formed in the TGS lattice will determine the growth rate and hence the crystal habit.

It is also worthwhile to describe the effect of the same impurity on different types of crystals. The growth kinetic data of TGS and KDP crystals grown in the

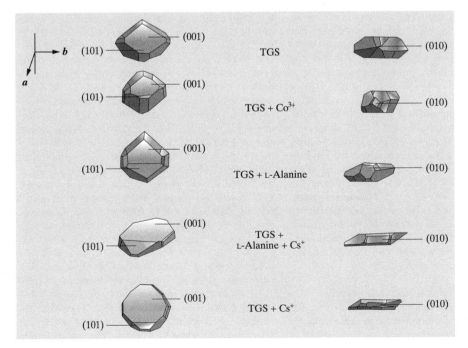

Fig. 17.18 Growth habits of doped TGS crystals

presence of Cr(III) ions are presented in Figs. 17.19 and 17.20. This data show the effect of impurity concentration on the growth rate of different faces of TGS and KDP crystals grown under constant, low supersaturation. In the case of KDP crystal, the mean growth rate along the [001] direction increases while along the [100] direction it remains almost constant with an increase in the concentration of Cr(III) in the solution/crystal with a slight fall below 7000 ppm. A similar type of effect of Fe(III) on growth rate of KDP crystal has been reported by *Owxzarek* and *Sangwal* [17.25]. $Cr_2(SO_4)_3$ molecules are considered to dissolve as an active complex such as $[Cr(H_2O)_2(OH)]^{2+}$, $[Cr(H_2O)_4(OH)_2]^+$ or $[Cr_2(SO_4)_2(H_2O)_7(OH)]^+$ and are assumed to become adsorbed on the crystal faces, thereby suppressing the growth rate. The impurities adsorbed on the surface of growing crystal at low supersaturation impede movements of steps by different mechanisms depending on the site of adsorption. Models of different types which describe the adsorption process and growth reduction have been reported in the literature [17.26, 27, 36, 93]. The models assume that the impurity species (ions, molecules or atoms) are adsorbed on the crystal surface into kinks, ledges, and terraces of growing surfaces. As soon as kinks and steps are occupied by impurity particles, there is a reduction in growth rates due to coverage of the crystal faces. This decrease in the growth rate can be explained on the basis of a model proposed by *Sangwal* and *Mielniczek-Brzoska* [17.32] based on their recent studies involving the Cu(II)–ammonium oxalate monohydrate crystal system. As shown in Fig. 17.19, the decrease in growth rate of (100) face of KDP crystals should be a kinetic effect involving a reduction in the value of the kinetic coefficient ($\beta = av \exp^{(-W/k_B T)}$), where a is the dimension of growth units perpendicular to the step, v is the frequency of vibration of molecules/atoms on the surface (s^{-1}), W is the activation energy for growth, k_B is the Boltzmann constant, and T is temperature (in Kelvin) for motion of steps on the surface. Above a certain critical concentration of impurity, there is no kinetic effect of impurity on growth kinetics. This may be due to the fact that all the active centers for crystallization are blocked, thus reducing the growth rate to zero. In our study, no growth of the {100} face was observed with more than 8000 ppm Cr(III) im-

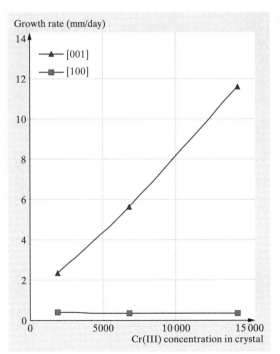

Fig. 17.19 Dependence of growth rate of the faces of KDP crystals on concentration of Cr(III)

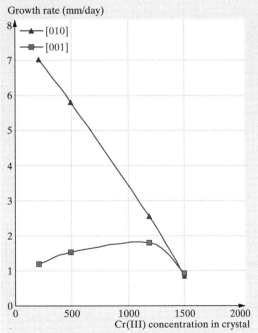

Fig. 17.20 Dependence of growth rate of the faces of TGS crystals on concentration of Cr(III)

purity in the KDP solution. An increase in growth rate along the [001] direction of KDP crystals may be caused by a decrease in the free energy of the face (thermodynamic effect); the surface energy decreases with an increase in impurity concentration, and hence increase in the growth rate. The above discussion suggests that the kinetic or thermodynamic effect depends on the structure of the crystal face, i.e., atomic arrangement, besides other factors.

Figure 17.20 shows that in the case of TGS crystals, the growth rate along the [010] direction decreases with an increase in the concentration of Cr(III) in the growth solution. This decrease in growth rate is due to the kinetic effect as explained above for the KDP crystal system. However, there is a slight increase in the growth rate along the [001] direction, with maxima around 1300 ppm Cr(III), and then there is a decrease. According to layer growth models, the consequence of a decrease in the edge free energy is an increase in the growth rate. Additionally, a decrease in the edge free energy may cause the growth mechanism to change. This effect of an initial increase followed by a subsequent decrease in growth rate with increasing concentration of impurity has been suggested by *Davey* [17.94], to opposite effects of thermodynamic and kinetic parameters. Furthermore, that ability of additives to form complexes with adventitious impurities present in a growth medium cannot be ruled out, as it can alter the atomic arrangement in crystal faces. To explain the effect of impurities on growth in more detail, one needs to collect more experimental data, including studies of the micromorphology of crystal surfaces as well as growth kinetics.

Effects of various organic dopants such as L-asparagine, L-tyrosine, L-cystine, guanidine, L-valine, and other dopants on morphology, growth, mechanical, and some physical properties of TGS have also been reported in the recent past [17.91, 95–99]. However, no explanation is given for change in the morphology of crystals by the authors of these publications.

17.5 Solution Growth of Triglycine Sulfate Crystals in Microgravity

The US National Aeronautics and Space Administration (NASA) has carried out about 115 Space Transportation Systems (STS) space flight missions (STS-1 to STS-127) from 1980 to the present day [17.100].

The authors were associated with two NASA missions called Spacelab-3 and the International Microgravity Laboratory (IML-1) on which single crystals of triglycine sulfate were grown from solution in microgravity for a period of 7 days aboard a Space Shuttle.

The general goal of the programs within NASA's Microgravity Research Division was to conduct basic and applied research under microgravity conditions (10^{-6} g) that would increase our understanding of fundamental physical, chemical, and biological processes specifically biotechnology, combustion science, fluid physics, fundamental physics, and materials science.

The microgravity environment of space provides a unique opportunity to further our understanding of various materials phenomena involving the molten, fluidic, and gaseous states by reducing or eliminating buoyancy-driven effects. Microgravity experiments in space are affected by residual microaccelerations on the spacecraft deriving from atmospheric drag, reaction control systems, momentum wheels, gravity gradients, crew involvement, and other disturbances. Mostly there is no actual suggestion by the scientific community as to the microgravity level required for their experiments. The general opinion is that microgravity will reduce the influence of convection, buoyancy, and sedimentation. Hardly any quantitative estimates have been made.

The anticipated results of microgravity materials science research range from establishing baselines for fundamental materials processes to generating results with more direct commercial significance. NASA's objectives for the microgravity materials science program include:

- Advancing our knowledge base for all classes of materials
- Designing and facilitating the execution of microgravity experiments that will help achieve this goal
- Determining road maps for future microgravity studies
- Contributing to NASA's Human Exploration and Development of Space enterprise
- Contributing to the national economy by developing enabling technologies valuable to the US private sector.

To accomplish these goals, the materials science program has tried to expand both its scientific scope and research community's involvement in microgravity research. Based on their requirements for experimental facilities most of the current materials science microgravity experiments can be divided into four general categories. The first category involves melt growth experiments, such as those used for processing multicomponent alloys from the liquid. The experiments in this category frequently require high temperatures and closed containers or crucibles to prevent elemental losses. The second group includes aqueous or solution growth experiments for materials such as triglycine sulfate and zeolite. These experiments usually require moderate to low temperatures. Hydrothermal processing of inorganic compounds and sol–gel processing also fit in this category. The third category of experiments involves vapor or gaseous environments, such as those used for growing mercury iodide or plasma processing. Unlike the first three categories that use containers for the parent materials and products, the fourth category involves processes and experiments that require containerless processing environments. Examples of these experiments include the formation of metallic and nonmetallic glasses during levitation melting and solidification, float-zone growth of crystals, and measurement of thermophysical properties such as diffusion coefficients and surface tension.

17.5.1 Rationale for Solution Crystal Growth in Space

In the microgravity environment of space, several physical phenomena taken for granted on Earth change dramatically. Convection in solution due to density differences is greatly reduced. Crystallization and solidification are two processes that can benefit from microgravity environment. As a part of the US National Aeronautics and Space Administration (NASA) microgravity and applications program, a study of TGS crystals growth from solution was carried out on Spacelab-3 (SL-3) and first International Microgravity Laboratory (IML-1) missions in 1985 and 1992, respectively. Crystals from solution are usually grown in a closed container of limited volume. Thus, any convection that is generated tends to lead to a circular to steady laminar convection, due to buoyancy. The density differences in the fluid can arise from both temperature and concentration variations. On Earth buoyancy-driven convection may cause microscopic gas/solution inclusions and fluctuating dopants incorporation and other defects in the crystals. Besides degrading pyroelectric device performance, the growth yield of useful crystals is also severely impacted due to incorporation of these types of defects. In a low-gravity environment, convection is greatly suppressed and diffusion becomes the predominant mechanism for thermal and mass transport. Thus, growth in microgravity can eliminate these problems and enhance our knowledge about the science of crystal growth.

17.5.2 Solution Crystal Growth Method in Space

Since the ground solution technique could not be used in the microgravity environment of space, the authors developed a new method known as the cooled sting technique to grow crystals in space from solution, as described below.

Cooled Sting Technique
As the conventional techniques of solution crystal growth cannot be used for growing crystals in space, a new technique was proposed and developed [17.100–102]. On Earth, in the absence of stirring, conventional techniques of solution crystal growth cause a lowering of concentration of the solution in the vicinity of the growing crystal, resulting in an upward flow of solution. At constant temperature this reduction in concentration would cause the growth rate to decrease rapidly. In a $1g$ environment, most solution growth techniques are directed towards increased convection mass transport by applying forced convection with very slow programmed cooling of a saturated solution. However, in the absence of convection, a change of temperature must move inward toward the crystal by conduction. The characteristic time for this to occur is $T = L^2 \rho C_p / k$, where L is the distance over which the heat must be conducted, ρ is the density, C_p is the heat capacity, and k is the thermal conductivity of solution. For water, it takes 48 min for a temperature change of 1 °C to be felt at a distance of 2 cm. This is too slow to keep a constant growth rate. So a unique technique was developed by the authors, which uses programmed cooling of the seed crystal itself. This is accomplished by using a cold finger (*sting*) in direct contact with the seed crystal, which allows temperature lowering in accordance with a predetermined polynomial [17.103, 104] for maintaining a supersaturated TGS solution near the surface of the crystal. Because of the L^2 dependence of T, it takes less time for a change of sting temperature to be transmitted through the growing crystal and to be felt at the surface. The construction of the ground-based cooled sting and solution growth apparatus [17.103, 104] are

illustrated in Figs. 17.21 and 17.22, respectively. In this case, crystals are grown by lowering the sting/seed and solution temperature, thereby creating a desired supersaturation.

Flight Hardware

The experiment in space utilizes the fluid experiments system (FES) and crystals are grown by a new technique developed by the authors called the *cooled sting technique* as described earlier [17.101–103]. This technique utilizes heat extraction from the seed crystal through a copper rod (sting), thereby creating the desired supersaturation near the growing crystal. The sting temperature profile follows a predetermined polynomial so as to obtain uniform growth. Figure 17.23 shows a detailed diagram of the flight cell with sting incorporated in the experimental module. The FES is an apparatus with the crystal growth cell as an integral part. It was developed by NASA and fabricated by TRW, CA.

The cell is designed to allow a variety of holographic diagnostics and real-time schlieren viewing of the crystal and the surrounding fluid. Schlieren images are transmitted down a link as black-and-white video to reveal flow patterns and variation in fluid density. Holograms that are recorded in space give 3-D information that leads to quantitative determination of concentration fields surrounding the crystal and motion of particles, if present, to determine g-jitters. The modified FES incorporates holographic tomography which enables the taking of optical data through the cell at multiple angles. During the SL-3 mission, two TGS crystals (named FES-2 and FES-3) were grown using (001)-oriented seed-type disc (as shown in Fig. 17.24). The objectives

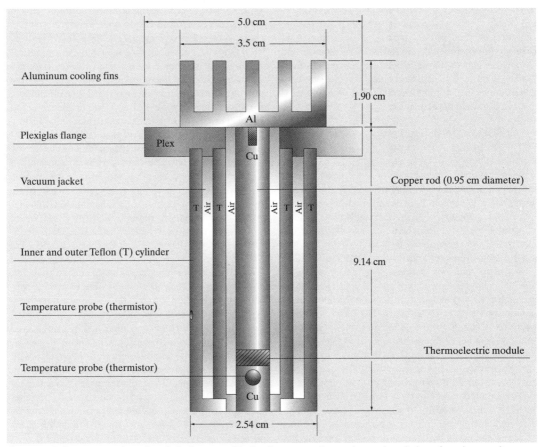

Fig. 17.21 Laboratory version of cooled sting assembly for the proposed crystal growth technique for microgravity

of the IML-1 flight experiments were: (a) to grow TGS crystals, (b) to perform holographic tomography of fluid field in the test cell in three dimensions, (c) to study fluid motion due to g-jitter by multiple-exposure holography of tracer particles (200, 400, and 600 μm), and (d) to study the influence of g-jitter on crystal quality and growth rate. One of the authors (R. B. Lal) was the principal investigator of the Spacelab-3 and IML-1 experiments. The coinvestigators were A. K. Batra, J. Trolinger, and W. R. Wilcox. During the IML-1 flight, due to serious hardware problems, only one TGS crystal was grown on a (010)-oriented seed crystal. The growth surface of seed crystal was a natural (010) face (unlike experiments performed in SL-3, in which processed seeds were used) cut from a polyhedral TGS crystal, with a thickness of about 3.5 mm. In TGS, the crystal growth rate is fast (maximum) in the [010] direction. On the ground, good-quality crystals are grown on (001)-oriented seed because growth on (010) face is nonuniform and multifaceted. Thus, it was important to investigate the growth on an (010)-oriented seed in the absence of buoyancy-driven convection,

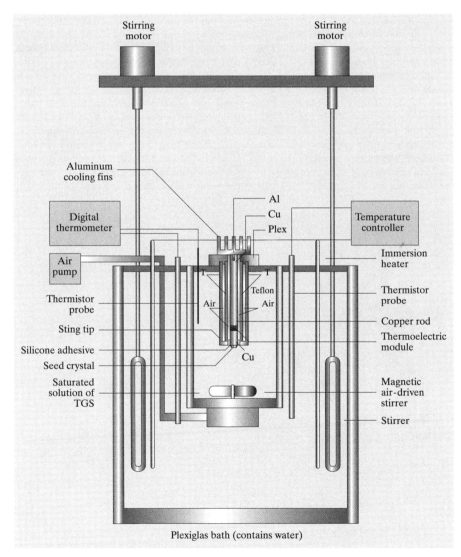

Fig. 17.22 Schematic diagram of the ground-based cooled sting solution growth apparatus

where growth is expected to be mainly diffusion controlled. This crystal was grown with undercooling of 4 °C for about 4 h. The growth rate was estimated to be about 1.6 mm/day and the quality of the grown crystal was substantially good. This can be attributed to a smooth transition from dissolution to growth in space experiment.

Flight Optical System

The fluids experiment system (FES) is a fully instrumented space flight chamber that can characterize the growth process through diagnostics of the crystal environment. Figure 17.25 shows the layout of optical system. Optical diagnostic instruments include two holographic cameras and a schlieren system, the output of which can be viewed in real time by television (TV) downlink. The optical and electronic instruments provide measures of solution concentration, temperature, convection, growth rate, and crystal properties during growth.

By recording light passing through the cell as well as light scattered from the crystal, holography provided diagnostics of the fluid through holographic interferometry, and particle diagnostics through three-dimensional particle imaging velocimetry. Figure 17.26 shows the layout of the optical system in which a 4 inch diameter, collimated He-Ne (Spectra Physics 107) laser beam passes through a double window into the crystal growth chamber, through the TGS solution, and across the surface of the crystal, finally emerging from a second set of windows. The beam then continues to the hologram plane, which is approximately 20 cm away, where it is mixed with a collimated reference wave on 70 mm-format roll film. The film is drawn flat on the platen by a vacuum in a unique film implementation of hologram recording for interferometry. In addition to the use of vacuum platens for recording and reconstruction, a special reconstruction process, necessary for holographic interferometry with film, described below, was developed to account for the imperfect optical quality of the film. A second holocamera views the crystal face directly from a lateral window [17.105]. Four types of holograms were produced, including singly exposed and multiply exposed holograms. The backlighting of the crystal was accomplished in two different ways, each with advantages and limitations; one method employs the direct laser beam and a second employs a diffuse beam, produced by inserting a diffuser into the object beam path before the beam enters the first cell window. The diffuse beam illuminates the field with

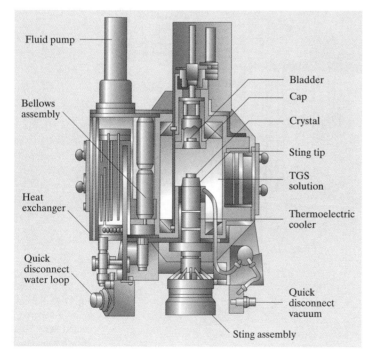

Fig. 17.23 Flight crystal growth cell designed and developed by NASA

many directions and is convenient for some types of viewing. However, such illumination is not useful for interferometry or schlieren in this system. Direct illumination is also used for interferometry and schlieren. With conventional optics, the direct illumination beam would provide a single illumination and viewing angle through the field.

Our previous experience in Spacelab-3 had taught us that more than one viewing angle is desirable. We achieved multiple viewing angles in IML-1 through the use of windows equipped with holographic optical elements (HOEs). The input window contains HOEs that convert the single input beam into three beams that pass over the crystal at angles of 0 and $\pm 23.5°$. The opposite window contains HOEs that redirect these beams to the recording film plane so that they can all be recorded and separated again during reconstruction.

Consequently, each recording comprises three superimposed, but independently viewable, holograms. The schlieren system is viewed by TV, allowing real-time viewing both by the crew and by the TV downlink. A primary use of the schlieren system is to view and judge the transition of the crystal from a dissolution phase to a growth phase since control of this transition is considered to be critical in producing a high-quality crystal. The knife-edge in the schlieren system was set so that, when the crystal was dissolving, light rays entering the resulting higher-refractive-index region above the crystal would be refracted in the direction of the crystal and be removed by the knife-edge, appearing dark in the image. When the transition from dissolution to growth occurred, the region immediately above the crystal would be depleted of solute, thus reducing the refractive index and causing the refracted rays to pass

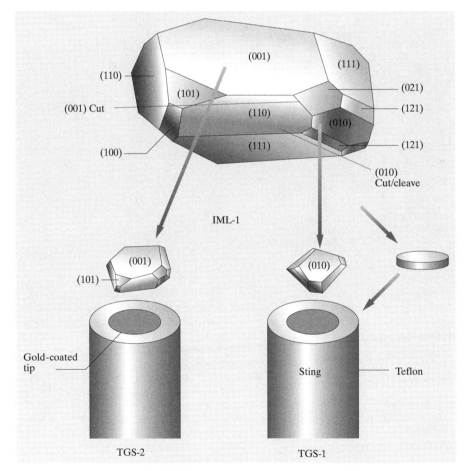

Fig. 17.24 TGS seed crystals used for growth runs on NASA's first International Microgravity Laboratory-1 (IML-1) mission

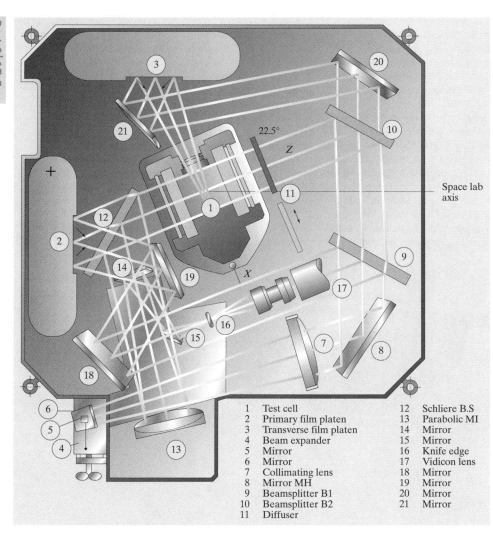

Fig. 17.25 Modified fluid experiment system (FES) optical system with various components

1	Test cell	12	Schliere B.S
2	Primary film platen	13	Parabolic MI
3	Transverse film platen	14	Mirror
4	Beam expander	15	Mirror
5	Mirror	16	Knife edge
6	Mirror	17	Vidicon lens
7	Collimating lens	18	Mirror
8	Mirror MH	19	Mirror
9	Beamsplitter B1	20	Mirror
10	Beamsplitter B2	21	Mirror
11	Diffuser		

above the knife edge, causing a bright region to appear above the crystal within the larger, dark region of higher concentration. The method proved to be an extremely sensitive way to identify the transition from dissolution to growth.

17.5.3 Results and Discussion

The flight TGS crystals were examined with a high-resolution monochromatic synchrotron x-radiation diffraction technique, both before and after slicing for the fabrication of infrared detectors to check the lattice regularity, identify inclusions and dislocations, draw inferences about growth mode and stability, and locate the interface between the seed and the new growth. The experiments were performed at the National Synchrotron Light Source at Brookhaven National Laboratory in collaboration with B. Steiner of the US National Institute of Standards and Technology (NIST). The performance of materials is determined by their structure; in this performance, irregularity typically plays a leading role. The growth of crystals in low-g has long been of interest because of the anticipation that reduction in gravitational forces would strongly affect crystal growth and

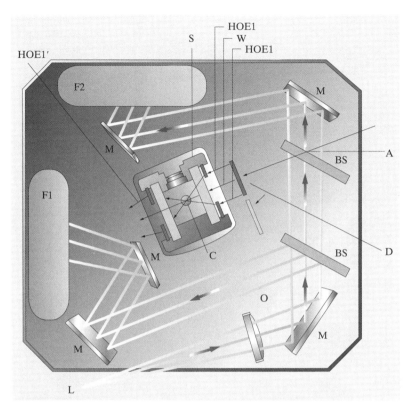

Fig. 17.26 Detailed optical layout for the fluid experiment system (FES) designed and developed by TRW. HOE 1′: note that ray emerges at an angle to simplify separation. A – Angle between optical axis and space shuttle axis, M – mirror, BS – beamsplitter, C – crystal, O – lens, D – removable diffusor, F1 – hologram 1, F2 – hologram 2, L – He-Ne laser underneath, S – side window, W – window

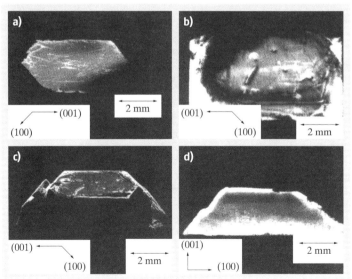

Fig. 17.27a–d High-resolution synchrotron x-ray radiation diffraction imaging of $1g$- and microgravity-grown TGS crystals. (**a**) Uncut flight seed crystal, (**b**) high-resolution ($\bar{1}10$) diffraction image of the interior of the IML-1 crystal, (**c**) space-grown crystal with polystyrene particles, (**d**) cut edge of the IML-1 crystal

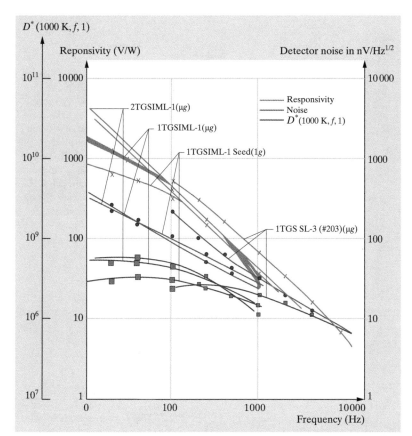

Fig. 17.28 Relevant parameters of infrared detectors fabricated from $1g$- and microgravity-grown TGS crystals

thereby the nature of resulting irregularities. Many factors affect crystal growth, and because these can interact strongly with one another, the understanding of the structural variation necessary for its effective exploitation has not been fully achieved. Gaining knowledge of irregularities in space- and Earth-grown crystals, developed in conjunction with an understanding of their genesis and detailed effects on properties, is an important challenge. Such knowledge is also expected to contribute to dramatically improving single-crystal production, both in space and on the Earth. The local acceptance angle for diffraction from the uncut flight TGS-1 crystal of 1–2 arcsec (Fig. 17.27) indicates extraordinary crystal quality [17.106]. Polystyrene particles that had been included in the space-grown material in IML-1 experiment are observed as small imperfections in Fig. 17.27b. Also, clearly distinguishable in Fig. 17.27b is the faceted growth mode. Two sets of edge dislocations in the seed, one [101] ori-

ented and the other [001] oriented, were noted as well in images taken in Laue geometry, but they appear not to have affected the space growth. Observation of the cut edge of the crystal (Fig. 17.27d) shows continuity between the seed at the top and space grown part. The demarcation between the seed and the space-grown material is indistinct. High-resolution imaging of terrestrial crystals has shown that the surface treatment of the seed crystal is critical to growth perfection. The ground control TGS crystals were of extremely high perfection. Slice next to many possible flight seeds were examined by high-resolution diffraction imaging. The selection of the flight seed was based on the perfection of the slice next to the seed crystal.

Infrared detectors from the flight and ground control crystals were fabricated at EDO/Barnes Engineering Division, Shelton, CT. The detectivity (D^*) and other parameters for these infrared detectors are shown in

Table 17.5 Detector characteristics of space-grown TGS crystals

Crystal (TGS)	Noise (nV/Hz$^{1/2}$)	Responsivity (V/W)	Detectivity (1000 K, f, 1) $\times 10^{-8}$	f (Hz)	Remarks
SL-3/FES2-TGS (μg grown)	320	510	0.99	100	Area = 0.3×3 mm^2, blackened
IML-1TGS seed (1g grown)	418	320	2	100	Area = 1×1 mm^2, no window
IML-1TGS (μg grown)	90	400	4.2	100	Area = 1×1 mm^2
IML-1TGS (1g grown)	98	420	4.5	100	Area = 1×1 mm^2
IML-1TGS (μg grown)	100	340	3	100	Area = 1×1 mm^2

Fig. 17.28. The detector characteristics are given in Table 17.5. The detectivity (D^*) for IR detectors fabricated from the IML-1 crystal shows an improvement over the ground-grown crystals and crystals grown on Spacelab-3.

The motion of three different-sized particles were mapped using techniques described elsewhere [17.105, 107]. The combined effects of fluid convection, particle interaction, residual gravity, Space Shuttle maneuvers, and g-jitters have been observed. However, the interferograms show several noteworthy features. When the crystal enters a growth phase, the solution in the region near the crystal is depleted of solute, thus reducing its refractive index below that of the surrounding fluid, creating a hemispherical cap of fringes over the crystal, as shown in Fig. 17.29. The stability of this cloud in the interferograms confirmed that the crystal was growing in a diffusion-controlled process. The cloud did show, however, that the process was not completely axisymmetric, due to equipment-related problems that were encountered during the mission.

To sum up, in spite of problems with the operation of the FES, two important objectives were attained in the IML-1 experiment: (a) a high-quality TGS crystal was grown, (b) the particle dynamics experiment was successful.

In spite of limited time and fast growth, the growth on the (010) face was substantially uniform over a period of 18 h. The growth on the (010) face on ground is mostly nonuniform. Polystyrene particles of three sizes that had been occluded by the growing TGS were observed as small imperfections in the grown crystal. The observations of the cut edge of the flight crystal (TGS-1) show continuity between the seed at the bottom and the space growth at the top, indicating a high degree of epitaxy of the space-grown material. The demarcation between the seed and the space-grown material is indistinct, indicating a smooth transition from dissolution to growth so that solvent inclusions are not formed between the seed and the grown layer. Experiments on Earth have shown that such inclusions tend to result in dislocations that propagate through subsequently grown material and degrade properties. The infrared detectors fabricated from the TGS-1 flight crystal show improved detectivity (D^*) compared with ground samples and even with detectors fabricated from crystals grown on Spacelab-3. The dielectric loss in the IML-1 crystal is lower than in ground crystals and in crystals grown in Spacelab-3.

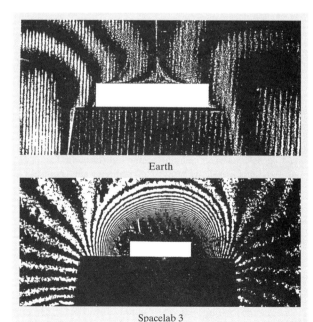

Fig. 17.29 Interferograms of concentration field in TGS solution on Earth and in microgravity onboard Spacelab-3

17.6 Protein Crystal Growth

The human body contains thousands of different proteins, which play essential roles in maintaining life. A protein's structure determines the specific role that it plays in the human body; however, researchers lack detailed knowledge about the structures of many proteins. The crystallization of proteins has three major applications: (1) structural biology and drug design, (2) bioseparations, and (3) controlled drug delivery. In the first application, protein crystals are used with the techniques of crystallography to ascertain the three-dimensional structure of the molecule. This structure is indispensable for correctly determining the often complex biological functions of these macromolecules. The design of drugs is related to this, and involves designing a molecule that can exactly fit into a binding site of a macromolecule and block its function in a disease pathway. Producing better-quality crystals will result in more accurate 3-D protein structure, which in turn means that the protein's biological function can be known more precisely, also resulting in improved drug design. The second application, bioseparations, refer to the downstream processing of the products of fermentation. Typically, the desired product of the fermentation process is a protein (e.g., insulin), which then needs to be separated from biomass. Crystallization is one of the commonly employed techniques for separation of protein. It has the advantage of being a benign separation process, that is, it does not cause the protein to unfold and lose its activity. The other application of protein, controlled drug delivery, is also very important. Most drugs are cleared by the body rapidly following administration, making it difficult to achieve a constant desired level over a period of time. When the drug is a protein such as insulin or α-interferon, administrating the drug in the crystalline form shows promise of achieving such controlled delivery. The challenge is to produce crystals of relatively uniform size so that dosage can be correctly prescribed.

17.6.1 Protein Crystal Growth Methods

Protein crystallization is inherently difficult because of the fragile nature of protein crystals. Proteins have irregularly shaped surfaces, which result in the formation of large channels within any protein crystal. Therefore, the noncovalent bonds that hold the lattice together must often be formed through several layers of solvent molecules. In addition, to overcoming the inherent fragility of protein crystals, successful production of x-ray-worthy crystals is dependent upon a number of environmental factors because so much variation exists among proteins, with each one requiring unique conditions for successful crystallization. Therefore, attempting to crystallize a protein without a proven protocol can be very tedious. Some factors that require consideration are protein purity, pH, concentration of protein, temperature, and the precipitants. In order to initiate crystallization the protein solution has to be brought to a thermodynamically unstable state of supersaturation. The solution can be brought back to the stable equilibrium state through precipitation of the protein, which is the most frequent process, or through crystallization. The supersaturation state can be achieved by several techniques: evaporation of solvent molecules, change of ionic strength, change of pH, change of temperature or change of some other parameter.

Two of the most commonly used methods for protein crystallization fall under the category of vapor diffusion [17.108–110]:

1. Hanging drop method
2. Sitting drop method.

Both of these methods entail the use of a droplet containing purified protein, buffer, and precipitants in higher concentration. Initially, the droplet of protein solution contains an insufficient concentration of precipitant for crystallization, but as water vaporizes from the drop and transfers to the reservoir, the precipitant concentration increases to a level optimal for crystallization. Since the system is in equilibrium, these optimum conditions are maintained until the crystallization is complete. Figures 17.30 and 17.31 depict

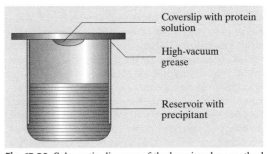

Fig. 17.30 Schematic diagram of the hanging drop method

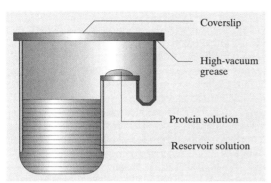

Fig. 17.31 Schematic diagram of the sitting drop method

the hanging drop and sitting drop systems, respectively. The hanging drop method differs from the sitting drop method in the vertical orientation of the protein solution drop within the system. It is important to mention that both methods require a closed system, that is, the system must be sealed off from outside using an airtight container. It is worth mentioning that the reservoir solution usually contains buffer and precipitant. The protein solution contains the same compounds, but in lower concentrations. The protein solution may also contain trace metals or ions necessary for precipitation; for instance, insulin is known to require trace amounts of zinc for crystallization.

17.6.2 Protein Crystal Growth Mechanisms

From the presence of well-defined facets on most protein crystals one can unambiguously conclude that growth occurs via the spreading of layers from growth step sources such as dislocations and 2-D nuclei. This has been confirmed on a molecular level. Ex situ electron microscopy observations have resolved individual growth steps on (010) and (110) faces of tetragonal lysozyme [17.111] that, contrary to recent claims [17.112, 113], are of monomolecular height. In situ atomic force microscopy of lysozyme has produced particularly instructive images of growth step generation at screw dislocations outcrops and of 2-D nucleation-induced islands.

Most recent atomic force microscopy observations on a larger number of other proteins and viruses [17.114] have reproduced the whole body of growth morphology and kinetics scenarios known for inorganic solution growth. These include layer spreading from dislocations and 2-D nuclei, interaction between growth steps from sources of different activities, and impediment of step propagation by foreign particles. Particle engulfment was observed to often result in dislocation formation. Crystallites that impinged on the interface became either epitaxially aligned with main crystal, or remained misaligned and caused various defects during further growth. There even appears to be some indication of kinetic roughening on certain facets of some proteins [17.115].

17.6.3 Protein Crystal Growth in Microgravity

The microgravity environment aboard spacecraft in low Earth orbit provides a convection- and sedimentation-free environment for the study and applications of fluid-based systems. With the advent of the US Space Shuttle, scientists had regular access to such environments and many experiments were initiated, including those in protein crystallization. After many trials it became clear that, for several proteins, crystallization in microgravity environment resulted in bigger and better crystals. In some instances, crystals that could not be crystallized at all on the ground were found to crystallize in space. Conversely, for numerous proteins the space environment was found to be no better or was worse than ground-based conditions. As a result of these observations, NASA has become one of the leading federal agencies in prompting and funding protein crystallization research. Efforts are directed at both utilizing the space environment to improve the crystallization of novel proteins and at fundamental studies of the causes (if any) of the improvement in protein crystals produced in microgravity. The results from flying more, and studying in more detail, have significantly altered attitudes towards space-based protein crystal growth. Persuasive explanations have emerged, and a strong theoretical model has emerged to explain why space-based growth is better.

Since the inception of protein crystal growth in microgravity research by *Littke* and *John* [17.116], several research groups have developed microgravity hardware and experiments [17.116–121]. Some of those are listed below:

1. Handheld protein crystallization apparatus for microgravity (HH-PCAM) [17.122]
2. Diffusion-controlled crystallization apparatus for microgravity (DCAM) [17.123]
3. High-density protein crystal growth system (HDPCG) [17.124]
4. Protein crystal growth facility (PCF) [17.125]

5. A multiuser facility-based protein crystallization apparatus for microgravity (PCAM) [17.122]
6. Advanced protein crystallization facility (APCF) [17.123].

Several thousand individual protein crystal growth experiments have been flown using the PCAM facility hardware aboard the US Space Shuttle. According to the developer of this facility, this hardware represents a pioneering development in design and deployment of space flight hardware based on disposable interface elements [17.122]. Furthermore, it has resulted in an ultrahigh-resolution structure and first example of neutron diffraction achieved as a result of protein crystal growth in microgravity [17.125–127]. Additionally, using this facility, fundamental differences in protein partitioning in microgravity have been documented, which represent the first direct experimental observation of the factors contributing to quality improvements in the growth of protein crystals in microgravity [17.128]. The other important hardware, referred to as the diffusion-controlled crystallization apparatus for microgravity (DCAM), utilizes the dialysis method and allows the equilibration rate of each individual experiment to be passively controlled from several days to several months. It is worth mentioning that precision control rate of supersaturation has routinely produced macrocrystals of size .5 mm to 1.25 cm for a variety of protein in this hardware. Analysis of serum albumin, ferritin, lysozyme, bacteriorhodopsin, and nucleosome core particles exhibit superior diffraction properties as compared with ground-based controls [17.123]. Further improvements of the hardware is ongoing for the International Space Station, where x-ray analysis can be done aboard the Space Station [17.124].

17.7 Concluding Remarks

Bulk high-quality single crystals are required for use in fabricating devices for various technological applications. Since crystal growth is a complicated process that depends on many parameters that can interact, the complete process is not well understood. This is one of the reasons to grow crystals in microgravity of space, to separate omnipresent convection on Earth and have only diffusion-controlled growth. The authors have attempted to give a comprehensive overview of the various problems encountered in the solution growth of single crystals on Earth and in space experiment based on their experience over almost three decades. The solutions of the various problems encountered during growth on ground and in spaceflight experiments are described. This chapter will serve as a foundation for those who desire to initiate a research program in the growth of bulk single crystals of technological importance that can be grown from low-temperature solution technique. A brief review of crystal growth fundamentals is presented, including key techniques for solution crystal growth such as solubility determination and the design of various crystal growth systems including mechanical and electronic crystal motor reciprocating arrangements. Three generations of modifications to solution crystallizers designed and fabricated in the laboratory and a crystallizer for space growth based on the cooled sting technique advanced by the authors are described. A number of solution-grown crystals grown at Alabama A&M University are shown. A detailed description of the crystal growth experiments on an important infrared material, triglycine sulfate, and the difficulties encountered in the space crystal growth experiment aboard the Space Shuttle in the NASA Spacelab-3 and first International Microgravity Laboratory (IML-1) missions are given. Since basic principles of solution growth technique are shared by protein crystal growth, basic techniques of protein crystal growth are briefly mentioned and efforts towards protein crystal growth in microgravity are also discussed.

References

17.1 H.A. Meirs, F. Isaac: The spontaneous crystallization, Proc. R. Soc. Lond. A **79**, 322–325 (1987)
17.2 K. Sangwal: Growth kinetics and surface morphology of crystals grown from solutions: recent observations and their interpretations, Prog. Cryst. Growth Charact. Mater. **36**(3), 163–248 (1998)
17.3 I.F. Nicolau: Growth kinetics of potassium-dihydrogen phosphate crystals in solution, Krist. Tech. **9**(11), 1255–1263 (1974)

17.4 R. Janssen-Van Rosmalen, P. Bennema, J. Garside: The influence of volume diffusion on crystal growth, J. Cryst. Growth **29**, 342–352 (1975)

17.5 A. Chianese: Growth and dissolution of sodium perborate in aqueous solutions by using RDC technique, J. Cryst. Growth **91**, 39–49 (1988)

17.6 K. Sangwal: On the mechanism of crystal growth from solutions, J. Cryst. Growth **192**, 200–214 (1998)

17.7 Y. Wang, X.L. Yu, D.I. Sun, S.T. Yin: Mass transport and growth kinetics related to the interface supersaturation oh lithium formate monohydrate, Cryst. Res. Technol. **36**(4/5), 441–448 (2001)

17.8 N. Zaitseva, L. Carman: Rapid growth of KDP-type crystals, Prog. Cryst. Growth Charact. Mater. **43**, 1–118 (2001)

17.9 I.V. Melikhov, L.B. Berliner: Crystallization of salts from supersaturated solutions; diffusion kinetics, J. Cryst. Growth **46**, 79–84 (1979)

17.10 P. Bennema: Sprial growth and surface roughening: development since Burton, Cabrera and Frank, J. Cryst. Growth **69**, 182–197 (1984)

17.11 X.Y. Liu, K. Malwa, K. Tsukamoto: Heterogeneous two dimension nucleation and growth kinetics, J. Chem. Phys. **106**(5), 1870–1879 (1997)

17.12 A.F. Izmailov, A.S. Myerson: Momentum and mass transfer in supersaturation solutions crystal growth from solution, J. Cryst. Growth **174**, 263–368 (1997)

17.13 M. Rak, K. Izdebski, A. Brozi: Kinetic Monte Carlo study of crystal growth from solution, Comput. Phys. Commun. **138**, 250–263 (2001)

17.14 F. Rosenberger, B. Mutaftschiev (Eds.): *Interfacial Aspects of Phase Transformation* (Reidel, Dordrecht 1982)

17.15 A.A. Chernov: Present day understanding of crystal growth from aqueous solutions, Prog. Cryst. Growth Charact. **26**, 121–151 (1993)

17.16 R. Mohan, A.S. Myerson: Growth kinetics: A thermodynamic approach, Chem. Eng. Sci. **57**, 4277–4285 (2002)

17.17 C.M. Pina, A. Putnis, J.M. Astilleros: The growth mechanisms of solid solutions crystallization from aqueous solutions, Chem. Geol. **204**, 145–161 (2004)

17.18 I. Sunagawa, K. Tsukamoto, K. Maiwa, K. Onuma: Growth and perfection of crystals from aqueous solution: Case studies on Barium nitrate and K-Alum, Prog. Cryst. Charact. **30**, 153–190 (1995)

17.19 W.R. Wilcox: Influence of convection on the growth of crystals from solution, J. Cryst. Growth **65**, 133–142 (1983)

17.20 I.V. Markov: *Crystal Growth for Beginners* (World Scientific, Hobohen 2004)

17.21 I. Sunagawa: *Crystals Growth, Morphology and Perfection* (Cambridge Univ. Press, Cambridge 2005)

17.22 T. Ogawa: A phenomenological analysis of crystal growth from solutions as an irreversible process, Jpn. J. Appl. Phys. **16**(5), 689–695 (1977)

17.23 S. Veintemillas-Verdaguer: Chemical aspects of the effect of impurities in crystal growth, Prog. Cryst. Growth Charact. **32**, 76–109 (1996)

17.24 K. Sangwal: Kinetics effects of impurities on the growth of single crystals from solutions, J. Cryst. Growth **203**, 197–212 (1999)

17.25 I. Owczarek, K. Sangwal: Effect of impurities on the growth of KDP crystals: Mechanism of adsorption on (101) faces, J. Cryst. Growth **102**, 574–580 (1990)

17.26 K. Sangwal: Effects of impurities on the crystal growth processes, Prog. Cryst. Charact. **32**, 3–43 (1996)

17.27 E. Kirkova, M. Djarova, B. Donkova: Inclusions of isomorphism impurities during crystallization from solutions, Prog. Cryst. Growth Charact. **32**, 111–134 (1996)

17.28 M. Rauls, K. Bartosch, M. Kind, S. Kuch, R. Lacmann, A. Mersmann: The influence of impurities on crystallization kinetics – A case study on ammonium sulfate, J. Cryst. Growth **312**, 116–128 (2000)

17.29 E. Mielniczek-Brzoska, K. Gielzak-Kocwin, K. Sangwal: Effects of Cu(II) ions on the growth of ammonium oxalate monohydrate crystals from aqueous solutions: Growth kinetics, segregation coefficient and characterization of incorporation sites, J. Cryst. Growth **212**, 532–542 (2000)

17.30 K. Sangwal, E. Mielniczek-Brzoska, J. Bore: Effect of Mn(II) ions on the growth of ammonium oxalate monohydrate crystals from aqueous solutions: Growth habit and surface morphology, Cryst. Res. Technol. **38**(2), 103–112 (2003)

17.31 N. Kubota, M. Yokota, J.W. Mullin: Supersaturation dependence of crystal growth in solutions in the presence of impurity, J. Cryst. Growth **182**, 86–94 (1997)

17.32 K. Sangwal, E. Mielniczek-Brzoska: On the effect of Cu(II) impurities on the growth kinetics of ammonium oxalate monohydrate crystals from aqueous solutions, Cryst. Res. Technol. **36**(8-10), 837–849 (2001)

17.33 K. Sangwal, E. Mielniczek-Brzoska, J. Bore: Study of segregation coefficient of cationic impurities in ammonium oxalate monohydrate crystals during growth from aqueous solutions, J. Cryst. Growth **244**, 183–193 (2002)

17.34 K. Sangwal, T. Palcznska: On the supersaturation and impurity concentration dependence of segregation coefficient in crystals grown from solutions, J. Cryst. Growth **212**, 522–531 (2002)

17.35 P.S. Delineshev: Growth of crystals in the presence of impurities, a hypotheses based on a kinetic approach, Cryst. Res. Technol. **33**(6), 891–897 (1998)

17.36 K. Sangwal: Kinetic effects of impurities on the growth of single crystals from solutions, J. Cryst. Growth **203**, 197–212 (1999)

17.37 A.S. Myerson, S.M. Jang: A comparison of binding energy an metastable zone width for adipic acid

17.38 D. Aquilano, M. Rubbo, G. Mantovani, G. Sgualdino, G. Vaccari: Equilibrium and growth forms of sucrose crystals in the {h0l} zone, J. Cryst. Growth **74**, 10–20 (1986)

17.39 D. Aquilano, M. Rubbo, G. Mantovani, G. Sgualdino, G. Vaccari: Equilibrium and growth forms of sucrose crystals in the {h0l} zone, J. Cryst. Growth **83**, 77–83 (1987)

17.40 A.A. Chernov, V.V. Sipyagin: Peculiarities in crystal growth aqueous solutions connected with their structure. In: *Current Topics in Materials Science*, Vol. 5, ed. by E. Kaldis (North-Holland, Amsterdam 1980) pp. 279–333

17.41 K. Byrappa, H. Klapper, T. Ohachi, R. Fornari (Eds.): *Crystal Growth of Technologically Important Electronic Materials* (Allied, New Delhi 2003)

17.42 M.D. Aggarwal, T. Gebre, A.K. Batra, M.E. Edwards, R.B. Lal, B.G. Penn, D.O. Frazier: Growth of nonlinear optical materials at Alabama University, Proc. SPIE **4813**, 52–65 (2002)

17.43 M.D. Aggarwal, W.S. Wang, K. Bhat, B.G. Penn, D.O. Frazier: Photonic crystals: crystal growth processing and physical properties. In: *Handbook of Advanced Electronic and Photonic Materials and Devices*, Vol. 9, ed. by H.S. Nalwa (Academic, New York 2001) pp. 193–228

17.44 H.K. Henisch: *Crystal Growth in Gels* (Dover, New York 1970)

17.45 A.F. Barton: *Handbook of Solubility Parameters and Other Cohesion Parameters* (CRC, Boca Raton 1983)

17.46 J. Novotny: A crystallizer for the investigation of conditions of growth of single crystals from solutions, Krist. Tech. **6**(3), 343–352 (1971)

17.47 M.D. Aggarwal, J. Choi, W.S. Wang, K. Bhat, R.B. Lal, A.D. Shields, B.G. Penn, D.O. Frazier: Solution growth of a novel nonlinear optical material: L-histidine tetrafluoroborate, J. Cryst. Growth **204**, 179–182 (1999)

17.48 A.K. Batra, S.C. Mathur: Reciprocating arrangement for solution crystal growth, Res. Ind. **29**, 20–22 (1984)

17.49 W.S. Wang, M.D. Aggarwal, J. Choi, K. Bhat, T. Gebre, A.D. Shields, B.G. Penn, D.O. Frazier: Solvent effects and polymorphic transformation of organic nonlinear optical crystal L-pyroglutamic acid in solution growth processes. I. Solvent effects and growth morphology, J. Cryst. Growth **198/199**, 578–582 (1999)

17.50 C. Owens, K. Bhat, W.S. Wang, A. Tan, M.D. Aggarwal, B.G. Penn, D.O. Frazier: Bulk growth of high quality nonlinear optical crystals of L-arginine tetrafluoroborate (L-AFB), J. Cryst. Growth **225**, 465–469 (2001)

17.51 R.B. Lal, H.W. Zhang, W.S. Wang, M.D. Aggarwal, H.W.H. Lee, B.G. Penn: Crystal growth and optical properties of 4-aminobenzophenone crystals for NLO applications, J. Cryst. Growth **174**, 393–397 (1997)

17.52 H.W. Zhang, A.K. Batra, R.B. Lal: Growth of large methyl-(2,4-dinitrophenyl)-aminopropanoate: 2-methyl-4-nitroaniline crystals for nonlinear applications, J. Cryst. Growth **137**, 141–144 (1994)

17.53 M.R. Simmons: An investigation of crystals matrices of single crystals doped with rare earth ions. M.Sc. Thesis (Alabama Univ., Alabama 2002)

17.54 J.-M. Chang, A.K. Batra, R.B. Lal: Growth and characterization of doped TGS crystals for infrared devices, Cryst. Growth Des. **2**(5), 431–435 (2002)

17.55 M.D. Aggarwal, R.B. Lal: Simple low-cost reciprocating crystallizer for solution crystal growth, Rev. Sci. Instrum. **54**, 772–773 (1983)

17.56 A.K. Batra, C.R. Carmichael-Owens, M. Simmons, M.D. Aggarwal, R.B. Lal: Design of a solution crystal growth crystallizer with versatile electronic reciprocal motion control for a crystal holder, Cryst. Res. Technol. **40**(8), 757–760 (2005)

17.57 A.K. Batra, M.D. Aggarwal, R.B. Lal: Growth and characterization of doped DTGS crystals for infrared sensing devices, Mat. Lett. **57**, 39–43 (2003)

17.58 R.B. Lal, M.D. Aggarwal: Reciprocating crystallizer: Automatic crystallizer grows crystals from aqueous solutions, NASA Tech. Briefs **8**, 419 (1984)

17.59 F. Jona, S. Shirane: *Ferroelectric Crystals* (Dover, New York 1993)

17.60 R.B. Lal, A.K. Batra: Growth and properties of triglycine (TGS) sulfate crystals: Review, Ferroelectrics **142**, 51–82 (1993)

17.61 E.A.D. White, J.D.C. Wood, V.M. Wood: The growth of large area, uniformly doped TGS crystals, J. Cryst. Growth **32**, 149–156 (1976)

17.62 B. Brezina, M. Havrankova, M. Vasa: Enhanced growth of non-polar {001} growth sectors of deuterated triglycine sulfate doped with L-alanine (LADTGS), Cryst. Res. Technol. **27**(1), 13–20 (1992)

17.63 S. Satapathy, S.K. Sharma, A.K. Karnal, V.K. Wadhawan: Effects of seed orientation on the growth of TGS crystals with large (010) facets needed for detector applications, J. Cryst. Growth **240**, 196–202 (2002)

17.64 M. Banan: Growth of pure and doped triglycine sulfate crystals for pyroelectric infrared detector applications. M.Sc. Thesis (Alabama A&M Univ., Alabama 1986)

17.65 D. Zhao-De: A new method of growth of ferroelectric crystal, Ferroelectrics **39**, 1237–1240 (1981)

17.66 F. Moravec, J. Novotny: Study on the growth of triglycine sulphate single crystals, Krist. Techn. **7**, 891–902 (1972)

17.67 W.K. Burton, N. Cabrera, F.C. Frank: The growth of crystals and the equilibrium structure of their surfaces, Philos. Trans. R. Soc. Lond. **243**, 299–358 (1951)

17.68 L.N. Rashkovich: Interferometric investigation of growth rate of {001} faces of triglycine sulfate at various supersaturation and temperature, Sov. Phys. Crystallogr. **28**, 454–458 (1983)

17.69 J. Novotny, F. Moravec, Z. Solc: The role of surface and volume diffusion in the growth of TGS single crystals, Czech. J. Phys. B **23**, 261–266 (1973)

17.70 J. Novotny, F. Moravec: Growth of TGS from slightly supersaturated solutions, J. Cryst. Growth **11**, 329–335 (1971)

17.71 D.A. Reiss, R.L. Kroes, E.E. Anderson: Growth kinetics of the (001) face of TGS below the ferroelectric transition temperature, J. Cryst. Growth **84**, 7–10 (1987)

17.72 M. Banan, R.B. Lal, A.K. Batra, M.D. Aggarwal: Effect of poling on the morphology and growth rate of TGS crystals, Cryst. Res. Technol. **24**(3), K53–K55 (1989)

17.73 F. Moravec, J. Novotny: A contribution to the study of the influence of impurities on the growth and some physical properties of TGS single crystals, Krist. Techn. **6**(3), 335–342 (1971)

17.74 R.V. Whipps, R.S. Cosier, L.K. Bye: Orthorhombic diglycine sulphate, J. Mater. Sci. **7**(12), 1476–1477 (1972)

17.75 E. Dominquez, R. Jimenez, J. Mendiola, E.J. Vivas: Diglycine sulphate – An interesting new dielectric crystal species, J. Mater. Sci. **70**(3), 363–364 (1972)

17.76 M.S. Tsedrik, V.N. Ulasen, G.A. Zaborovski: Growing TGS and TGSe single crystals at various pH of solution, Krist. Techn. **10**(1), 49–54 (1975)

17.77 E.J. Weidmann, E.A.D. White, V.M. Wood: Induced growth anisotropy in TGS crystals, J. Mater. Sci. Lett. **7**, 719–720 (1972)

17.78 L. Szczepanska: Growth investigations of single crystal sulphates containing molecular glycine groups, Krist. Techn. **11**, 265–271 (1976)

17.79 L. Prokopova, J. Novotny, Z. Micka, V. Malina: Growth of triglycine sulphate crystals doped by cadmium phosphate, Cryst. Res. Technol. **36**, 1189–1195 (2001)

17.80 G.R. Pandya, D.D. Vyas: On growth and morphological studies of TGS single crystals, Cryst. Res. Technol. **16**, 1353–1358 (1981)

17.81 R.B. Lal, S. Etminan, A.K. Batra: Effect of simultaneous organic and inorganic dopants on the characteristics of triglycine sulfate crystals, Proc. 9th IEEE Int. Symp. Appl. Ferroelectr. (1994) pp. 695–697

17.82 M. Banan, A.K. Batra, R.B. Lal: Growth and morphology of triglycine sulphate (TGS) crystals, J. Mater. Sci. Lett. **8**, 1348–1349 (1989)

17.83 N. Nakatani: Ferroelectric domain structure and internal bias field in DL-α-alanine-doped triglycine sulfate, Jpn. J. Appl. Phys. **30**(12A), 3445–3449 (1991)

17.84 J. Eisner: The physical properties of TGS single crystals, grown from aqueous TGS solutions containing aniline, Phys. Status Solidi **43**, K1–K4 (1977)

17.85 F. Moravec, J. Novotny, J. Strajblova: Single crystals of triglycine sulphate containing palladium, Czech J. Phys. **23**, 855–862 (1977)

17.86 A.S. Sidorkin, A.M. Kostsov: Exoelectron emission from a ferroelectric crystal of triglycine sulfate with defects, Sov. Phys. Solid State **33**(8), 1383–1384 (1991)

17.87 L. Yang, A.K. Batra, R.B. Lal: Growth and characterization of TGS crystals grown by cooled sting technique, Ferroelectrics **118**(1-4), 85–89 (1991)

17.88 N. Nakatani: Ferroelectric domain structure and internal bias field in DL-α-alanine doped triglycine sulfate, Jpn. J. Appl. Phys. **30**, 3445–3449 (1991)

17.89 B. Brezina, M. Havrankova: Orientation of structure and crystals of TGS and TGS doped with D, al or L, al, Cryst. Res. Technol. **20**, 781–786 (1985)

17.90 G.M. Loiacono, J.P. Dougherty: *Final Technical Report* (Night Vision and Electro-optics Laboratories, Fort Belvoir 1978), Contract No. DAAK70-77-C-0098

17.91 G. Arunmozhi, E. De Matos Gomes, J. Ribeiro: Dielectric properties of L-asparagine doped TGS (Asp-TGS) crystals, Ferroelectrics **295**, 87–95 (2003)

17.92 S. Seif, K. Bhat, A.K. Batra, M.D. Aggarwal, R. B.Lal: Effect of Cr(III) impurity on the growth kinetics of potassium dihydrogen phosphate and triglycine sulfate grown from aqueous solutions, Mater. Lett. **58**, 991–994 (2004)

17.93 V.A. Kuznetov, T.M. Okhrimenko, M. Rak: Growth promoting effect of organic impurities on growth kinetics of KAP and KDP crystals, J. Cryst. Growth **193**, 164–173 (1998)

17.94 R.J. Davey: *Industrial Crystallization*, Vol. 78, ed. by E.J. de Jong, S.J. Jancic (North-Holand, Amsterdam 1979) pp. 169–198

17.95 O.W. Wang, C.S. Fang: Investigation of the solution status of TGS and ATGSP crystals, Cryst. Res. Technol. **27**, 245–251 (1992)

17.96 K. Meera, R. Muralindharan, A.K. Tripathi, P. Ramasamy: Growth and characterization of l-threonine, dl-threonine, l-methionine admixtured TGS crystals, J. Cryst. Growth **263**, 524–531 (2004)

17.97 G. Su, Y. He, H. Yao, Z. Shi, Q. Wu: A new pyroelectric crystal L-lysine-doped THS (LLTGS), J. Cryst. Growth **209**, 220–222 (2000)

17.98 J. Novotny, J. Zelinkay, F. Moravec: Broadband infrared detectors on the basis of PATGS/Pt(IV) single crystals, Sens. Actuators A **119**, 300–304 (2005)

17.99 S. Kalainathan, M.B. Margaret, T. Trusan: Morphological changes of L-asparagine doped TGS crystal, Cryst. Eng. **5**, 71–78 (2002)

17.100 http://www.nasa.gov/missions/index.html (last accessed August 9, 2009)

17.101 R.B. Lal, R.L. Kroes: Solution growth of crystal on Spacelab-3, Proc. 24th AIAA Sci. Meet. (Reno 1985)

17.102 R.B. Lal, M.D. Aggarwal, R.L. Kroes, W.R. Wilcox: A new technique of solution crystal growth, Phys. Status Solidi (a) **80**, 547–551 (1983)

17.103 A.K. Batra, R.B. Lal, M.D. Aggarwal: Electrical properties of TGS crystals grown by new technique, J. Mater. Sci. Lett. **4**, 1425–1427 (1985)

17.104 R.B. Lal, M.D. Aggarwal, A.K. Batra, R.L. Kroes: Solution growth of crystals in zero-gravity, final technical report (NASA, Washington 1987), NASA contract number NAS8–32945

17.105 R.B. Lal: Solution growth of crystals in low-gravity, final technical report (1995), NASA contract number NAS8–36634

17.106 B. Steiner, R. Dobbyn, D. Black, H. Burdette, M. Kuriyama, R. Spal, L. van den Berg, A. Fripp, R. Simchick, R.B. Lal, A.K. Batra, D. Mathiesen, B. Ditchek: High resolution diffracting imaging of crystals grown, Microgravity and closely related terrestrial crystals (NIST, Gaithersburg 1991), Technical Note 1287

17.107 R.B. Lal, A.K. Batra, J.D. Trolinger, W.R. Wilcox: TGS crystal growth experiment on the first international microgravity laboratory (IML-1), Microgravity Q. **4**(3), 186–198 (1994)

17.108 D.E. McRee: *Practical Protein Crystallography* (Academic Press, San Diego 1993) pp. 1–23

17.109 G. Rhodes: *Crystallography Made Crystal Clear* (Academic, San Diego 1993) pp. 8–10

17.110 G. Rhodes: *Crystallography Made Crystal Clear* (Academic, San Diego 1993) pp. 29–38

17.111 S.D. Durbin, G. Feher: Studies of crystal growth mechanisms of proteins by electron microscopy, J. Mol. Biol. **212**, 763–774 (1990)

17.112 E. Forsythe, M.L. Pusey: The effects of temperature and NaCl concentration on tetragonal lysozyme face growth rates, J. Cryst. Growth **139**, 89–94 (1994)

17.113 A. Nadaraja, E.L. Forsythe, M.L. Pusey: The averaged face growth rates of lysozyme crystals: The effect of temperature, J. Cryst. Growth **151**, 163–172 (1995)

17.114 W. Littke, C. John: Protein single crystal growth under microgravity, Science **225**, 203–204 (1984)

17.115 F. Rosenberger: Protein crystallization, J. Cryst. Growth **166**, 40–54 (1996)

17.116 W. Littke, C. John: Protein single crystal growth under microgravity, J. Cryst. Growth **76**, 663–672 (1986)

17.117 L.J. DeLucas, F.L. Suddath, R.S. Snyder, R. Naumann, M.B. Broom, M.L. Pusey, V. Yost, B. Herren, D.C. Carter, B. Nelson, E.J. Meehan, A. McPherson, C.E. Bugg: Preliminary investigations of protein crystal growth using the space shuttle, J. Cryst. Growth **76**, 681–693 (1986)

17.118 L.J. DeLucas, C.D. Smith, H.W. Smith, V.K. Senadhi, S.E. Senadhi, S.E. Ealick, D.C. Carter, R.S. Snyder, P.C. Weber, F.R. Salemme, D.H. Ohlendorf, H.M. Einspahr, L.L. Clancy, M.A. Navia, B.M. Mc-Keever, T.L. Nagabhushan, G. Nelson, A. McPherson, S. Koszelak, G. Taylor, D. Stammers, K. Powell, G. Darby, C.E. Bugg: Protein crystal growth in microgravity, Science **246**, 651–653 (1989)

17.119 S. Simic-Stefani, M. Kawaji, H.U. Hu: G-jitter induced motion of a protein crystal under microgravity, J. Cryst. Growth **294**, 373–384 (2006)

17.120 A. McPherson: Virus and protein crystal growth on earth and in microgravity, J. Phys. D **26**, B104–B112 (1993)

17.121 D.C. Carter, T.E. Dowling: Protein crystal growth apparatus for microgravity, US Patent 5643540 (1997)

17.122 D.C. Carter, B. Wright, T. Miller, J. Chapman, P. Twigg, K. Keeling, K. Moody, M. White, J. Click, J.R. Ruble, J.X. Ho, L. Adcock-Downey, T. Dowling, C.-H. Chang, P. Ala, J. Rose, B.C. Wang, J.-P. Declercq, C. Evrard, J. Rosenberg, J.-P. Wery, D. Clawson, M. Wardell, W. Stallings, A. Stevens: PCAM: A multi-user facility-based protein crystallization apparatus for microgravity, J. Cryst. Growth **196**, 610–622 (1999)

17.123 D.C. Carter, B. Wright, T. Miller, J. Chapman, P. Twigg, K. Keeling, K. Moody, M. White, J. Click, J.R. Ruble, J.X. Ho, L. Adcock-Downey, G. Bunick, J. Harp: Diffusion-controlled crystallization apparatus for microgravity (DCAM): Flight and ground-based applications, J. Cryst. Growth **196**, 602–609 (1999)

17.124 L.J. DeLucas, K.M. Moore, M.M. Long, R. Rouleau, T. Bray, W. Crysel, L. Weise: Protein crystal growth in space, past and future, J. Cryst. Growth **237–239**, 1646–1650 (2002)

17.125 J.P. Declercq, C. Evrard, D.C. Carter, B.S. Wright, G. Etienne, J. Parello: A crystal of a typical EF-hand protein grown under microgravity diffracts x-rays beyond 0.9 resolution, J. Cryst. Growth **196**, 595–601 (1999)

17.126 S. Gorti, E.L. Forsythe, M.L. Pusey: Measurable characteristics of lysozyme crystal growth, Acta Cryst. D **61**, 837–843 (2005)

17.127 http://science.nasa.gov/ssl/msad/pcg/ (last accessed August 9, 2009)

17.128 D.C. Carter, K. Lim, J.X. Ho, B.S. Wright, P.D. Twigg, T.Y. Miller, J. Chapman, K. Keeling, J. Ruble, P.G. Vekilov, B.R. Thomas, F. Rosenberger, A.A. Chernov: Lower dimmer impurity incorporation may result in higher perfection of HEWL crystals grown in microgravity: a case study, J. Cryst. Growth **196**, 623–637 (1999)

18. Hydrothermal Growth of Polyscale Crystals

Kullaiah Byrappa

In this chapter, the importance of the hydrothermal technique for growth of polyscale crystals is discussed with reference to its efficiency in synthesizing high-quality crystals of various sizes for modern technological applications. The historical development of the hydrothermal technique is briefly discussed, to show its evolution over time. Also some of the important types of apparatus used in routine hydrothermal research, including the continuous production of nanosize crystals, are discussed. The latest trends in the hydrothermal growth of crystals, such as thermodynamic modeling and understanding of the solution chemistry, are elucidated with appropriate examples. The growth of some selected bulk, fine, and nanosized crystals of current technological significance, such as quartz, aluminum and gallium berlinites, calcite, gemstones, rare-earth vanadates, electroceramic titanates, and carbon polymorphs, is discussed in detail. Future trends in the hydrothermal technique, required to meet the challenges of fast-growing demand for materials in various technological fields, are described. At the end of this chapter, an Appendix 18.A containing a more or less complete list of the characteristic families of crystals synthesized by the hydrothermal technique is given with the solvent and pressure–temperature (PT) conditions used in their synthesis.

Crystals are the unacknowledged pillars of modern technology owing to their ever-increasing applications in various technologies such as electronics, microelectronics, magnetics, optics, nonlinear optics, photonics, optoelectronics, magnetoelectronics, biomedicine, biophotonics, biotechnology, nanotechnology, etc. Accordingly the size and quality of crystals control their application potential, as the properties also vary greatly with size, i. e., from bulk crystals to nanocrystals, due to the quantization effect. Hence, a new terminology (*polyscale crystals*) has become more appropriate in recent years for contributions like this devoted to the hydrothermal growth of crystals of different compositions and sizes. When the hydrothermal technique was initiated in the mid 19th century, the focus was essentially on mineral synthesis, also in the form of bulk single crystals. During World War II the importance of the hydrothermal technique was realized with the tremendous success in the growth of larger-size quartz crystals, and the focus shifted to the design of different types of autoclaves which could hold fluids under high-pressure and high-temperature conditions over a longer period. A greater variety of crystals hitherto unknown or without natural counterparts was synthesized during the 1960s and 1970s. Although there was a slight decline in the popularity of the hydrothermal technique during the 1980s, it has now picked up as one of the best methods to grow not only bulk crystals, but also fine and nanocrystals with desired shape, size, and properties. Hence, there has been a sudden surge in activities related to hydrothermal research in the last decade, because of the high-quality crystals of different sizes that can be grown well under hydrothermal conditions. This is further supported by the overwhelming success in the field of thermodynamic modeling and solution chemistry under hydrothermal conditions, which have drastically reduced the PT conditions required for crystal growth.

18.1	History of Hydrothermal Growth of Crystals	603
18.2	Thermodynamic Basis of the Hydrothermal Growth of Crystals ..	606
	18.2.1 Hydrodynamic Principles of the Hydrothermal Growth of Crystals	606

18.2.2 Thermodynamic Modeling of the Hydrothermal Growth of Crystals 608			18.4 **Hydrothermal Growth of Some Selected Crystals** 620	
18.2.3 Solutions, Solubility, and Kinetics of Crystallization under Hydrothermal Conditions 610			18.4.1 Quartz......................... 620	
			18.4.2 Aluminum and Gallium Berlinites . 625	
			18.4.3 Calcite 628	
			18.4.4 Gemstones 629	
18.3 **Apparatus Used in the Hydrothermal Growth of Crystals** .. 615			18.4.5 Rare-Earth Vanadates................. 633	
			18.5 **Hydrothermal Growth of Fine Crystals** 634	
18.3.1 Morey Autoclave........................ 617			18.6 **Hydrothermal Growth of Nanocrystals** 637	
18.3.2 Tuttle–Roy Cold-Cone Seal Autoclaves............................... 617			18.7 **Concluding Remarks** 640	
18.3.3 General-Purpose Autoclaves and Others 618			18.A **Appendix**... 641	
			References 646	

The term *hydrothermal* is purely of geological origin. It was first used by the British geologist *Sir Roderick Murchison* (1792–1871) to describe the action of water at elevated temperature and pressure, in bringing about changes in the Earth's crust leading to the formation of various rocks and minerals. It is well known that the largest single crystal formed in nature (a beryl crystal of > 1000 kg) and some of the largest quantities of single crystals created by man in one experimental run (quartz crystals of several 1000 kg) are both of hydrothermal origin [18.4]. The technique is very important for its technological efficiency in developing high-quality crystals of different sizes. The term *polyscale* is relatively new and refers to different sizes of crystals, covering bulk single crystals, small crystals, and micrometer to nanosize crystals [18.5]. This concept becomes more relevant with the progress achieved in nanotechnology, wherein the quantization size effect explains the changes in the physical properties of the crystalline materials with size. The hydrothermal technique is becoming one of the most important tools for advanced materials processing, particularly owing to its advantages in the processing of nanostructural materials for a wide variety of technological applications such as electronics, optoelectronics, catalysis, ceramics, magnetic data storage, biomedicine, biophotonics, etc. The hydrothermal technique greatly helps in processing monodispersed and highly homogeneous nanocrystals [18.6]. The hydrothermal technique refers to any

Table 18.1 Different terminologies used in the hydrothermal technique

Conventional hydrothermal	Solvothermal	Supercritical hydrothermal	Related terminologies	Multi-energy hydrothermal
Aqueous solvent; Refers to conditions above atmospheric temperature and pressure; Suitable for high quality; Bulk, fine nanocrystals [18.1]	Nonaqueous solvents; Low to high temperature conditions; Suitable for good quality; Bulk, fine nanocrystals [18.1, 2]	Critical to supercritical conditions; Both aqueous and nonaqueous solvents; Suitable for fine and nanocrystals; Rapid [18.3]	Ammonothermal, glycothermal, lyothermal, alcothermal, carbonothermal, etc. depending upon the specific solvent used; Spray pyrolysis; Suitable for fine to nanocrystals and thin films [18.1, 2]	Hydrothermal in combination with extra energy likesuch as microwave, electrochemical, sonar, mechanochemical, biomolecular, sol–gel, etc. Extremely efficient for thin films, fine to nanocrystals; Epitaxy, etc. Very fast processing [18.2]

homogeneous (nanocrystals) or heterogeneous (bulk single crystals) reaction in the presence of aqueous solvents or mineralizers under high-pressure and high-temperature conditions to dissolve and recrystallize (recover) materials that are relatively insoluble under ordinary conditions. *Byrappa* and *Yoshimura* define *hydrothermal* as any homogeneous or heterogeneous chemical reaction in the presence of a solvent "(whether aqueous or nonaqueous) above the room temperature and at pressure greater than 1 atm in a closed system" [18.1, 2]. However, chemists prefer to use the term *solvothermal*, meaning "any chemical reaction in the presence of a nonaqueous solvent or solvent in supercritical or near-supercritical conditions". Table 18.1 gives different terminologies used by different specialists for hydrothermal technology.

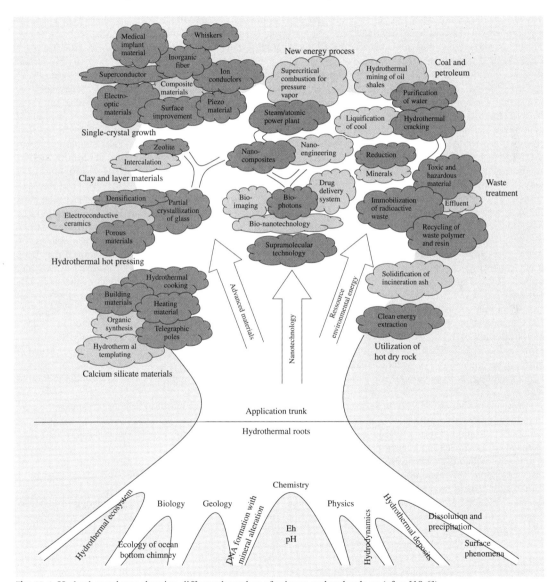

Fig. 18.1 Hydrothermal tree showing different branches of science and technology (after [18.6])

In recent years, the addition of energy into the hydrothermal process, such as in hydrothermal-electrochemical, hydrothermal-mechanochemical, hydrothermal-microwave, hydrothermal-sonar, hydrothermal-sol–gel, hydrothermal-biomolecular, etc., methods has made the process extremely effective and fast. It also leads us to a new concept of chemistry at the speed of light [18.2, 6]. However, this new process is only suitable for the preparation of fine to nanosize crystals, and thin films with high crystallinity and desired properties. In recent years, capping agents, organic molecules, surfactants, etc., have become popular for use in crystal growth in order to achieve growth in the desired crystallographic direction and stunt growth in the undesired crystallographic directions under hydrothermal conditions. Such developments have made the hydrothermal technique unique in terms of its ability to fabricate functional products with in situ control over their growth. All these recent developments in the last 10 years have completely changed the concept of hydrothermal growth of crystals. Earlier, researchers had always considered hydrothermal growth of crystals to be a high-temperature and high-pressure technique that was very expensive and slow in terms of growth rate. However, a new generation of hydrothermal researchers now consider it to be a simple technique that requires mild to low temperature/pressure, which is not expensive, and which is fast with a maximum product yield. Thus hydrothermal technique is being popularly employed by physicists, chemists, ceramists, hydrometallurgists, material scientists, biologists, engineers, geologists, technologists, and so on. Figure 18.1 shows the various branches of science either emerging from the hydrothermal technique or closely linked with it. One could firmly say that this family tree will keep expanding its branches and roots in the years to come.

Here, the author uses the term *hydrothermal* only in a broad sense covering a set of all the above-mentioned variations given in Table 18.1 to describe chemical reactions taking place in the presence of a solvent – aqueous or nonaqueous – under subcritical or supercritical conditions in a closed system. Similarly, the hydrothermal method has been widely accepted since the 1960s and practically all inorganic species, from native elements to the most complex silicates, germinates, phosphates, and others, have been obtained by this method. The technique is being employed on an industrial scale to prepare bulk to nanosize crystals for piezoelectric, optoelectronic, magnetic, ceramic, photonic, etc., applications. It offers several advantages over the conventional techniques of crystal growth in terms of purity, homogeneity, crystal symmetry, metastable-phase formation, reproducibility, lower crystallization temperature, single-step processing, simple equipment, lower energy requirements, fast reaction times, desired polymorphic modifications, growth of ultralow-solubility materials, etc.. For example, it is the only method that has been successfully employed to produce large-size single crystals of α-quartz on an industrial scale. Similarly compounds with elements in oxidation states that are difficult to obtain (especially transitional-metal compounds) by other ordinary methods can be synthesized well under hydrothermal conditions, for example, ferromagnetic chromium(IV) oxide. The synthesis of metastable phases such as subiodides of tellurium (Te_2I) can be carried out more easily under hydrothermal conditions [18.7]. The scope of the present chapter has been limited to the hydrothermal growth of crystals – bulk single crystals to selected nanocrystals – instead of describing the application of the entire hydrothermal technology that deals with the hydrothermal reactions, hydrothermal treatment of various organic and inorganic materials including recycling, frequently employed on a large scale. Also it is impossible to discuss the growth of each and every crystalline compound available in the literature, as they exceeds several hundred; the focus is limited to selected crystals of current interest and also some exceptionally major works of the past. The following aspects have been covered in the present chapter:

1. The history of hydrothermal growth of crystals
2. The thermodynamic basis of the hydrothermal growth of crystals
3. Hydrothermal apparatus and safety measures to be adopted
4. Hydrothermal growth of selected crystals:
   a) Bulk crystals
   b) Small crystals
   c) Nanocrystals
5. Future trends in the hydrothermal growth of crystals.

The theory of hydrothermal growth of crystals dealing with the growth mechanism is a recent subject; hence more emphasis is placed on the experimental aspects of crystal growth under hydrothermal conditions. However, the growth mechanism is briefly discussed for those compounds where this is available in the literature.

18.1 History of Hydrothermal Growth of Crystals

The history of hydrothermal growth of crystals has been elaborated in several works by *Byrappa* and co-workers [18.1, 2, 6, 7]. Here, the history of the hydrothermal growth of crystals is discussed briefly. The hydrothermal technique was initiated by Earth scientists during the middle of the 19th century to understand the genesis of rocks and minerals by simulating the natural conditions existing under the Earth's crust and crystallizing them in the laboratory. The first hydrothermal experiment was carried out by *Schafthaul* in 1845 to synthesize quartz crystals upon transformation of freshly precipitated silicic acid in Papin's digester [18.8]. Followed by this there were quite a few works by French and German mineralogists on the hydrothermal synthesis of minerals [18.9–12]. However, the size of the crystals obtained in general did not exceed thousands or hundreds of a millimeter. Thus, it was well known that the majority of the early hydrothermal experiments carried out during the 1840s to the early 1900s mainly dealt with fine to nanocrystalline products, which were discarded as failures due to the lack of sophisticated electron microscopic techniques available during that time to observe such small-sized products. Many times, when the bulk crystals or single crystals of several millimeter sizes were not obtained in the products, the experiments were considered as failures and the materials were washed away. Perhaps the greatest contribution during the 19th century in the field of hydrothermal synthesis of crystals was by De Senarmount, the founder of hydrothermal synthesis in geoscience. He synthesized mineral carbonates, sulfates, sulfides, and fluorides using glass liners in autoclaves [18.11]. *Bunsen* first used thick-walled glass tubes to contain high-temperature high-pressure liquids and prepare strontium and barium carbonates [18.9]. Although *Saint-Claire Deville* attempted to transform bauxite into corundum under hydrothermal conditions using NaOH as a mineralizer, the experimental results were not definite; perhaps he was the first to use a mineralizer other than water [18.13]. *Von Chrustschoff* first proposed the noble-metal lining of autoclaves to prevent corrosion [18.14]. Before this the glass tubes that were used were frequently attacked under hydrothermal conditions, and earlier researchers do not mention anything about the precautions taken in this regard. With the introduction of steel autoclaves and noble-metal linings, the tendency to reach higher-pressure/temperature conditions began. However, the majority of works up to 1880 continue to pertain to quartz, feldspar, and related silicates. *Hannay* claimed to have synthesized artificial diamond by the hydrothermal technique [18.15]. Similarly, *Moissan* also claimed to have synthesized diamond as large as 0.5 mm artificially from charcoal [18.16]. Though, the success of these experiments was treated as dubious, they certainly provided a further stimulus for hydrothermal research, particularly the development of high-pressure techniques. Perhaps the first large crystals obtained using this technique in the 19th century were by *Friedel* and *Sarasin* (1881), who synthesized hydrated potassium silicate, 2–3 mm long [18.17]. Then, in 1891 *Friedel* obtained corundum crystals by heating a solution of NaOH with excess Al_2O_3 at a high temperature for that time: 530–535 °C [18.18].

Towards the end of the 19th century, Spezia from the Torino Academy of Science began his classical work on the seeded growth of quartz. His contribution to the field of hydrothermal growth of crystals is remembered even today. *Spezia* (1896) found that plates of quartz kept at 27 °C for several months with water under a pressure of 1750–1850 atm did not lose their weight and also showed no etch figures; thus, he concluded that pressure alone has no influence on the solubility of quartz [18.19]. *Spezia* studied the solubility of quartz in such great detail that the growth of bulk single crystals of quartz became possible [18.20]. So before the end of the 19th century, a large number of minerals had been synthesized, and experiments had been carried out on a wide variety of geological phenomenon ranging from the origin of ore deposits to the origin of meteorites. According to *Morey* and *Niggli*, around 150 mineral species including diamond had been synthesized by 1900 [18.21]. *Morey* quotes a horrible experience of one of the earliest hydrothermal experiments in which, after reacting a mixture of colloidal silicon, colloidal ferric hydroxide, colloidal ferrous oxide, lime water, magnesium hydroxide, and potassium hydroxide in glass for 3 months at 550 °C, some fine crystals of hornblende were obtained, which had no bearing on the petrogenesis or the phase equilibria of the mineral system [18.22]. This was definitely due to lack of knowledge in areas of solvent chemistry, thermodynamics, kinetics, and compound solubility.

The entire activity on the hydrothermal research was concentrated in Europe, and there was no activity in North America or Asia, including Japan, China, India, and Taiwan, which are included today in the top ten countries actively engaged in hydrothermal research.

Perhaps the first published work from North American on hydrothermal research was by *Barus*, who essentially worked on the impregnation of glass with water to such an extent that it melted below 200 °C by using steel autoclaves [18.23]. Following this, Allen published his classic work on the growth of quartz crystals 2 mm long using steel autoclaves with copper sealing. The first commercialization of the hydrothermal technique took place in the early 20th century to leach bauxite, an aluminum ore, by Bayer's process. However, this is not related to crystal growth. The establishment of the Geophysical Laboratory at the Carnegie Institute of Washington in 1907 probably marked the most important milestone in the history of hydrothermal research. The credit goes to pioneers such as Bridgman, Cohen, Morey, Niggli, Fenner, and Bowen in the early 20th century, who changed the scenario of hydrothermal research. They carried out an impressive amount of basic research, along with their European counterparts. However, the total research effort was small and the study passed into a period of dormancy except for phase equilibria studies in some systems relevant to natural systems. This was connected with the need for materials with a combination of high strength and corrosion resistance at high temperatures. Although, a great deal of research was carried out during the 20th century, the facilities for large-scale hydrothermal research before the end of World War II were virtually nonexistent, with the exception of at the University of Chicago and Harvard University. This situation changed dramatically with the development of test-tube-type pressure vessels by Tuttle, later modified by Roy. These test-tube-type pressure vessels are some of the most versatile autoclaves used worldwide today; also popularly known as batch reactors, that could hold temperatures up to 1150 °C at lower pressures and pressure up to 10 kbar at lower temperatures (the modified titanium zirconium molybdenum (TZM) autoclaves).

The actual impetus for hydrothermal growth of crystals began during World War II with the growth of quartz. Here, in contrast to the very slow growth rate achieved by earlier workers such as *De Senarmount* and *Spezia*, captured German reports show that *Nacken*, using natural α-quartz as seed crystals and vitreous silica as the nutrient, had grown quartz crystals in an isothermal system and succeeded in obtaining a large quartz crystal from a small seed [18.25, 26]. *Nacken* (1884–1971) worked on the synthesis of various crystals from 1916 onwards, but left this field. In 1927 or 1928, he started working only on the hydrothermal growth of quartz crystals. On Nacken's work, Sawyer writes (cited by Bertaut and Pauthenet, 1957) that, "... Nacken made quartz crystals of 1″ diameter by using hydrothermal method and the conditions are given as ..." followed by some biographical data. Similarly, Nacken's emphasis on quartz growth has also been documented by Sawyer. Almost at the same time, Nacken made emerald single crystals by the hydrothermal method and also beryl or corundum crystals for watch bearings. He prepared a large number of synthetic emeralds by using a trace of chromium to produce color and could obtain hexagonal prisms of emerald weighing about 0.2 mg in a few days [18.27]. Figure 18.2 shows the first large-size manmade quartz crystals obtained by *Nacken* [18.24]. At almost this time, Russian researchers were producing bulk quartz crystals using old discarded cannon barrels as autoclaves. Unfortunately, these works were not published. However, the author of this chapter had an opportunity to interact on several occasions with the late *Prof. Shternberg*, from the Institute of Crystallography, Academy of Sciences (erstwhile USSR), who was mainly responsible for such great and unpublished Russian work [18.28].

With the publication of Nacken's work in 1950, several laboratories around the world began working on large-scale production of quartz crystals. During the 1960s, interest in the hydrothermal method was boosted in connection with the industrial growth of good-quality quartz single crystals and the successful growth of most complex inorganic compounds. Perhaps this is the beginning of the growth of crystals which did not have natural analogues. During the 1970s, there was a question about the search for and growth of hitherto unknown compounds of photosemiconductors, ferromagnets, lasers, and piezo- and ferroelectrics,

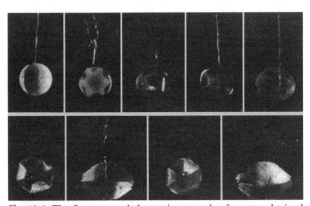

Fig. 18.2 The first manmade large-size crystals of quartz, obtained by *Nacken* [18.24]

and the hydrothermal method attracted great attention. Several established laboratories in the world began to study systematically various aspects of the hydrothermal growth of crystals such as the physicochemical principles, kinetics, designing new apparatus, growing new compounds, and so on. Specific aspects of the hydrothermal method as a modeling tool to understand the natural processes of mineral formation changed dramatically into an important method characteristic for inorganic chemistry. During the 1980s, a new sealing method for large-size autoclaves was designed, viz. Grey-Loc sealing, which facilitates the construction of very large-size autoclaves with a volume of 50000 l [18.29]. Toyo Electric Co., Japan, houses the world's largest autoclave for the growth of quartz crystals.

Towards the end of the 1970s, on the whole, the hydrothermal field experienced a declining trend for two reasons: there was no major scope for further work on the growth of large-size single crystals of quartz on the one hand, and on the other hand, large-scale attempts to grow larger crystals of other compounds investigated during the 1960s and 1970s failed miserably. It was unanimously decided that the hydrothermal technique was not suitable for the growth of large crystals other than quartz. The focus at this time was on Czochralski, molecular chemical vapor deposition (MOCVD), and molecular-beam epitaxy (MBE). This is mainly connected to the general approach of the hydrothermal researchers to grow large single crystals without looking into the hydrothermal solvent chemistry and the kinetics of the crystallization process. However, the Nobel Symposium organized by the Swedish Academy of Sciences, during September 17–21, 1979, on *The Chemistry and Geochemistry of Solutions at High Temperatures and Pressures* is remembered as an eye-opener. The presence of pioneers in the field of hydrothermal physical chemistry such as *Franck, Seeward, Helgeson, Pitzer*, and so on, drew the attention of hydrothermal crystal growers and a new trend was set to look into the hydrothermal solvent chemistry and the physical chemistry of the hydrothermal systems [18.30–33]. Following this, *Prof. Somiya*, Japan, organized the first ever International Symposium on Hydrothermal Reactions, in 1982, which was attended largely by specialists from different branches of science such as physical chemistry, inorganic chemistry, solid-state physics, material scientists, organic chemists, hydrometallurgists, hydrothermal engineers, etc. [18.34]. This was the dawn of modern hydrothermal research, and since then new avenues in the field of hydrothermal research are being explored. As evident from the recent literature data, the hydrothermal technique is one of the most efficient techniques for the growth of high-quality crystals of GaN, ZnO, GaPO$_4$, etc. as well as traditional quartz crystals. Obviously there is a surge in the activities related to hydrothermal growth of crystals. This is greatly supported by developments in thermodynamic modeling and also understanding of the fluid dynamics in autoclaves through simulation.

There is a great difference between the hydrothermal research carried out during the previous century and the early 21st century. During the mid-20th century the hydrothermal technique was at its peak and the focus was mainly on the high-temperature and high-pressure regime of crystal growth, because of lack of knowledge on the solubility of several compounds and also on the selection of an appropriate solvent. The First International Hydrothermal Symposium (1982) held at the Tokyo Institute of Technology, Japan, brought together specialists from the interdisciplinary branches of science [18.34]. Since then knowledge on the physical chemistry and the PVT relationship in hydrothermal systems has greatly improved, which helped in drastically reducing the temperature and pressure conditions required for crystal growth. Similarly solvothermal and supercritical processing have been developed, which use a variety of other solvents such as organic and organometallic complexes in the synthesis, thereby moving this technique towards "green" chemistry. Table 18.2 presents trends in the hydrothermal growth of crystals, taking the hydrothermal

Table 18.2 Current trends in hydrothermal technology [18.5]

Crystal	Earlier works	Present Author
Li$_2$B$_4$O$_7$	$T = 500$–$700\,°\mathrm{C}$	$T = 240\,°\mathrm{C}$
	$p = 500$–$1500\,\mathrm{bar}$	$p \leq 100\,\mathrm{bar}$
Li$_3$B$_5$O$_8$(OH)$_2$	$T = 450\,°\mathrm{C}$	$T = 240\,°\mathrm{C}$
	$p = 1000\,\mathrm{bar}$	$p = 80\,\mathrm{bar}$
NaR(WO$_4$)$_2$	$T = 700$–$900\,°\mathrm{C}$	$T = 200\,°\mathrm{C}$
R = La, Ce, Nd	$p = 2000$–$3000\,\mathrm{bar}$	$p \leq 100\,\mathrm{bar}$
R:MVO$_4$	Melting point	$T = 100\,°\mathrm{C}$
R = Nd, Eu, Tm;	$> 1800\,°\mathrm{C}$	$p \leq 30\,\mathrm{bar}$
M = Y, Gd		
LaPO$_4$	Synthesized at	$T < 120\,°\mathrm{C}$
	$> 1200\,°\mathrm{C}$	$p < 40\,\mathrm{bar}$
Zoisite	$T = 500\,°\mathrm{C}$	$T = 250\,°\mathrm{C}$
	$p = 5\,\mathrm{kbar}$	$p \leq 100\,\mathrm{bar}$

technique towards green technology for sustained human development, since it consumes less energy, no or little solid/liquid waste, no recovery treatment, no hazardous processing materials, high selectivity, a closed processing system, etc. [18.5]. Hydrothermal chemistry has to be understood precisely in order to grow crystals under soft and environmentally benign conditions.

18.2 Thermodynamic Basis of the Hydrothermal Growth of Crystals

The thermodynamic basis of the hydrothermal growth of crystals is perhaps the least explored aspect in the literature. As such, there is no major review or book dealing with this aspect, although some significant developments have occurred in this area since the Nobel Symposium organized by the Royal Swedish Academy of Sciences in 1978, followed by the First International Symposium on Hydrothermal Reactions, organized by the Tokyo Institute of Technology in 1982, helped in setting a new trend in hydrothermal technology by attracting physical chemists in large numbers. Hydrothermal physical chemistry today has enriched our knowledge greatly through a proper understanding of hydrothermal solution chemistry. The behavior of the solvent under hydrothermal conditions dealing with aspects such as structure at critical, supercritical, and subcritical conditions, dielectric constant, pH variation, viscosity, coefficient of expansion, density, etc., are to be understood with respect to pressure and temperature. Similarly, thermodynamic studies have yielded rich information on the behavior of solutions at various pressure–temperature conditions. Since most hydrothermal crystal growth experiments are carried out under conditions with temperature gradients in standard autoclaves, growth occurs due to the recrystallization of the solid substance, including dissolution in the liquid phase and convective mass transfer of the dissolved part of the substance to the growth zone or seed, and also through the dissolution of the mixture of the nutrient components with the help of their convective mass transport into the growth zone and interaction of the dissolved components on the seed surface. However, there are several macro- and microprocesses occurring at the interface boundary of the solution and the crystal/seed. Therefore, the composition and concentration of the solution, and the temperature, pressure, and hydrodynamic conditions and surface contact of the phases are some of the basic physical and chemical parameters that determine the regime and rate of dissolution of the nutrient, the mass transport, and the possibility of the formation of new phases.

18.2.1 Hydrodynamic Principles of the Hydrothermal Growth of Crystals

The hydrodynamic principles of growth of crystals under hydrothermal conditions have been studied by several groups since the 1990s using numerical modeling to understand the fluid flow in an autoclave. There are quite a few publications on this aspect. In order to grow good-quality crystals with minimum possible defects, one has to understand the flow dynamics. *Laudise* and *Nielsen* were perhaps the first to draw the attention of hydrothermal crystal growers to the thermal conditions during growth, which lead to strong buoyancy-driven flow carrying nutrient from the lower dissolution chamber to the upper growth chamber [18.35]. Then *Klipov* and *Shmakov* studied the shape and morphology of the surfaces, the growth rate, the macroscopic defects, and the inclusion density of hydrothermally grown quartz crystals, and found a strong correlation with the fluid flow around the crystals [18.36]. *Ezersky* et al. have studied the hydrodynamics under hydrothermal conditions using a shadowgraph technique [18.37, 38]. This helped the authors to understand the spatiotemporal structure of hydrothermal waves in Marangoni convection. *Roux* et al. have developed both two-dimensional (2-D) axisymmetric and three-dimensional (3-D) models for hydrothermal crystal growth, by focusing on the bulk flow pattern in an autoclave [18.39]. Their 2-D axisymmetric model gave an unrealistic nonaxisymmetric flow pattern, while their 3-D model considers a square cross-section for the container instead of a circular or cylindrical one actually used in hydrothermal experiments. However, the results obtained were based on very low Rayleigh numbers (up to 6×10^4), corresponding to very small autoclaves, which cannot be applied to the larger autoclaves used for bulk or commercial crystal growth. *Chen* et al. [18.40, 41] have studied the flow dynamics by using a numerical model with a higher Rayleigh numbers and developed a comprehensive 3-D model for the hydrothermal growth of crystals. The significant aspect of their work is the introduction of a fluid-superposed porous raw material bed in the lower dissolution chamber of the autoclave. This model predicts a strong 3-D flow in the autoclave and a strong temperature fluctuation. However, the best works to date on the numerical modeling of flow pattern dynamics comes from *Evans* and group [18.42–45]. Their 3-D modeling is based on industrial-scale hydrothermal autoclaves with various aspect ratios with nonuniform heating conditions. The nonuniform heating is introduced on the surface of the lower dissolving chamber and the upper growing chamber of an autoclave with or without a baffle at the middle height. It was found that the circumferentially nonuniform surface temperature has dramatic effects on the fluid flow and therefore the temperature distribution in the bulk fluid. With a temperature deviation, the flow is three dimensional. When only the dissolving chamber is subjected to circumferentially nonuniform heating, a baffle is essential to create a uniform growth environment in the growth chamber. It is evident from their work that, in order to obtain high-quality single crystals, wall-temperature control on the growth chamber wall is more important than on the dissolving chamber wall. Figure 18.3 shows the velocity distributions and temperature contours for an autoclave without a baffle and with a 15% baffle, respectively. The velocity field is much stronger in the case without a baffle than that with a baffle. The numerical study of the effects of various baffles (with 2–25% opening) on the fluid flow and temperature fields have been investigated by these authors. Accordingly a smaller opening leads to a weaker flow field and a more uniform temperature profile. A multihole baffle establishes a more uniform temperature in the upper chamber than does a single-hole baffle of the same area opening. The number of holes in the multihole baffle has significant effects, while the hole arrangements affect the thermal condition only near the baffle.

The flow pattern, with a high flow velocity in the upper growing chamber, could have a significant negative effect on solute transport from the solution to the seed crystals and therefore on the crystal growth rate and quality. The strong flow also leads to strong distortion of the temperature field. A nonuniform temperature field in the growth region is not desirable for growing uniform high-quality crystals. A significant development in the field of numerical modeling of heat transfer processes and flow fields under hydrothermal conditions is the work related to the crystal growth of beryl, $AlPO_4$, and $GaPO_4$ [18.46, 47]. These studies have yielded some insight into the flow pattern under hydrothermal con-

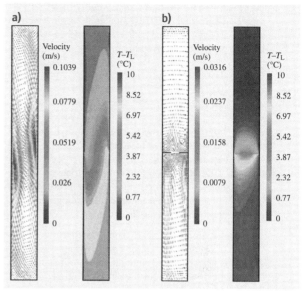

Fig. 18.3a,b Velocity pattern and isotherms in the symmetry plane. $\Delta T = 10\,°C$. On the lower chamber $\Delta T = 0.1\,°C$ only. (**a**) Without baffle; and (**b**) with a 15% area baffle opening at the median plane

ditions. For example, the influence of a rotating solid crystal or a seed crystal inside the autoclave has also been considered in such numerical modeling.

However, most of these models on numerical flow patterns are based on several assumptions and are applied mainly to simple or single-phase systems. Also, the nature of the fluid is important and the flow dynamics will vary with the type of fluid; and 3-D simulation models of the flow in the autoclave show that the turbulent flow is not in fact 3-D. The effect of the simultaneous existence of the crystals and the raw materials has not been considered in most of these models. Thus there is an incomplete picture on the whole with respect to the flow dynamics. A lot more experimental and theoretical investigations are needed to clarify and complete the description of the hydrothermal flow dynamics.

During the late 1980s to early 1990s a good number of publications appeared on the physicochemical foundation of crystal growth under hydrothermal conditions. Balitsky and Bublikova carried out detailed physicochemical investigations on malachite bulk crystal growth. Similarly Kuznetsov has reviewed the physical chemistry of hydrothermal crystal growth of II–VI compounds. Further, Popolitov has reviewed the physical chemistry of the hydrothermal growth of tellurium dioxide crystals. The reader can get more valuable information in the book *Hydrothermal Growth of Crystals* [18.4]. However, all these studies are again based on several assumptions and indirect approaches with less bearing on the thermodynamic modeling principles.

18.2.2 Thermodynamic Modeling of the Hydrothermal Growth of Crystals

A key limitation to the conventional hydrothermal method has been the need for time-consuming empirical trial-and-error methods as a means of process development. Currently, research is being focused on the development of an overall rational-engineering-based approach that will speed up process development. The rational approach involves four steps:

1. Computation of thermodynamic equilibria as a function of chemical processing variables
2. Generation of equilibrium diagrams to map the process variable space for the phases of interest
3. Design of hydrothermal experiments to test and validate the computed diagrams
4. Utilization of the processing variables to explore opportunities for controlling reactions and crystallization kinetics.

Hydrothermal crystallization is only one of the areas where our fundamental understanding of hydrothermal kinetics is lacking due to the absence of data related to the intermediate phases forming in solution. Thus our fundamental understanding of hydrothermal crystallization kinetics is in the early stage, although the importance of studies of the kinetics of crystallization was realized with the commercialization of the synthesis of zeolites during the 1950s and 1960s. In the absence of predictive models, we must empirically define the fundamental role of temperature, pressure, precursor, and time on the crystallization kinetics of various compounds. Insight into this would enable us to understand how to control the formation of solution species, solid phases, and the rate of their formation. In recent years, thermochemical modeling of chemical reactions under hydrothermal conditions has become very popular. The resulting thermochemical computation data helps in the intelligent engineering of the hydrothermal processing of advanced materials. The modeling can be successfully applied to very complex aqueous electrolyte and nonaqueous systems over wide ranges of temperature and concentration and is widely used in both industry and academy. For example, OLI Systems Inc., USA, provides software for such thermochemical modeling, and using such a package, aqueous systems can be studied within the temperature range $-50-300\,°C$, pressure ranging from 0 to 1500 bar, and concentration $0-30\,M$ in molal ionic strength; for nonaqueous systems the temperature range covered is $0-1200\,°C$ and pressure from 0 to 1500 bar, with species concentration from 0 to 1.0 mol fraction.

Such a rational approach has been used quite successfully to predict the optimal synthesis conditions for controlling phase purity, particle size, size distribution, and particle morphology of lead zirconium titanates (PZT), hydroxyapatite (HA), and other related systems [18.48–50]. The software algorithm considers the standard state properties of all system species as well as a comprehensive activity coefficient model for the solute species. Table 18.3 gives an example of thermodynamic calculations and the yield of solid and liquid species outflows at $T = 298$ K, $p = 1$ atm, $I = 0.049$ M, and pH = 12.4.

Using such a modeling approach, theoretical stability field diagrams (also popularly known as yield diagrams) are constructed to obtain 100% yield. Assuming that the product is phase-pure, the yield Y can be expressed as

$$Y_i = 100\frac{m_i^{ip} - m_i^{eq}}{m_i^{ip}}\%, \quad (18.1)$$

where m^{ip} and m^{eq} are the input and equilibrium molal concentrations, respectively, and the subscript i indicates the designated atom. Figures 18.4 and 18.5 show stability field diagrams for the PZT and HA systems.

From Fig. 18.4 it is observed that the region shaded represents 99% yield of PZT, although the PZT forms within a wide range of KOH and Ti concentrations. The figure clearly illustrates the region where all the solute species transform towards 100% product yield. Simi-

Table 18.3 Thermodynamic calculations for the HAp system

Species name	Inflows (mol)	Outflows Liquid/mol	Solid/mol
H_2O	55.51	55.51	
$Ca(OH)_2$	0.1	7.2×10^{-6}	
CaO			
Ca^{2+}		1.5×10^{-2}	
$Ca(OH)^+$		4×10^{-3}	
H^+		4.45×10^{-13}	
OH^-		3.41×10^{-2}	
Total	55.61	55.56	8.1×10^{-2}

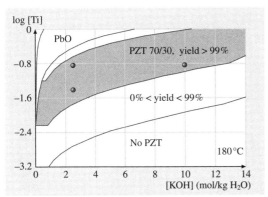

Fig. 18.4 Calculated stability field diagram for the PZT system at 180 °C with KOH as the mineralizer (after [18.48])

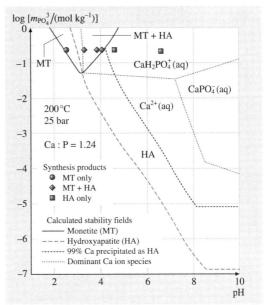

Fig. 18.5 Calculated stability field diagram for the HAp system at 200 °C and 25 bar with Ca : P ratio of 1.24 (after [18.50])

larly, from Fig. 18.5, it is observed that all the Ca species participate in the reaction to form HA, thus leading to 100% yield of HA in the region denoted by a black square. Thick dotted lines indicate the boundary above which 99% Ca precipitates as HA. The other regions mark mixed-phase precipitation such as hydroxyapatite, monatite, and other calcium phosphate phases.

Such thermodynamic studies help to intelligently engineer the hydrothermal process and also to obtain a maximum yield for a given system. This area of research has great potential for application in crystal growth, including of nanocrystals. However, in the majority of cases, the conventional and nonconventional (composition diagrams) phase diagrams are still constructed based on phase equilibria studies carried out under hydrothermal conditions for systems of interest. Nonconventional phase diagrams (composition diagrams) plotted for equilibrium conditions are popular, especially among Russian workers, and are known as NC diagrams or TC diagrams, where N stands for the nutrient composition, T stands for experimental temperature, and C stands for solvent concentration. There are hundreds of such diagrams in the literature. Figure 18.6 represents an NC diagram for the system Na_2O–RE_2O_3–SiO_2–H_2O, showing the distribution of silicon–oxygen radical groups in the rare-earth silicates under hydrothermal conditions with fixed pressure and temperature conditions [18.51]. The solid lines indicate the regions of monophase crystallization, and dashed lines indicate the beginning of crystallization of the excess component. Similarly there is one more type of composition diagrams used routinely in crystal growth to select the conditions of crystallization of various phases in a given system. Figure 18.7 represents the concentration versus temperature diagram for the system K_2O–La_2O_3–P_2O_5–H_2O [18.52]. This dia-

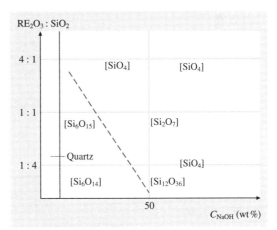

Fig. 18.6 NC diagram for the system Na_2O–RE_2O_3–SiO_2–H_2O giving the distribution of silicon–oxygen radical groups in the rare-earth silicates under hydrothermal conditions with fixed pressure and temperature conditions (after [18.51])

gram helps in selecting the experimental temperature for the growth of rare-earth orthophosphate crystals. These nonconventional phase diagrams are relatively easy to obtain and are highly useful for the growth of single crystals as they clearly depict the growth conditions.

In addition to such unconventional phase diagrams, researchers frequently use standard phase diagrams such that shown in Fig. 18.8 for the potassium titanyl phosphate (KTP) system [18.53]. This type of phase diagram occurs in the thousands for various compounds, and is commonly used in crystal growth. Such phase diagrams give in general the phase boundaries and the phase formation within a given system under fixed PT conditions, and help to select the conditions for crystal growth.

18.2.3 Solutions, Solubility, and Kinetics of Crystallization under Hydrothermal Conditions

This is one of the most important aspects of the hydrothermal growth of crystals. Initial failures in the hydrothermal growth of a specific compound are usually the result of lack of proper data on the type of solvents, the solubility, and the solvent–solute interaction. A hydrothermal solution is generally considered as a thermodynamically ideal one, yet in the case of strong and specific interaction between the solute and the solvent, or among the components of the soluble substance in them, significant deviations from Raoult's law occur. Consequently, real hydrothermal solutions differ from ideal solutions and their understanding requires knowledge of the influence of the solvent in

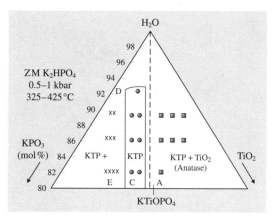

Fig. 18.8 Phase diagram for the KTP system (after [18.53])

the process of dissolution and crystallization of various compounds. Obviously, as shown in most of the experiments, the type of solvent and its concentration determine a specific hydrothermal process and its important characteristics such as the solubility of the starting materials, the quantity of the phases, their composition, and the output of the phases, kinetics, and growth mechanism of single crystals.

At the moment there is no theory which can explain and estimate solubility in real solutions. However many of the problems connected with solubility can be explained on the basis of overall physicochemical principles or laws.

In some cases, it is always better to use the empirical rule that solubility becomes high in solvents with higher dielectric constant (ε) and for types of chemical bond which are close to those of the solute substance. Deviation from this takes place when specific interactions between the solid substance and solvent occur.

The synthesis and recrystallization of a specific compound and the growth of single crystals on the seed are all carried out using different solvents on the basis of physicochemical considerations.

The following conditions are adopted in selecting the most suitable mineralizers:

1. Congruence of the dissolution of the test compounds
2. A fairly sharp change in the solubility of the compounds with changing temperature or pressure
3. A specific quantitative value of the absolute solubility of the compound being crystallized
4. The formation of readily soluble mobile complexes in the solution

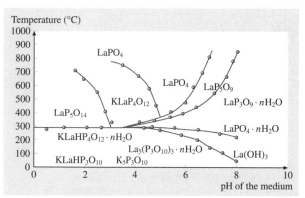

Fig. 18.7 Concentration versus temperature diagram for the system K_2O–La_2O_3–P_2O_5–H_2O (after [18.52])

5. A specific redox potential of the medium, ensuring the existence of ions of the required valence.

Additionally the solvent should have the desired viscosity, insignificant toxicity, and very weak corrosion activity with respect to the apparatus. These factors fulfill the requirements of the hydrothermal mineralizer (solvent) in addition to determining the values of solubility of the compound under investigation. Some of the basic properties of the hydrothermal medium, such as viscosity, dielectric constant, compressibility, and coefficient of expansion, are discussed here briefly in the context of crystal growth. Since diffusion is inversely proportional to solvent viscosity, one can expect very rapid diffusion in hydrothermal growth. This leads to the growth of perfect single crystals with well-developed morphology. We can expect higher growth rates, a narrower diffusion zone close to the growing interface, and less likelihood of constitutional supercooling and dendritic growth. It is thus no wonder that quartz growth rates as high as 2.5 mm/day without faults or dendritic growth have been observed [18.54].

Let us consider briefly the role of water as a solvent in the hydrothermal growth of crystals. Water is always the major component of hydrothermal solutions, with various chemical compositions in the laboratory and nature. All solutions used in hydrothermal experiments vary from one another in their properties, their ability to dissolve and crystallize, and the nature of the linking between water and electrolyte. Moreover, the properties of each solution depend upon physicochemical aspects and the structure of the pure water. The formation of associates and complexes in the aqueous solutions of electrolytes is possible because of the presence of structural water, i.e., water molecules with directional hydrogen bonding. Several chemical changes arise from changes in ionic dissociation of the solution. Therefore, it is better to understand the characteristics of the pure water under hydrothermal conditions. The hydrogen ions show an influence on the solubility of various compounds under hydrothermal conditions. Figure 18.9 shows the viscosity of water as a function of density and temperature [18.55]. It has been demonstrated that the mineralizer solutions (typically 1 M NaOH, Na_2CO_3, NH_4F, K_2HPO_4, etc.) have properties quite close to those of water. For 1 M NaOH at room temperature, $\eta_{solution}/\eta_{water} = 1.25$ [18.56] and one can expect that the viscosity of hydrothermal solutions can be as much as two orders of magnitude lower than *ordinary* solutions. The mobility of molecules and ions in the supercritical range is much higher than under normal conditions. Also electrolytes which are completely dissociated under normal conditions tend to associate with rising temperature [18.57].

Hasted et al. (1948) observed that the dielectric constant of electrolyte solutions (ε) can be regarded as a linear function of molarity up to 1–2 M, depending on the electrolyte [18.58]. *Franck* [18.55] discussed ionization under hydrothermal conditions and made careful and complete conductivity studies, showing that the conductance of hydrothermal solutions remains high despite the decrease in ε, because this effect is more than compensated for by an increase in the ion mobility brought about by the decreased viscosity under hydrothermal conditions. Figure 18.10 shows the dielectric constant of water [18.59]. Thermodynamic and transport properties of supercritical water are remarkably different from those of ambient water. Supercritical water is unique as a medium for chemical processes. The solubility of nonpolar species increases, whereas that of ionic and polar compounds decreases as a result of the drop of the solvent polarity, and the molecular mobility increases due to a decrease in the solvent viscosity (η). Drastic changes of ionic hydration are brought about by the decrease in the dielectric constant (ε) and density (ρ). Largely as a consequence of the dramatic decrease in the dielectric constant of water with increasing temperature at constant pressure and/or de-

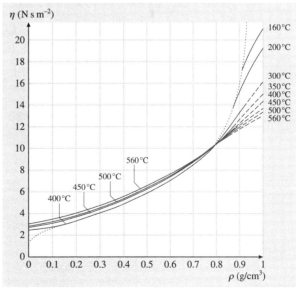

Fig. 18.9 Viscosity of water as a function of density and temperature (after [18.55])

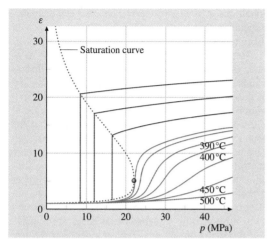

Fig. 18.10 Dielectric constant of water around the critical point (after [18.59])

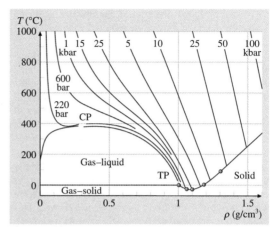

Fig. 18.11 Temperature–density diagram of water with pressure as a parameter (after [18.60])

creasing pressure at constant temperature, *completely* dissociated electrolytes at low temperatures and pressures may become highly associated in the supercritical region.

Water is an environmentally safe material and cheaper than other solvents, and it can act as a catalyst for the transformation of desired materials by tuning temperature and pressure. Studies of the structure dynamics and reactivity of supercritical water have led to a better understanding of how complicated compounds are adaptively evolved from simple ones in hydrothermal reactions. It is important to elucidate how the structure and dynamics of supercritical water are controlled by intermolecular interactions (ρ) as well as kinetic energies (T). This is a new frontier of solution chemistry. The PVT data for water up to 1000 °C and 10 kbar is known accurately enough (to within 1% error) [18.60, 62, 63]. At very high-PT conditions (1000 °C and 100 kbar), water is completely dissociated into H_3O^+ and OH^-, behaving like a molten salt, and has a higher density of the order of 1.7–1.9 g/cm^3. Figure 18.11 shows the temperature–density diagram of water, with pressure as a parameter [18.60].

In hydrothermal crystal growth, the PVT diagram of water proposed by *Kennedy* is very important even today for reasons of simplicity (Fig. 18.12), although there are several recent diagrams available with greater accuracy [18.61]. Usually, in most routine hydrothermal experiments, the pressure prevailing under the working conditions is determined by the degree of filling and the temperature. When concentrated solutions are used, the critical temperature can be several hundred degrees above that of pure water. The critical temperatures are not known for the, usually complex, solutions at hand; hence one cannot distinguish between sub- and supercritical systems for reactions below 800 °C. Although the temperature in the growth zone and the actual vapor pressure are not known to 100% accuracy, the PVT diagram of Kennedy is routinely used by most hydrothermal crystal growers. The PVT relations for several other systems, such as $AlPO_4$ and SiO_2, have been reviewed in [18.52, 64, 65].

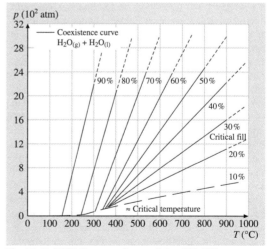

Fig. 18.12 PVT diagram of water (after [18.61])

Solubility

Solubility is one of the most important aspects of hydrothermal crystal growth. The early failures in the growth of bulk crystals using the hydrothermal technique were mainly due to the lack of solubility data on those compounds. This topic was almost neglected until the works of Spezia, who studied the solubility of quartz in detail and gave a new direction to hydrothermal growth of crystals. During the 1960s, new methods of investigating the solubility and the new experimental setup for the determination of solubility under hydrothermal conditions were proposed [18.66, 67]. There are two basic methods of determining the solubility under hydrothermal conditions: (a) the sampling technique, and (b) the weight loss method [18.68, 69]. Both methods have their merits and demerits. In many cases the solubility obeys the van't Hoff equation and is linearly dependent on solution density [18.70]. The ratio of solubility to mineralizer concentration often gives clues to the species present [18.71]. Some dedicated equipment have been designed and fabricated for solubility studies under hydrothermal conditions [18.72]. Herein we describe the different types of solubility under hydrothermal conditions by selecting appropriate examples such as quartz, berlinite, gallium orthophosphate, and potassium titanyl phosphate, because of their contrasting solubilities. As mentioned earlier (Sect. 18.1) the earliest hydrothermal researchers used only water as the solvent. Today we use a great variety of mineralizers, both aqueous and nonaqueous solvents, and also mixed solvents in many cases. However, *Hannay* and *Hogarth* used alcohol as the solvent [18.73], although that work did not draw the attention of scientists until recently. Although water can be a good solvent under very high-pressure and high-temperature conditions, the use of acids, salts, bases, and mixtures of them considerably reduces the PT conditions required for crystallization. The commonly used nonaqueous solvents are NH_3, HF, HCl, HBr, Br_2, S_2Cl_2, S_2Br_2, $SeBr_2$, $H_2S + N(C_2H_5)_3$, NH_4Cl, C_2H_5OH, CS_2, CCl_4, C_6H_6, CH_3NH_2, etc. In the last 15 years, the growth conditions for a variety of crystals have been further reduced with the use of some high-molar acid solvents such as HCl, H_2SO_4, H_3PO_4, HNO_3, $HCOOH$, etc. which has reduced the growth temperature to below 300 °C. This reduction in the growth temperature considerably reduces the working pressure, which in turn helps in the use of simple apparatus. Even silica autoclaves can be used, which facilitates direct visibility of the growth medium just as in any other low-temperature solution growth. This has greatly attracted the attention of crystal growers in the last decade, and a great variety of compounds which hitherto were produced only at high temperature and pressure have been today obtained at lower-PT conditions. Hence the mineralizer plays an important role in the hydrothermal growth of crystals and has given a new perspective to the method.

Solubility of Quartz. The first systematic study of the solubility of quartz was carried out by *Spezia*, which opened up the trend for the growth of bulk crystals on the whole using the hydrothermal method [18.20]. The solubility of quartz in pure water was found to be too low for crystal growth (0.1–0.3 wt %), but the solubility could be markedly increased by the addition of OH^-, Cl^-, F^-, Br^-, I^-, and acid media, which act as mineralizers. For example, the reactions

$$SiO_2 + 2OH^- \rightarrow SiO_3^{2-} + H_2O,$$
$$2OH^- + 3SiO_2 \rightarrow Si_3O_7^{2-} + H_2O,$$

show the formation of various complexes or species during the hydrothermal crystallization of quartz. *Hosaka* and *Taki* have used Raman spectra to identify and quantify such species [18.74].

In pure aqueous solutions (even at 400 °C and 25 000 psi), the solubility of quartz is too low to allow growth to take place in any reasonable time. Alkaline additions, such as $NaOH$, Na_2CO_3, KOH, and K_2CO_3, are all effective as mineralizers in this pressure and temperature range. However, higher molarity of the alkaline solutions introduce other alkali silicates along with quartz. Therefore, a moderate molarity and especially mixed mineralizers are more effective to achieve higher growth rate. *Laudise* and *Ballman* have measured the solubility of quartz in 0.5 M NaOH as a function of temperature [18.75] and even today the classical work carried out by Laudise and group remains the standard one as far as quartz growth under hydrothermal conditions is considered. Figure 18.13 shows the solubility curve for quartz in 0.5 M NaOH as a function of temperature and percentage fill [18.75]. An important result of solubility determinations is the delineation of the pressure, temperature, and composition regions where the temperature coefficient of solubility is negative. These regions are to be avoided in the growth of quartz, since it requires a different setup to avoid the loss of seed crystals. In recent years, for quartz crystal growth, mixed solvents are used in most cases.

Solubility of Berlinite. The solubility of berlinite was first determined by *John* and *Kordes* in orthophosphoric

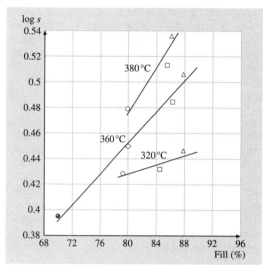

Fig. 18.13 Quartz solubility dependence on percentage fill (after [18.75])

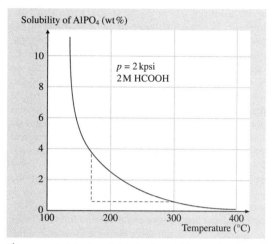

Fig. 18.14 Solubility of AlPO$_4$ (in wt%) as a function of temperature at a pressure of 2 kpsi, in 2 M HCOOH solution (after [18.78])

acid above 300 °C and was found to be positive [18.76]. Subsequently *Stanley* reported a negative solubility for berlinite in 6.1 M H$_3$PO$_4$ [18.77]. The solubility of AlPO$_4$ varies widely with the type of solvent used. Some authors claim that solubility of AlPO$_4$ in HCl is similar to that in H$_3$PO$_4$. The most important difference is the higher solubility at comparable mineralizer concentration. The authors of [18.78] have studied the solubility of AlPO$_4$ in some new solvents such as HCOOH, NH$_4$Cl, Na$_2$CO$_3$, NH$_4$H$_2$PO$_4$, NaF, KF, and LiF. The solubility of AlPO$_4$ (in wt%) as a function of temperature and at a pressure of 2 kpsi, in 2 M HCOOH solution is shown in Fig. 18.14.

Solubility of Gallium Orthophosphate. Gallium orthophosphate has been studied extensively for the past 12 years owing to its excellent piezoelectric properties, which are much better than those of conventional α-quartz. The solubility of gallium orthophosphate is quite interesting, although it is isostructural and isoelectronic with α-quartz and berlinite. Several studies exist on the solubility of GaPO$_4$ [18.79–82]. GaPO$_4$ shows a negative coefficient of solubility, like berlinite, with some significant differences between the two. Figure 18.15 shows a more precise solubility curve for GaPO$_4$ crystals [18.82]. The solubility of another very important crystal, ZnO, has been discussed in detail by M. J. Callahan et al. in this Handbook.

Kinetics of Crystallization under Hydrothermal Conditions

Studies related to the kinetics of crystallization under hydrothermal conditions began during the 1950s and 1960s with the commercialization of the synthesis of zeolites, and the large-scale growth of bulk crystals of quartz. Subsequently, several other crystals, such as GaPO$_4$, AlPO$_4$, malachite, ZnO, calcite, and ruby, were studied in detail. The anisotropy in the rate of growth of different faces during growth on the seed, if understood

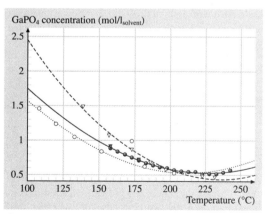

Fig. 18.15 Solubility curve of GaPO$_4$ under hydrothermal conditions in 15 M phosphoric acid: △ and ▽ correspond to the crystallization data; ○ and ● correspond to the dissolution data (△, ○ after [18.79]; ▽, ● after [18.80])

clearly, can allow the orientation of the seeds in such a way that the fastest growth can be achieved within a short time without compromising crystal quality. The crystallization kinetics of ZnO crystals under hydrothermal conditions have been discussed in this Handbook by the other authors (*Callahan* et al., Chap. 19). However, the crystallization kinetics are discussed in general for one or two selected compounds in this section. Various general laws characterizing the relationship between the anisotropy of the rates of growth of the faces and the chemical nature of the solvent have been studied by several workers.

The high activation energies of the growth of the faces, combined with other factors – the anisotropy of the rates of growth of the faces, the marked dependence of the rate of crystallization on the composition of the solution, etc. – provide strong evidence in support of the suggestion that, in hydrothermal crystallization under conditions of excess mass transfer, a primary role is played by surface processes taking place directly at the crystal–solution interface. It should be remembered that the activation energy of diffusion in solutions usually does not exceed 4–5 kcal/mol [18.83], while the activation energies of dissolution rarely exceed ≈ 10 kcal/mol. These values are much lower than the activation energies of growth, indicating that diffusion in the solution and dissolution of the charge do not limit the rate of crystallization with increase in the concentration of the solution. The rate of crystallization can increase in two ways. For crystals which do not contain components of the solvent, the rate increases sharply at low concentrations and remains practically unchanged at high concentrations. Similarly, in some cases, an increase in the rate of growth of the faces with increase in pressure has been observed. Pressure apparently does not have any significant direct effect on the rate of growth crystals, but it may have an influence through other parameters: mass transfer and solubility.

Kuznetsov reported the effects of temperature, seed orientation, filling, and temperature gradient (ΔT) on the growth kinetics of corundum crystals [18.84]. The activation energies calculated were 32 kcal/mol for the (1011) face and 17.5 kcal/mol for the (1120) face. *Kaneko* and *Imoto* have investigated the effects of pressure, temperature, time, and Ba:Ti ratio on the kinetics of a hydrothermal reaction between barium hydroxide and titania gels to produce barium titanate powders [18.85]. *Ovramenko* et al. conducted kinetics studies to compute the activation energy (E_a) of 21 kJ/mol [18.86]. In contrast, *Hertl* calculated an activation energy of 105.5 kJ/mol [18.87]. *Riman* and group have investigated the crystallization kinetics of various perovskite-type oxides using the computation modeling under hydrothermal conditions, and were able to construct speciation and yield diagrams [18.88, 89].

Thermodynamic modeling, solubility studies, and kinetics of crystallization under hydrothermal conditions is still an attractive field of research in the hydrothermal growth of crystals. During the 1980s and 1990s, external energy such as microwave, sonar, mechanochemical, and electrochemical, was employed to enhance the crystallization kinetics under hydrothermal conditions, especially in the preparation of fine crystals; and now in the last half-decade, the processes have been better understood to prepare crystalline particles with desired shape and size [18.90, 91].

18.3 Apparatus Used in the Hydrothermal Growth of Crystals

Designing an ideal hydrothermal apparatus, popularly known as an autoclave, is the most difficult task and perhaps impossible to define, because each project has different objectives and tolerances. Obviously several autoclave designs exist to meet requirements for a specific objective such as the study of phase equilibria, solubility, growth of bulk or small crystals, continuous production of nanocrystals, etc. However, an ideal hydrothermal autoclave should have the following characteristics irrespective of the goal:

1. Inertness to acids, bases, and oxidizing agents
2. Ease of assembly and disassembly
3. Sufficient length to obtain a desired temperature gradient
4. Leakproof, with unlimited resistance to the required temperature and pressure
5. Rugged enough to bear high-pressure and high-temperature experiments for long duration, so that no machining or treatment is needed after each experimental run.

The most commonly used autoclaves in hydrothermal research are listed in Table 18.4. The last four types listed are useful for the preparation of nanocrystals and powders under extreme conditions. When selecting a suitable autoclave, the first and foremost parameter is the experimental temperature and pressure conditions and its corrosion resistance in that pressure temperature range for a given solvent or hydrothermal fluid. If the reaction is taking place directly in the vessel, the corrosion resistance is of course a prime factor in the choice of autoclave material. The most successful materials are corrosion-resistant high-strength alloys, such as 300-series (austenitic) stainless steel, iron, nickel, cobalt-based superalloys, and titanium and its alloys. However, the hydrothermal experimenter should pay special attention to systems containing hydrogen, nickel, etc., which can be dangerous under high-PT conditions in the presence of some solvents. Therefore, selection of autoclaves, PT conditions, and the chemical media under hydrothermal conditions are extremely important to know before running actual experiments. Glass autoclaves were used in the 19th century for lower-PT conditions. In recent days in some laboratories thick quartz tubes are used for lower-temperature experiments [18.92]. Autoclaves are usually provided with liners made of various materials depending upon the type and purpose of the crystals to be grown. For example, quartz growth can be carried out in regular stainless-steel autoclaves without any lining or liners. If high-quality and high-purity quartz crystals are desired then platinum or gold liners should be used. Even glass, copper, silver, Teflon, etc., tubes can be used, depending upon the PT conditions and the hydrothermal media. Autoclave lining is not new and in fact began

Table 18.4 Autoclaves

Type	Characteristic data
Pyrex tube 5 mm i.d., 2 mm wall thickness	6 bar at 250 °C
Quartz tube 5 mm i.d., 2 mm wall thickness	6 bar at 300 °C
Flat plate seal, Morey type	400 bar at 400 °C
Welded Walker-Buehler closure 2600 bar at 350 °C	2000 bar at 480 °C
Delta ring, unsupported area	2300 bar at 400 °C
Modified Bridgman, unsupported area	3700 bar at 500 °C
Full Bridgman, unsupported area	3700 bar at 750 °C
Cold-cone seal, Tuttle–Roy type	5000 bar at 750 °C
Piston cylinder	40 kbar, 1000 °C
Belt apparatus	100 kbar, > 1500 °C
Opposed anvil	200 kbar, > 1500 °C
Opposed diamond anvil	Up to 500 kbar > 2000 °C

Table 18.5 Materials used as reactor linings

Material	T (°C)	Solutions	Remarks
Titanium	550	Chlorides Hydroxides Sulfates Sulfides	Corrosion in >25% NaOH solution in >10% NH_4Cl solution (at 400 °C)
Armco iron	450	Hydroxides	Gradual oxidation producing magnetite
Silver	600	Hydroxides	Gradual recrystallization and partial dissolution
Platinum	700	Hydroxides Chlorides Sulfates	Blackening in chlorides in the presence of sulfur ions Partial dissolution in hydroxides
Teflon	300	Chlorides Hydroxides	Poor thermal conduction
Tantalum	500	Chlorides	Begins to corrode in 78% NH_4Cl solution
Pyrex	300	Chlorides	
Copper	450	Hydroxides	Corrosion reduced in the presence of fluoride ions and organic compounds
Graphite	450	Sulfates	Pyrolytic graphite most suitable for linings
Nickel	300	Hydroxides	
Quartz	300	Chlorides	
Gold	700	Hydroxides Sulfates	Partial dissolution in hydroxides

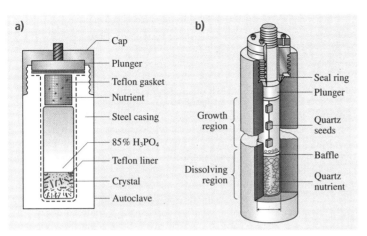

Fig. 18.16a,b The internal liner assembly. (a) Berlinite with negative temperature coefficient of solubility has the seed crystals (growth zone) at the bottom and nutrient (dissolution zone) at the top [18.78]. (b) Quartz with positive temperature coefficient of solubility has the nutrient (dissolution zone) at the bottom and the seed crystals (growth zone) in the top (courtesy of *Laudise* [18.94])

during the 19th century. Table 18.5 gives a list of the materials used as linings [18.93].

The internal assembly of the liners in hydrothermal experiments varies with the type of compounds to be grown and their solubility. For example, quartz and berlinite have contrasting solubility, and accordingly their crystal growth insists upon a completely different experimental setup, as shown in Fig. 18.16.

Here, only the standard designs used in hydrothermal research related to phase equilibria studies, solubility, kinetics, and crystal growth will be discussed, since the number of autoclave designs available in the literature exceeds more than 100.

18.3.1 Morey Autoclave

This autoclave was designed in 1913 by *Morey* with a simple sealing gasket; its volume is generally 25–100 ml [18.21]. It is widely used in hydrothermal research and a cross-section of a typical Morey autoclave is shown in Fig. 18.17. The usual dimensions of are 10–20 cm length and 2.5 cm inner diameter. The closure is made by a Bridgman unsupported area seal gasket made of copper, silver or teflon. Therefore it is also called a flat plate closure autoclave. The autoclave generates an autogeneous pressure depending on the degree of filling, the fluid, and the temperature. The autoclave is limited to $\approx 450\,°C$ and 2 kbar in routine use. In later versions, the pressure can be directly measured and adjusted during an experiment by providing an axial hole through the closure nut, but this sometimes causes compositional changes as the material is transported to the cooler region. A thermocouple is inserted in the well close to the sample and the vessel is placed inside a suitable furnace so that the entire Morey autoclave and closure lie within the element of the furnace. At the end of the run the vessel is quenched in a jet of air followed by dipping in water and the closure seal is broken.

18.3.2 Tuttle–Roy Cold-Cone Seal Autoclaves

This consists of a longer vessel in which the open end and seal are outside the furnace, hence the term *cold-cone seal* (although in fact the seal is far from being cold). Pressure is transmitted to the sample, which is contained in a sealed capsule, through a hole in the clo-

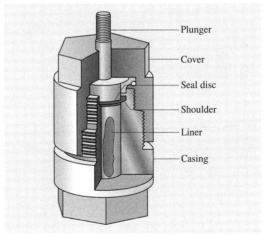

Fig. 18.17 Cross-section of a typical Morey autoclave [18.21]

sure. The capsules are normally made of noble metals (platinum, gold or silver). These vessels may be operated closure up, closure down, or horizontally. The ratio of the vessel diameter to the wall diameter determines the strength of the vessel. In most standard Tuttle vessels this ratio is 4%. Using steallite 25, experiments can be carried out at 900 °C and 1 kbar or 750 °C and 3 kbar for long-term use (from hours to weeks). Roy and Tuttle made many hydrothermal experimental runs for several months in the late 1950s. Moreover, the reaction could be *quenched* by lowering the furnace quickly and surrounding the bomb with a container of water [18.95]. Figure 18.18 shows the cross section of a Tuttle–Roy autoclave [18.95].

Around 1950, *Roy* and *Osborn* [18.96] also designed a simple universal pressure intensifier for compressing (virtually) all gases (H_2, N_2, CO_2, and NH_3) or liquids such as H_2O from the compressed gas tank pressures of about 100–200 bar to 5 kbar. From 1950 onwards *test-tube racks* of hydrothermal vessels including such compressors were used worldwide for hydrothermal research, mainly in the geochemical community. Figure 18.19 shows the advertisement of Leco Company, Tem-Press Research (Bellefonte, USA), which was set up by Tuttle, Roy, and Licastro to make these key tools for hydrothermal research available worldwide. This continues to the present day, mainly due to the overriding convenience and simplicity of the design.

These autoclaves are highly useful for hydrothermal researchers to carry out phase equilibria studies, solubility, high-temperature/pressure synthesis, nanocrystals synthesis, etc. Today they are also called batch reactors by several researchers and are popularly used for nanomaterials synthesis. The advent of new materials, particulary molybdenum-based alloys, has extended the temperature capabilities of cold-seal vessels to about 1150 °C at pressure above 4 kbar (so-called TZM vessels). The prospects of new refractory alloys extending up to this range are encouraging. These cold-seal vessels are safe, inexpensive, simple, and operationally routine. Unlike the Morey type of autoclaves, in the Tuttle cold-seal vessel the pressure is built up by an external pump and the pressure medium is usually distilled water with a little glycol added to inhibit corrosion. For higher fluid pressure, a hydraulic intensifier or air-driven pump is used with argon or nitrogen as the pressure medium. Normally, several vessels run from a single pump, where each vessel is connected to a high-pressure line by a valve and T-junction. The (two-way) valve isolates the vessels from the line. Each vessel is connected to a separate Bourdon gauge with a safety glass dial and blowout block. An autocontroller is normally used. The types of pressure tubing, valves, and fittings used depend on the operating pressures. Normally stainless-steel alloy is used up to 13 kbar, and size is recommended to be 1.4 inch outer diameter and 1/16 inch inner diameter. Although there has been rather little evolution in the basic design of cold-seal pressure vessels, marked changes have occurred in the controllers and pumps.

18.3.3 General-Purpose Autoclaves and Others

Other autoclaves, such as welded closures and modified Bridgman autoclaves, were used as the workhorses during the 1950s to 1980s; now their usage has reduced significantly because handling of them is quite cumbersome. However, these autoclaves contributed enormously to the development of both the hydrothermal technique itself and the growth of bulk crystals.

For the synthesis of crystals under hydrothermal conditions, these days very simple general-purpose autoclaves are used. Such autoclaves are commercially

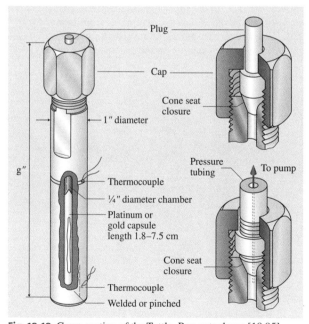

Fig. 18.18 Cross-section of the Tuttle–Roy autoclaves [18.95]

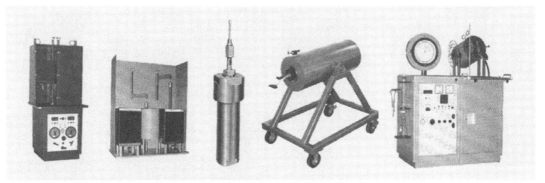

Fig. 18.19 Various test-tube-type cold-cone closure autoclaves (after [18.95])

available from many sources. Also several laboratories fabricate such autoclaves in their laboratories. Figure 18.20 shows the general-purpose autoclave used at the author's laboratory.

Many other designs are available today to meet the specific requirements, such as stirred autoclaves (Fig. 18.21), which are very useful to stir the mixture during the hydrothermal synthesis, to extract the contents from inside the reactor during the experiments, and also to pump in the desired gas into the reactor at any given time. The internal pressure can be read directly. The products can be quenched readily by the circulation of chilled water through the cooling coils running inside the autoclave. These additional advantages have helped greatly to understand the hydrothermal crystallization mechanism, kinetics, and the metastable phases more precisely.

In the last two decades the use of continuous-flow reactors has become very popular to synthesize small, micrometer-sized crystals to nanocrystals with a high degree of control over the shape and size of the crystals.

Fig. 18.20 General-purpose autoclaves popularly used for hydrothermal synthesis

These reactors operate at or above the supercritical conditions of the solvents – both aqueous and nonaqueous in nature. The method is also popularly known as the supercritical fluid (SCF) technology. With the invention of *green chemistry* in the early 1990s, there was a surge in the popularity of SCF technology. Today SCF technology has replaced organic solvents for the fabrication of a number of nanocrystals. Also it is emerging as an alternate to most existing techniques for designing and crystallizing new drug molecules.

Several designs and variations are available today for continuous-flow reactors to generate small to nanocrystals within the shortest possible time. The solubility of the materials has no major role to play in these flow systems. The use of capping agents and surfactants helps to control the size and shape of the nanocrystals and also to modify the surface characteristics of the crystals produced. By this method a great variety of crystals starting from metal oxides to silicates, vanadates, phosphates, germanates, titanates, etc., are produced for a wide range of modern technological applications within a few seconds. Hence, this technique is highly suitable for the commercial production of nanocrystals. The technique has been reviewed extensively and the reader can find additional information in several works [18.97–99]. Figure 18.22 shows a continuous-flow semi-pilot-scale reactor used in the crystallization of nanocrystals.

There are many more reactor designs for special purposes, such as rocking autoclaves, PVT apparatus, multichamber autoclaves, fluid sampling autoclaves, micro-autoclaves, autoclaves for visual examination, hydrothermal hot-pressing, vertical autoclaves, continuous-flow reactors, hydrothermal-electrochemical autoclaves, autoclaves for solubility measurements, autoclaves for kinetic study, pendulum autoclave, hor-

620   Part C | Solution Growth of Crystals

Fig. 18.21 Commercially available stirred autoclaves [18.100]

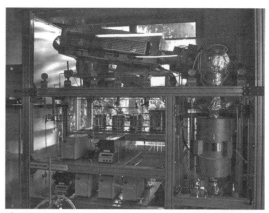

Fig. 18.22 The semi-pilot plant scale continuous-flow reactor used in the crystallization of nanocrystals [18.6]

izontal autoclaves for controlled diffusion study, and so on. The reader can find details of the construction and working mechanism of these autoclaves from the source [18.1].

Safety and maintenance of autoclaves is the prime factor one has to bear in mind while carrying out experiments under hydrothermal conditions. It is estimated that for a $100\,\mathrm{cm}^3$ vessel at $20\,000\,\mathrm{psi}$, the stored energy is about $15\,000\,\mathrm{lbf/in}^2$. Figure 18.23 shows the autoclaves ruptured because of pressure surges.

Hydrothermal solutions – either acidic or alkaline – at high temperatures are hazardous to human beings if the autoclave explodes. Therefore, the vessels should have rupture discs calibrated to burst above a given pressure. Such rupture discs are commercially available for various ranges of bursting pressure. The most important arrangement is that provision should be made for venting the live volatiles in the event of rupture. Proper shielding of the autoclave should be given to divert the corrosive volatiles away from personnel. In the case of a large autoclave, the vessels are to be placed with proper shielding in a pit.

18.4 Hydrothermal Growth of Some Selected Crystals

As mentioned earlier, in this chapter only the growth of some selected crystals of current technological significance will be discussed. However, a more or less complete list of the compounds obtained under hydrothermal conditions is given in the form of a table at the end of this chapter, as Appendix 18.A.

18.4.1 Quartz

Advances in the hydrothermal method of growing crystals have been linked very closely with the progress achieved in the field of quartz growth ever since the technique was discovered. Quartz (SiO_2) exists in both crystalline and amorphous forms in nature. The crystalline form has over 22 polymorphic modifications. Among them, α-quartz is the most important one, which transforms into β-quartz above $573\,°\mathrm{C}$. In order to grow this low-temperature modification of quartz, only the hydrothermal technique is suitable. α-quartz, as a piezoelectric crystal, has the ability to convert electric waves into mechanical waves and vice versa. Because of this property α-quartz is widely used in the electronics industries. In recent years, quartz for *tuning forks* has become essential for timing functions in electronic watches and in timing circuits for computers and telecommunications.

The principal source of electronics-grade natural quartz is Brazil. Today, the electronics industries are largely inclined to use synthetic quartz, because natural quartz crystals are generally irregular in shape, automatic cutting is cumbersome, and the yield is low. Many countries in the world, such as the USA, Japan, the UK, Germany, Russia, Poland, China, Belgium, and Taiwan, have entered the world market. Over $3000\,\mathrm{t}$ of quartz is produced annually for a variety of applications. The production of high-quality defect-free quartz has a large potential market among all electronic materials. Japan alone produces more than 50% of the world's production, followed by the USA and China. More recently several other Japanese companies such as Kyocera, Shin-Etsu, Asahi Glass Co., etc. have started producing quartz for highly specialized applications. Perhaps Toyo is the largest single company in the world producing quartz. About 50% of the quartz produced goes into devices in the automotive industries, 30% of production goes into frequency control-devices, and the remaining 20% goes into optical devices. The total cost involved in the production of quartz is:

End users: ≈ US$ 1 000 000 000 000
Equipment: ≈ US$ 5 000 000 000
Components: ≈ US$ 100 000 000.

Thus, quartz takes the first place in value and quantity of single-crystal piezoelectric materials produced. Much of the research related to quartz today is confined to industry, while only scant publications emerge from research laboratories on new applications of quartz and the preparation of nanoscale silica particles. Otherwise, it is mainly research and development (R&D)-based industries that carry out research on the production of high-quality quartz and its applications, and these results are not published.

After the successful growth of quartz (Fig. 18.2) by Nacken in the 1940s, *Walker* synthesized large-size crystals of quartz (Fig. 18.24), much bigger than the size of the crystals produced by Nacken [18.101]. Subsequently major research activity was initiated at AT&T Bell Labs, and the most important contributions to the quartz growth came from the workers such as Ballman, Laudise, Kolb, and Shtenberg during the 1970s. Figure 18.25 shows the quartz crystals obtained at AT&T Bell Labs during the 1970s, the biggest crystals of that period. Each autoclave could produce around 70 kg of quartz crystals during the 1970s. In contrast, today in Japan, in one experimental run, about 4000–4500 kg of quartz is produced using the world's largest autoclave with a volume of 5000 l. Figure 18.26 shows the growth of quartz crystals in the world's largest autoclave at Toyo Communications Ltd., Japan. The early success in producing commercial quartz crystals in Japan during the post-war period paved the way for the entry of several others into the field. Some of the prominent contributors were Professor Noda, Doimon, Kiyoora, Dr. Itoh, and Dr. Taki. Today Japan is the world's largest producer of synthetic quartz for a variety of applications, followed by China.

In recent years mixed solvents are being used in the growth of quartz. High-quality quartz crystals have been obtained in NaCl and KCl solutions, NaOH, and

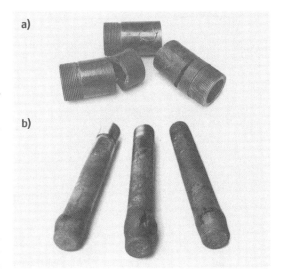

Fig. 18.23a,b Ruptured autoclaves. (a) General-purpose autoclaves, (b) Tuttle–Roy autoclaves

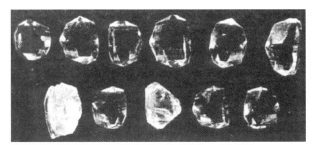

Fig. 18.24 Quartz crystals obtained by *Walker* [18.101]

Na_2CO_3 solutions (10%) [18.102, 103]. It has been observed that the high-frequency applications of α-quartz require sheets with small thickness, of the same order of size as the defects (such as inclusions, etch pits, and dislocations). Thus the pressure can promote new solvents viable for hydrothermal growth, especially by reducing their concentrations. The recent experimental results have shown the following enthalpy values calculated for α-quartz

$$\Delta H_T = 2395 \pm 5 \, \text{cal/mol}, \quad \text{for NaOH (1 M)},$$
$$200 \, \text{MPa} \leq p \leq 350 \, \text{MPa};$$
$$\Delta H_T = 4001 \pm 2 \, \text{cal/mol}, \quad \text{for Na}_2\text{CO}_3 \, (1 \, \text{M}),$$
$$150 \, \text{MPa} \leq p \leq 350 \, \text{MPa}.$$

In the growth of α-quartz, available nutrient material, such as small-particle-size α-quartz, silica glass, high-

In most experiments the percentage opening of the baffle is 20%, although lower values are used by several workers. However, the actual percentage opening of the baffle area and its geometry are not disclosed, especially by commercial growers.

The optimum growth conditions for synthesis of quartz based on the work in Bell Laboratories are [18.105, 106]:

Dissolution temperature:	425 °C
Growth temperature:	375 °C
Pressure:	15 000–25 000 psi.
Mineralizer concentration:	0.5–1.0 M NaOH
Temperature gradient (ΔT):	50 °C
% fill:	78–85%
Growth rate in (0001):	1.0–1.25 mm/day.

The quality of the grown crystals is also a function of the seed orientation and its quality. Strained seeds generally produce strained growth regions [18.107]. The seeds are polished to a very fine finish before use. Most high-quality crystals are grown using seeds with the surface perpendicular to the z-direction, since this region has the lowest aluminum concentration. Though most quartz production consists of y *bar* crystals, that is, small crystals ($z = 20$–25 mm, 64 mm seed) capable of several y bar per crystal, *pure z bars* are also produced, representing 10–20% of production. However, for medium- and high-quality grades we notice a rise in demand for crystals of very large dimensions and upper–medium quality, especially in the USA, for manufacturing wafers used in surface wave applications [18.108]. Earlier, most seeds used were natural quartz cut in a definite orientation, but in recent years this practice is only used when a high-quality crystal is desired.

The growth of quartz crystals has been understood precisely with reference to the growth temperature, temperature gradient, percentage fill, solubility, percentage of baffle opening, orientation and nature of seed, and type of nutrient. Also many kinetic studies have been carried out [18.35, 109]. Figure 18.27 shows the growth rate of quartz as a function of seed orientation [18.109]. The solubility of quartz has been studied as a function of temperature, and the growth rate as a function of seed orientation and percentage fill. Figure 18.28 shows hydrothermally grown quartz crystals.

The type of crystal to be grown depends on the application, as different properties are required in each case. For optical use, high uniformity, low strain, and low inclusion counts are needed, since all of these can affect transparency. For surface acoustic wave de-

Fig. 18.25 Quartz crystals obtained at AT&T Bell Labs during the 1970s [18.102]

quality silica sand or silica gel, is placed in a liner made of iron or silver with a suitable baffle and a frame holding the seed plates. A mineralizer solution with a definite molarity is poured into the liner to achieve the required percentage fill. The increased solubility in the presence of mineralizer increases the supersaturation without spontaneous nucleation and consequently allows more rapid growth rates on the seeds. Figure 18.16b shows a cross-section of a modified Bridgman autoclave used in the growth of quartz crystals.

The commercial autoclaves used have 10 inch inner diameter, and are 10 ft long unlined. These autoclaves can work at conditions up to 30 000 psi and 400 °C. Most of these experiments are carried out for 25–90 days to obtain full-size crystals of 4 cm in the z-direction and 12.5–15 cm in the y-direction. The temperature gradient is varied according to the nutrient used. About 1 N NaOH or Na_2CO_3 is the most commonly used mineralizer. The solubility change with temperature is smaller in NaOH and slightly larger in Na_2CO_3. The temperature of the autoclave at the nutrient zone is usually kept at 355–369 °C and, in the growth zone, kept at 350 °C. The addition of lithium improves the growth rate, and small amounts of Li salts are routinely added to the solution [18.104].

The solubility is also, to some extent, a function of increasing pressure. The pressure is controlled by the percentage fill of the autoclave and is usually about 80% for hydroxyl mineralizer (20 000 psi internal pressure).

Fig. 18.26 Growth of quartz crystals in the world's largest autoclave at Toyo Communications Ltd., Japan (courtesy of S. Taki). Dissolution temperature: 425 °C; growth temperature: 375 °C; pressure: 15 000–25 000 psi; mineralizer concentration: 0.5–1.0 M NaOH; temperature gradient (ΔT): 50 °C; percentage fill: 78–85%; growth rate of (0001): 1.0–1.25 mm/day

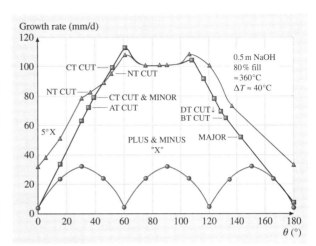

Fig. 18.27 Growth rate of quartz as a function of seed orientation [18.109]

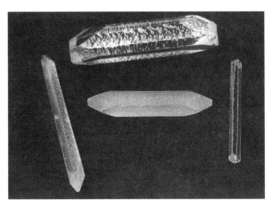

Fig. 18.28 Hydrothermally grown quartz crystals

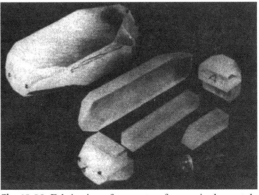

Fig. 18.29 Fabrication of resonators from a single crystal

vices, large pieces that can take a very high-quality surface finish are needed. The quality of the material required for resonators used in time and frequency devices varies with the application. The more precise the need, the more stringent the requirements. For most applications, a truly high-quality material is not needed. For high-precision uses, such as in navigational devices and satellites, a very high-quality material must be used. Most recent research on quartz growth is for improved resonator performance, which requires the growth of high-quality low-dislocation quartz. Figure 18.29 shows the fabrication of resonators from a single crystal.

Several criteria are used to evaluate the quality of quartz crystals. The most commonly used criterion is the Q value or quality factor, which is a measure of the acoustic loss of the material. It is important for a resonator to have high electrical Q value and superior frequency–temperature characteristics. In a piezoelectric resonator, electrical energy and mechanical energy are interconvertible. In such a case, Q is expressed as

$$Q = \frac{[X]}{R}, \tag{18.2}$$

where X is the inductive or capacitive reactance at resonance and R is the resistance.

The acoustic Q for natural α-quartz crystal varies in the range 1 to 3×10^6, while for synthetic quartz crystals the value drops to 2×10^5 to 1×10^6. Thus in the last two decades, the main objective among quartz crystal growers has been to improve Q, which in turn leads to a low concentration of physicochemical and structural defects.

The growth rate is determined by the ratio of increase in thickness of the seed and the duration of the run. The growth rate along the main crystallographic axes R_c is determined by dividing the thickness of the layer grown on the seed, $(h_1 - h_0)/2$, by the run duration t, and is expressed in millimeters per day. It has also been known for quite some time that, qualitatively, acoustic Q is inversely proportional to growth rate and directly related to chemical impurities. The internal friction (the inverse of the mechanical Q) is dependent on the growth rate of synthetic quartz crystals. *Martin* and *Armington* have studied the effect of growth rate on the aluminum content at low growth rates [18.110]. Al^{3+} and H^+ are the most deleterious impurities in quartz used for frequency and timing applications, and they also influence the growth rates. Similarly the percentage fill and pressure strongly influence the growth rate of quartz.

For any crystal growth, seed crystals play a vital role in controlling the quality, growth rate, and morphology of the grown crystals. Usually, in the case of quartz for higher-frequency applications, a smaller x-axis dimension is needed. z-Growth material is desired for resonators, as it has been shown that this material has about an order of magnitude lower aluminum concentration. In the majority of crystal growth experiments, in general the thickness of the seed crystal is usually $1-2$ mm. Until recently, it was necessary to use natural seeds for the preparation of low-dislocation crystals. However, most crystals grown from synthetic seeds contain large numbers of dislocations, of the order of several hundred per square centimeters. These results in the formation of etch channels in the resonator, which weaken it mechanically and can be a problem when electronic devices are deposited on the surface. When a seed perpendicular to the z-axis (but from the x-growth region) is used, the dislocation density can be reduced to below ten and sometimes to zero dislocations per square centimeter [18.111]. The purity of the crystals produced using seeds cut from the x-region is as good as that from crystals produced from the usual z-seed.

In the growth of quartz crystals, the nutrient has a profound effect on the quality of the grown crystal.

It is well known that most of the impurities and in turn the defects in the grown crystals come from the nutrient and only to some extent from the autoclave walls. Nutrient selection is one of the attractive research topics in crystal growth on the whole. Even in the 19th century, researchers tried the use of several varieties of nutrient to obtain quartz under hydrothermal conditions. There are several reasons for using varieties of nutrients, including: to obtain high-purity crystal, to obtain higher solubility in the desired PT conditions for a given solvent, to achieve higher growth rate, to obtain higher yield, and to lower the PT conditions for growth. In the case of quartz, to obtain high-quality crystals, even α-cristobalite has been used as a nutrient [18.112].

Tables 18.6 and 18.7 show the distribution of impurities in different sectors of synthetic quartz. Hence, attention has to be paid to the solvent, nutrient, seed orientation, and seed quality in order to obtain high-quality crystals, as well as the experimental PT conditions.

In the growth of α-quartz for high-frequency device applications (24–100 MHz or more), the existence of defects, either physicochemical or structural, in synthetic quartz crystals leads to critical modifications of devices. Low-defect, high-purity synthetic quartz should have the characteristics given in Table 18.8 [18.113]. Sweeping is one of the most popularly used techniques in recent years to enhance the performance of quartz resonators. Sweeping, or solid-state electrolysis or electrodiffusion, is generally performed under vacuum, air, hydrogen or a desired atmosphere. During sweeping, the crystal is placed in an electric field and heated. Then, migration of the impurities and some modifications are induced within the

Table 18.6 Impurities in different sectors in synthetic quartz

Sector	Al	Na	Li
Z	5	1	0.5
+X	31	9	5
−X	122	40	5
S	85	26	16

Table 18.7 Dislocation density in different seeds

Seed	Mineralizer	Etch channel density
Z	Hydroxide	253
Z	Carbonate	247
X+	Hydroxide	1
X+ (reused)	Hydroxide	14

Table 18.8 Desirable and achieved parameters in synthetic quartz [18.113]

Parameter	Desirable	So far achieved
Etch channel density	$< 10/\text{cm}^2$	< 86
Inclusion density	$< 10/\text{bar}$	
Impurity concentration (ppb)		
Al	< 200	700
Li	< 300	300
Na	< 500	1640
K	< 40	300
Fe	< 100	1800
Q (3500/3800 cm^{-1})	$< 2.5 \times 10^6$	$< 2.5 \times 10^6$
Strain	None	Variable
Fringe distorion	< 0.05 RMS	Variable

crystal. Sweeping reduces the formation of etch tunnels. The effect of sweeping is to remove lithium and sodium deposited interstitially in the lattice during the growth. These ions are usually trapped along an angstrom-wide tunnel, which is parallel to the z-axis in the quartz crystal lattice. These ions, in an interstitial position, interact with substitutional aluminum impurities in the lattice to form Al-Li centers, which have been shown to be weakly bonded and are the cause of low radiation tolerance in a resonator. In the sweeping process, these are replaced by Al–OH or Al–hole centers, which have a much higher radiation tolerance. The sweeping is carried out at 500–550 °C usually for a period of 7 days or even more to remove alkalies. However, it should be remembered that sweeping cannot become a routine process even if the same experimental conditions are always applied.

There are several other techniques employed for the fabrication of piezoelectric high- to ultrahigh-frequency devices based on quartz resonators. The important ones are: chemical polishing, ion beam etching (IBE). The reverse thermodynamic relations are employed for the refinement of chemical polishing. Several solvents such as HF and NH$_4$HF$_2$, NaOH, KOH · xH$_2$O, and NaOH · xH$_2$O are employed [18.114].

Similarly, the industrial chemical etching process is specially dedicated for large thickness removals without damaging the blank surface texture. This is most useful for frequency applications of quartz, because the mechanical grinding and lapping introduce surface stresses. Fluoride media is the most popular for this type of chemical polishing [18.115]. Recently, *Cambon* et al. tried industrial chemical etching successfully in the temperature range 150–180 °C using concentrated NaOH solvents [18.116]. During the chemical etching process, several factors influence the process: kinetics, etching temperature, etching time, plate orientation, SiO$_2$ concentration, solvent concentration, wafer carrier geometry, and so on. By using this process about 3200 quartz plates can be processed in a single run. The resonators manufactured by this process have demonstrated a high level of performance, even higher than that of plates obtained by mechanical means.

In recent years there has been growing interest in morphological variations, growth-rate monitoring using various seed orientations, and computer simulation for the precise calculation of growth rates of various faces [18.117–119]. These studies greatly help in understanding the growth technology for commercial production of defect-free and economic quartz crystals.

The growth of bulk single crystals of quartz is still an attractive field of research, especially the growth of defect-free quartz crystals for space-grade applications. Understanding of solubility and the growth mechanism is still incomplete, although laser Raman spectroscopy and other advanced techniques such as computer simulation are yielding rich information about how to intelligently engineer the growth processes, thereby minimizing growth defects.

The preparation of quartz for other applications, such as photomask substrates for lithography, poly-Si thin-film transistor liquid-crystal displays (TFT-LCDs), and high-purity and high-precision polishing devices, is a new trend in science. For example, Shin-Etsu Chemical (Japan), Asahi Glass Co. (Japan), NDK America (USA), and APC International Ltd. (USA) are engaged in such research and development, whereas companies such as Sawyer Research Inc., General Electric (GE), Philips, and Toyo Communication are engaged in production of bulk crystals of quartz for oscillators, frequency devices, etc.

18.4.2 Aluminum and Gallium Berlinites

Berlinite (AlPO$_4$) and gallium berlinite (GaPO$_4$) are the two most important crystals that are isoelectronic and isostructural with α-quartz. These compounds are used to replace quartz in electronic devices because of their larger mechanical coupling factors than α-quartz, and because their resonant frequency is nearly independent of temperature for certain orientations [18.120–122]. Table 18.9 gives a comparison of some piezoelectric characteristics of these materials [18.123]. The growth of berlinite and gallium berlinite does not differ much between the two materials because they both have a negative temperature coefficient of solubility, and hence a similar experimental arrangement is used for both. During the 1980s the growth of berlinite was studied extensively, and it was considered at one time as a replacement for α-quartz [18.124]. However, results were not encouraging for bulk crystal growth of berlinite on a par with that of α-quartz. Although the growth of gallium berlinite began sometime during the late 1980s, momentum picked up during 1995. It is important to discuss the growth of these crystals from a scientific point of view in order to learn about the hydrothermal technique in general, as these crystals – although isoelectronic and isostructural with α-quartz – have a completely contrasting experimental setup for their crystal growth.

John and *Kordes* were the first to grow berlinite crystals successfully, followed by *Stanley* using completely different conditions [18.76, 77]. The greatest disadvantage in the growth of these berlinites is their negative temperature coefficient of solubility, which demands a special experimental setup. Although much progress has been achieved in the growth of these berlinite crystals, our knowledge on the growth of AlPO$_4$ and GaPO$_4$ is still comparable to that of quartz some 50 years ago, particularly with reference to size, crystal perfection, reproducibility, etc. The highly corrosive nature of the solvent is the major hindrance in the growth of bulk crystals of these berlinites.

The solubility of these berlinites has been discussed earlier. Although several reports have appeared on solubility measurements as a function of various parameters, there is still no unanimity in these results.

Berlinites could be best grown using the hydrothermal technique. α-Berlinite is stable up to 584 °C, and below 150 °C the hydrates and AlPO$_4 \cdot$ H$_2$O are stable. Therefore, the growth temperatures should be greater than 150 °C but less than 584 °C. However, solubility and PVT studies have clearly shown that crystal growth must be carried out at less than 300 °C because of the reverse solubility. Several versions of the hydrothermal growth of berlinite single crystals have been tried to suit the solubility. In contrast to aluminum berlinite, gallium berlinite has no phase transitions, and therefore some researchers have even synthesized this crystal using the flux growth method [18.125]. However, the quality of these flux-grown gallium berlinite crystals is far inferior to that of hydrothermally grown crystals. In the hydrothermal growth of these berlinites, usually acid mineralizers are used at various concentrations. Mixed acid mineralizers such as HCl, H$_3$PO$_4$, H$_2$SO$_4$, HCOOH, etc. are much more widely used in practice. However, other mineralizers such as NaOH, Na$_3$PO$_4$, Na$_2$HPO$_4$, NaHCO$_3$, NaF, KF, LiF, NH$_4$HF$_2$, NaCl, Na$_2$CO$_3$, etc. have also been tried. More details can be found in [18.126].

Small crystals of both aluminum berlinite and gallium berlinite can be obtained through hydrothermal reactions of H$_3$PO$_4$ with various chemicals containing Al and Ga [18.94, 126, 127]

$$M_2O_3 + 2H_3PO_4 \rightarrow 2MPO_4 + 3H_2O .$$

These fine crystals of berlinite can in turn be used as the nutrient to grow bigger bulk single crystals. Since the solubility of these compounds is negative, the seed is placed in the hotter zone in the bottom, while nutrient is placed in the upper cooler region. The most important modifications suggested from time to time in the growth of berlinite crystals are:

1. Crystal growth by slow heating method
2. Crystal growth by composite gradient method
3. Crystal growth by temperature gradient
4. Growth on seeds.

The crystal growth by slow heating method was mainly used for improving the quality of seeds and to enhance the size of the crystal, with conditions similar to those used for nucleation. A major drawback of this method is the impossibility of obtaining crystals of sufficient size in one operation due to the limited quantity of metal phosphate available in the solution. Usually Morey autoclaves are used to carry out crystal growth by the slow heating method ($T \leq 250$ °C; $p = 100-1000$ atm). The size of the crystals obtained through spontaneous nucleation is usually 0.1–4 mm, depending upon the experimental conditions.

The crystal growth by composite gradient method gives more flexibility in adjusting the growth rates and avoids the preliminary work of crystal growth or nutrient preparation.

Table 18.9 Comparison of some piezoelectric characteristics (* AT cut)

Parameter	Quartz	Berlinite	Gallium berlinite
Coupling coefficient K (%)*	8.5	11.0	> 16.0
Surtension coefficient Q*	3×10^6	10^6	$> 5 \times 10^4$
α–β phase transition (°C)	573	584	No

The crystal growth with reverse temperature gradient method is much more versatile for these compounds in general. In this approach, the nutrient (either crystalline powder or fine grains of AlPO$_4$ usually obtained from the slow heating method) is kept in a gasket at the upper portion of the autoclave, which is cooler than the bottom of the liner that is kept at slightly higher temperature, forming the crystallization zone (Fig. 18.16a).

The crystal growth on seeds method uses spontaneously nucleated seeds and oriented seeds cut from grown crystals, usually mounted on a platinum frame placed in the bottom (hotter) region of an autoclave, and the nutrient (≈ 60 mesh particle size prepared by other methods) is placed above in a platinum gasket in the upper region (or in a Teflon gasket). Thus, the seeds are in the warmer, supersaturated region at the bottom of the autoclave and the temperature gradient achieved by cooling the top allows proper convection.

Figure 18.30 shows aluminum berlinite crystals grown by the hydrothermal method. The orientation of the seed is important, as in the case of quartz, because the growth rate and crystal quality depend upon the seed orientation. The following growth rates are shown for seeds whose faces are indicated below:

(0001) (basal plane): 0.25–0.50 mm/day
(102$\bar{1}$) (x-cut): 0.23–0.30 mm/day
(10$\bar{1}$0) (y-cut): 0.12–0.15 mm/day
(01$\bar{1}$1) (minor rhombohedral face): 0.12 mm/day
(10$\bar{1}$1) (major rhombohedral face): 0.15 mm/day.

The relative growth rates are in general agreement with the morphology of the equilibrium form, as judged from observing spontaneously nucleated crystals which are bounded by small prism faces and terminated by minor and major rhombohedral faces. The dissolution rate of aluminophosphate glass charge is three times greater than that of the crystalline charge [18.128].

The morphology of spontaneously crystallized aluminum berlinite crystals varies greatly, depending upon the type of solvent, its concentration, and the experimental temperature. The most commonly observed morphologies are hexagonal, rhombohedral, rods, needles, and equidimensional crystals. Impurities also play a prominent role in controlling the crystal morphology. In the growth of aluminum berlinite the typical experimental conditions used are:

Growth temperature: 240 °C
Pressure: 15–35 MPa
Solvents: Mixed acid mineralizers
Nutrient: Powdered nutrient, preferably aluminum phosphate glass
Filling: 80%
ΔT: 5 °C $< T <$ 30 °C
Growth rate: 0.35 to 0.50 mm/day in H$_3$PO$_4$
0.35 to 0.45 mm/day in HCl
0.25 to 0.35 mm/day in HNO$_3$
0.2 to 0.3 mm/day in H$_2$SO$_4$.

It is interesting to note that the berlinite crystals show internal structural defects on x-ray Lang topography. No fundamental differences were found by x-ray topography between the horizontal gradient, vertical gradient, and slow heating methods. The two principal imperfections readily observable in berlinite are crevice flawing and cracks. Although crystals appear quite transparent from the outside, they contain many internal defects. A combination of optimum growth conditions, reducing the water content, and using heat treatment to control the water distribution could be of great interest for improving the piezoelectric device potential of aluminum berlinite.

During the 1980s and early 1990s work on aluminum berlinite was at its peak, and then activity suddenly dropped, because of several difficulties associated with its growth as bulk crystal. However, crystal growth of gallium berlinite, which began during the late 1980s, is still an attractive field, and several research groups are actively engaged in bulk crystal growth of gallium berlinite. The solubility and crystal growth kinetics of gallium phosphate have been extensively studied by various groups [18.129–131] and basically resemble those of aluminum berlinite. The necessary conditions for crystal growth are:

1. Solubility of GaPO$_4$ nutrient is sufficient for crystal growth.
2. Convection of saturated fluids is intensive and not varying during the crystal growth cycle.
3. No formation of spontaneous crystallites.
4. Saturation of an initial solvent slowly appears, simultaneously. with the end of seed etching (very soft restart conditions).

The typical growth conditions are:

Experimental temperature: 280 °C
Experimental pressure: 8 MPa
Mineralizer: H$_3$PO$_4$ + H$_2$SO$_4$
(pH = 3 − 4 under ambient conditions).

The biggest disadvantage is that, in the case of seeded growth under retrograde solubility, it is often accompanied by precipitation of spontaneous crystals or degradation of growth surfaces because of the irreversible character of the mass transportation (high or low relation of ΔS to growth kinetic parameters).

Figure 18.31 shows gallium berlinite crystals obtained under hydrothermal conditions. The commonly appearing faces are $(10\bar{1}1)$, $(01\bar{1}1)$, $(10\bar{1}0)$, and $(10\bar{1}2)$. The growth rates along various crystallographic directions decrease in the sequence

$$\gamma(0001) \gg \gamma(01\bar{1}2) > \gamma(10\bar{1}0) > \gamma(01\bar{1}1)$$
$$> \gamma(10\bar{1}2) > \gamma(10\bar{1}1).$$

The common twins observed in $GaPO_4$ obey the Dauphiné, Brazil, and Laydolt's twinning laws.

The crystallization process, dissolution process, rate of crystallization (i.e., the amount of $GaPO_4$ crystal formed per unit time), action of solvent, its concentration and temperature regime, etc. have been studied in detail for $GaPO_4$ crystal growth by several workers [18.127, 132, 133]. Also characterization of defects has been carried out in detail [18.134–136].

Although the first stage of $GaPO_4$ crystal growth began with epitaxy of $GaPO_4$ on $AlPO_4$ seeds, the subject still remains popular. Several substrates including quartz have been used to grow $GaPO_4$ crystal.

If the size of the $GaPO_4$ crystals grown by the hydrothermal technique is increased sufficiently, it would become the best piezoelectric material replacing quartz.

18.4.3 Calcite

Calcite (Ca_2CO_3) is an important carbonate mineral. Pure and optically clear calcite is called Iceland spar. Calcite single crystals form an important optical material owing to its large birefringence and transparency over a wide range of wavelengths. These properties make it a significant material for polarized devices such as optical isolators and Q-switches. It also exhibits antiferromagnetic properties. Although there are many deposits of calcite in the world, the optically clear quality calcite, Iceland spar, has been depleted in recent years, leading to its shortage in nature. However, the demand for optically clear calcite single crystals is increasing greatly with the development of laser devices such as the optical isolator [18.137, 138]. Many researchers have tried to grow calcite single crystals at relatively low temperatures, since calcite crystals dissociate to form CaO and CO_2 above $900\,°C$ under atmospheric pressure. Various solvents have been used for hydrothermal growth of calcite, but none of them has been found to be the best to grow large single crystals.

The solubility of calcite is very interesting. The solubility is positive for increasing CO_2 pressure and negative for increasing temperature. This behavior of calcite has posed a real problem in the search for a suitable solvent to optimize the growth rate. Despite a large number of reports concerning calcite growth, hydrothermal reactions of calcite in chloride and other chloride solutions have appeared, because natural calcite crystals are formed in both chloride and carbonate hydrothermal solutions and these chloride solutions are almost analogous to natural carbonate thermal springs. *Ikornikova* has done extensive work on aspects of solubility, designing an apparatus to grow calcite crystals with changing CO_2 concentration as the pressure is reduced at constant temperature, crystal growth kinetics, and mechanism [18.139]. The following groups have carried out extensive studies on the solubility of calcite:

1. Ikornikova, Russia, during the 1960s and 1970s, using chloride solutions
2. Belt, USA, during the 1970s using carbonate solutions

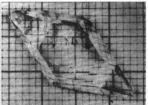

Fig. 18.30 Hydrothermally grown aluminum berlinite crystals (courtesy of Dr. Y. Toudic)

Fig. 18.31 Characteristic photographs of $GaPO_4$ crystals obtained by the hydrothermal method (courtesy of Prof. L. N. Demianets)

3. Hirano, Japan, during the 1980s and early 1990s using nitrate solutions
4. Kodaira, Japan, during the 1990s using H_2O/CO_2
5. Yamasaki, Japan, during the 1990s using ammonium acetate and other organic solvents.

Each of the last three groups claims superiority of their solvents over the others. Figure 18.32 shows solubility curves for calcite in different solvents. *Yanagisawa* et al. have studied the effect of pH of CH_3COONH_4 on the growth of calcite [18.142]. When as-cleaved seeds are used, the calcite crystals obtained are usually opaque or not transparent and the surfaces show a greater degree of defects such as growth hillocks, a lot of small secondary grown crystals, and strains. In contrast, crystals grown on etched seed crystals are usually transparent and their surfaces are very smooth. Figure 18.33 shows a characteristic photograph of calcite single crystal grown in NH_4NO_3 solution using as-cleaved and etched seeds. Earlier experiments on calcite single-crystal growth were carried out under higher-PT conditions. For example, *Kinloch* et al. carried out calcite crystal growth at temperature greater than 435 °C with 1.72 kbar pressure using platinum liners and achieved about 50 µm/day growth rate for the {1011} face [18.137]. However, in recent years crystal growth has been carried out using lower-PT conditions using Teflon liners (below 250 °C), and a growth rate of > 120 µm/day has been achieved. Also several types of nutrients such as natural limestone, natural Iceland spar, reagent-grade carbonates, etc., have been attempted in recent years to obtain high-quality calcite single crystals [18.143].

There are several works in the literature on the growth of other carbonates such as $MnCO_3$ (rhodochrosite), $FeCO_3$ (siderite), $CdCO_3$ (otavit), $NiCO_3$, etc., using alkali chloride and carbonate solutions in the temperature range 300–500 °C and pressure up to 1 kbar with a temperature gradient [18.139].

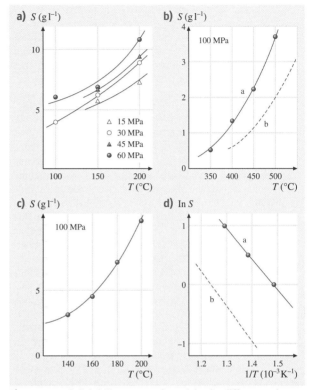

Fig. 18.32a–d Solubility curves for calcite in different solvents: (a) in 1 M NH_4NO_3; (b) in 1.5 M $Ca(NO_3)_2$ (*a*), 3 M NaCl (*b*); (c) in 3 M NH_4NO_3; (d) in 1.5 M $Ca(NO_3)_2$ (*a*), 3 M NaCl (*b*) (after [18.140, 141])

Fig. 18.33 Characteristic photographs of calcite single crystal grown in NH_4NO_3 solution

18.4.4 Gemstones

The hydrothermal method has been popularly used from the third quarter of the 19th century for the growth of gemstones such as corundum, ruby, emerald, and garnet [18.15, 16, 18]. During the 1960s and 1970s work on hydrothermal growth of gemstones reached its peak, when successful and high-quality crystals of emerald, colored quartz, amethyst, ruby, corundum, malachite, silicon nitride, garnet, etc. were obtained. However, today, like quartz, much of the research on the hydrothermal growth of gemstones is commercial and hence few publications appear from research laboratories. However, some publications appear periodically on the growth of new gemstones, for example, zeosite. As the sources of natural colored gemstones continue to shrink and the population of the world continues to grow, one could expect increased demand for materials such as synthetic emerald (tempered, however, by occasional reverses in the world economy). Here, in order to understand the hydrothermal growth of crystals in general, knowledge on the growth of gemstones is also essential. Therefore, a few selected gemstones grown using this hydrothermal technique will be discussed below.

Corundum

Corundum (Al_2O_3) is stable under hydrothermal conditions at $T > 400\,°C$ [18.146, 147] and is soluble in alkali and carbonate solutions. The hydrothermal growth of corundum is achieved using a metastable-phase technique because of its very low solubility. Gibbsite ($Al(OH)_3$) is used as a nutrient, α-Al_2O_3 as a seed, and an alkaline solution as a solvent. A higher growth rate is obtained in KOH or K_2CO_3 solution than in NaOH or Na_2CO_3 solution. The solubility is lower in KOH or K_2CO_3, but the interaction between the impurity absorbed beforehand onto the crystal surface and K^+ ions existing in the solution acts as a factor to augment the growth rate. As the pH of solution is lowered, thickness of the growth layer increases along the c-axis. The growth of corundum is extremely low in 6 N HCl. Kashkurov et al. have studied the growth of large corundum crystals at pressures of up to 2000 atm and temperatures of up to $550\,°C$ in alkali solutions of various concentrations. Crystals weighing up to 1 kg were prepared by the hydrothermal technique, and the imperfect state of these crystals was apparently associated with internal stresses and the mosaic structure of seed crystals, which were prepared by Varneuil technique [18.148]. Figure 18.34 shows the solubility of corundum in potassium carbonate aqueous solutions.

The ideal experimental conditions for the growth of good-quality corundum crystals are:

Temperature: $480\,°C$
Pressure: $1.33\,\text{kbar}$
Mineralizer: $4\,\text{M}\,K_2CO_3$
Fill: 85%
ΔT: $30\,°C$.

Using these experimental conditions colorless and transparent crystals with (211), (311), and (111) faces were obtained. Increasing the temperature yields a higher growth rate in the temperature interval $400–500\,°C$.

Ruby and Sapphire

Ruby and sapphire crystals are grown using the hydrothermal method, with appropriate dopants added to experimental conditions similar to those described for corundum crystal growth. Sapphire was probably the second material after quartz to be grown in any size by the hydrothermal method. Similarly large ruby crystals could be grown using carbonate solutions. The solubilities of sapphire and ruby are the same as that of corundum and increase with temperature. Oxides of chromium and iron taken in the nutrient along with Na_2CO_3 solution strongly influence the crystallization of these two varieties of crystals than K_2CO_3 solution. However, when the concentration of these components in the nutrient is $> 1.6\%$ in 10% Na_2CO_3 solution, crystals barely grow. Figure 18.35 shows the growth rate of the (1011) face versus Cr_2O_3 concentration in the initial charge (10% Na_2CO_3 solution, $550\,°C$, autoclave 60% filled). In carbonate and bicarbonate solutions, chromic oxide and aluminum oxide have substantially different solubilities and rates of dissolution.

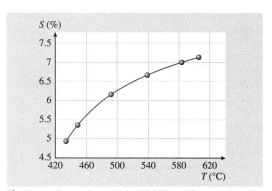

Fig. 18.34 Dependence of solubility of corundum in K_2CO_3 aqueous solutions (after [18.144])

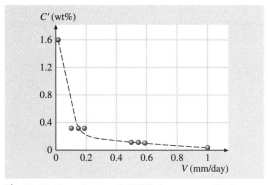

Fig. 18.35 Growth rate of (1011) face versus Cr_2O_3 concentration in the initial charge (after [18.145])

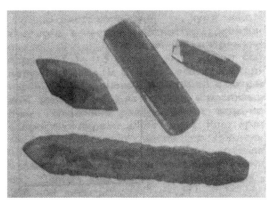

Fig. 18.36 Hydrothermally grown ruby crystals (courtesy of Prof. V. S. Balitskii)

Monchamp et al. have carried out very large-scale growth of sapphire and ruby with considerable success [18.149, 150]. Figure 18.36 shows hydrothermally grown ruby crystal [18.151]. Green crystals containing iron have been grown by carrying out the crystallization in welded liners directly, and essentially pure colorless crystals have been prepared by the use of silver tubes [18.150]. The overall features are similar to those for the growth of corundum crystals.

Emerald

Among the other gemstones grown popularly under hydrothermal conditions, emerald is one of the most important ones. It is also used as an effective tunable laser medium. The presence of Cr^{+3} in beryl ($Be_3Al_2(SiO_3)_6$) gives a green-colored emerald. The bluish-green variety is known as aquamarine. Gems that are greenish/yellow to iron-yellow and honey-yellow are called golden beryl. The rose-colored variety is called morganite or vorobyevite. Alkali elements such as Na^+, Li^+, and Cs^+ are sometimes present, replacing the beryllium at 0.25–5%. Artificial growth of beryl crystals began in the previous century itself. *Hautefeuille* and *Perrey* were the first to synthesize beryl crystals artificially [18.152]. Probably the earliest best work on hydrothermal growth of emerald was by Nacken. He made larger numbers of synthetic emeralds, using a trace of chromium to produce the color. Hexagonal prisms weighing about 0.2 g were grown in a few days. Several workers have reviewed the hydrothermal growth of emerald [18.153, 154]. Lechleitner of Innsbruck, Austria, released the first and not completely satisfactory product into the market during the 1960s and 1970s [18.155–157]. He used faceted beryl gemstones as the seed and grew a thin layer of hydrothermal emerald on the surface, which was subsequently given a light polish (not necessarily on all faces). Such stones were marketed for a short time under the names *Emerita* and *Symerald*. The first completely synthetic hydrothermal emerald was put on the market by the Linde Division of the Union Carbide Corporation in 1965. The first patent on beryl revealed that it could be grown in a neutral to alkali medium (pH of 7–12.5) using mineralizers such as alkali or ammonium halides (e.g., $NH_4F + NH_4OH$ or KF), and that Fe, Ni, or Nd could be used as dopants [18.158]. The second patent dealt with the analogous growth of beryl in an acid medium (pH 0.2–4.5), using similar halide mineralizers but with an acid reaction (e.g., $8N\ NH_4Cl$) or with extra acid added [18.159]. Typical emerald growth conditions included a pressure of 10 000–20 000 psi at temperatures of 500–600 °C resulting from a 62% fill. A small temperature gradient of 10–25 °C was employed. The Al was supplied from gibbsite $Al(OH)_3$, Be from $Be(OH)_2$, Si from crushed crystals of quartz, and Cr from $CrCl_3 \cdot 6H_2O$. Growth rates as high as 0.33 mm/day could be attained. The pressure vessel was lined with gold.

Biron Mineral (Pvt.) Ltd., Australia, produces commercially popular Biron synthetic emerald, but the growth conditions remain unknown. However, chlorine, chromium, vanadium, some water, and also tiny gold crystals have been detected in these crystals [18.160, 161]. Today Russia produces the largest quantities of synthetic emerald in the world. Also there are several recent publications from Russian laboratories on synthetic emerald. This activity began in 1965 and the growth conditions used are:

Growth temperature: 590–620 °C
Temperature gradient: 45 °C 20–100 °C 70–130 °C
Pressure: 790–1481 atm
Solvent: Acidic solution or flourine bearing solutions of complex components
Nutrient: Natural beryl or oxides of Be, Al, and Si
Vessel: Stainless steel (200–800 ml)
Seed: Plate cut parallel to (5510).

Cr^{3+}, Fe^{2+}, Fe^{3+}, Ni^{3+}, and Cu^{2+} have been detected through chemical analysis as color-associated transition elements. Fe^{2+} and Cu^{2+} ions were assigned to the tetrahedral sites, and Ni^{3+}, Cr^{3+}, and Fe^{3+} ions to the octahedral sites [18.162]. Evolution of ions

Fig. 18.37 As-grown crystals of regency emerald

from the inner wall of the autoclave is possible. It is, therefore, unknown which ions were added during the process of growth. The length of the crystal is 6.1 cm, width 1 cm, and thickness 0.7 cm. The solubility of emerald has been restudied recently using H_2SO_4 aqueous solution [18.163]. Also there are several works related to defect formation and zoning in emerald crystals obtained under hydrothermal conditions [18.164]. Figure 18.37 shows hydrothermally prepared regency emerald.

Colored Quartz

Throughout history, quartz has been the common chameleon of gemstones, standing in for more expensive gemstones ranging from diamond to jade. However, the incredible variety of quartz is now beginning to be appreciated for its own sake. Purple to violet amethyst and yellow to orange citrine are jewelry staples that continue to increase in popularity.

Ametrine combines the appeal of both amethyst and citrine as well as both the purple and yellow in one bicolored gemstone. Other major colors include the brownish smoky quartz, pink rose quartz, and colorless rock crystal quartz. Different colors and types of chalcedony, including agate, bloodstone, chrysoprase, and black onyx, have grown in popularity with growing appreciation for carved gemstones and art cutting and carving. Unusual quartz specialties such as drossy quartz, with its surface covered by tiny sparking crystals, and rutilated quartz, which has a landscape of shining gold needles inside, are adding variety and nature's artistry to unusual one-of-a-kind jewelry. The hydrothermal growth of colored quartz is a major activity today owing to its commercial value. No colored synthetic quartz producing company has ever published the exact recipe of the growth of colored quartz.

The synthesis of amethyst was first described by *Tsinobar* and *Chentsova* [18.165]. After the growth of quartz on positive (r) and negative (z) seeds in an iron-containing alkaline solution, an amethystine color is produced by irradiation. The color intensity is high from the (z) to the (r) directions. This is due to the difference of growth rates between the positive and negative rhombohedra. This phenomenon has been exploited by a large group of researchers in growing a wide range of bicolored, color-zoned quartz crystals, which are very attractive as gemstones, for example, ametrine (amethyst-citrine) [18.166]. In alkaline solutions the growth rate of the positive rhombohedron is lower than that of the negative rhombohedron. So, the content of entrapped iron is higher in the positive than in the negative rhombohedron. This color inhomogeneity with the amount of entrapped iron is distinctly discerned at areas where Dauphiné twin quartz exists. A brown or green color is emitted from the basal surface of crystals, which is caused by nontrapping of structural iron in the basal surface and also a cut surface close to the basal surface [18.167]. The credit for hydrothermal growth of gem-quality colored quartz goes to pioneers such as Tsinobar, Nassau, and Balitsky. At present Balitsky is still very active in this area of research.

Purple-colored quartz has been synthesized under the following conditions [18.168]:

Growth temperature: 400–450 °C
ΔT: 25–50 °C
Pressure: 774–1260 atm
Solvent: 10 wt % NaCl; 10 wt % KCl,
Dopant: Fe in the form of a metal; α-Fe_2O_3 in the form of a powder
Vessel: Platinum-lined (16.5 ml)
Irradiation: $2-15 \times 10^6$ R (Röntgen) (^{60}Co)

When the amount of impurities exceeds a certain level the growth is inhibited and pits or crevice flaws are induced in the crystal. Also, the type of dopant and its concentration controls the degree of growth inhibition. Faint-pink-colored quartz can be obtained in the presence of iron and titanium ions. The type and concentration of impurity ions used by the industries producing these colored quartz crystals commercially are not exactly known. However, these synthetic colored quartz crystals are even called poor man's diamond, emerald, ruby, etc., because of their beauty and closeness to those expensive natural gemstones. Readers can find more extensive information on this subject in [18.151, 169].

Zoisite

The commercial name for zoisite is tanzanite, a calcium aluminum silicate hydroxide ($Ca_2Al_3Si_3O_{12}(OH)$) having a very high commercial value that poses a real challenge to crystal growers to grow it as bulk single crystals. There are several groups throughout the world working on this, but the PT conditions involved are extreme. The special chatoyancy phenomena with vitreous luster observed in this crystal has made it a highly valuable gemstone. There are several imitations on the market as well.

Byrappa et al. have reported the synthesis of zoisite using mild hydrothermal conditions with sillimanite (Al_2SiO_5), calcium carbonate ($CaCO_3$), and quartz (SiO_2) in the ratio $3:4:3$ [18.170]. Sillimanite gel was prepared using commercially available corundum (Al_2O_3) and quartz (SiO_2) gels. The advantages of gels as starting material is that numerous and complex phases can be intimately mixed. The experiments were carried out in general-purpose autoclaves using Teflon liners in the temperature range $10-250\,°C$ with pressure of $60-80\,bar$ in a desired solvent. The experimental temperature was raised slowly at the rate of $20\,°C/h$. Several mineralizers such as HCl, CH_3COOH, C_2H_5OH, glycol, methanol, NaOH, etc. have been tried, but only HCOOH and n-butanol were found to be the most suitable mineralizers to synthesize zoisite crystal. The probable reaction for zoisite synthesis when HCOOH is used as a mineralizer is as follows:

$$3Al_2SiO_5 + 4CaCO_3 + 3SiO_2 + HCOOH$$
$$\rightarrow 2Ca_2Al_3Si_3O_{12}(OH) + 5CO_2 \ .$$

The usual pH conditions to obtain zoisite is between 1.6 and 1.8. The crystals were obtained through spontaneous nucleation in the size range $0.5-2\,mm$.

18.4.5 Rare-Earth Vanadates

R:MVO$_4$ (where R = Nd, Er, Eu; M = Y, Gd) form an important group of highly efficient laser-diode pumped microlasers, efficient phosphors, polarizer materials, and low-threshold laser hosts [18.171–174]. These crystals offer many advantages over the conventional Nd:YAG crystal, with larger absorption coefficient and gain cross section. Rare-earth vanadates are known as high-melting materials ($> 1800\,°C$) with low solubility, and obviously their synthesis by any technique usually requires higher-temperature conditions. In spite of its excellent physical properties, high-tech applications have not been realized due to crystal growth difficulties. One of the major problems encountered in the growth of RVO$_4$ crystals is the presence of oxygen imperfections (color centers and inclusions), which are introduced during the crystal growth processes. Although YVO$_4$ melts congruently [18.175], vanadium oxides vaporize incongruently, causing changes in Y/V ratio and oxygen stoichiometry in the melt. These undesired effects could generate additional phases and oxygen defects in the YVO$_4$ crystals grown, especially from the melt [18.176]. Efforts to eliminate these defects did not yield significant success with the flux and melt techniques. The instability of pentavalent vanadium at higher temperatures and the loss of oxygen through surface encrustation by the reaction of the melt with the crucible material further complicate the growth processes. In order to overcome most of the difficulties encountered in the melt and high-temperature solution techniques, the hydrothermal technique has been proposed as a solution [18.177–181]. Since the experiments are carried out in a closed system, the loss of oxygen can be readily prevented. The experiments are usually carried out in the temperature range of $240-400\,°C$ at pressure of $40\,bar$ to $1\,kbar$. The starting materials (oxides of rare earths and V_2O_5 with a desired solvent in a particular concentration) are held in Teflon or platinum liners depending upon the PT conditions of the experiments, and the Morey or Tuttle type of autoclaves are used. The solubility of rare-earth vanadates is shown in Fig. 18.38 and is found to be negative. It increases with the concentration of the solvent. Both acid and basic mineralizers are used in the growth of these vanadates. However, acid mineralizers are more effective, especially the mixed acid mineralizers ($HCl + HNO_3$ in a particular molar ratio).

The morphology of these crystals can be tuned to a desired shape using appropriate starting materials and

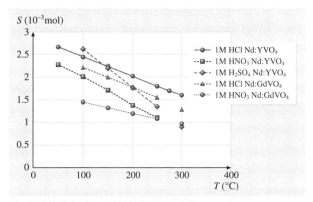

Fig. 18.38 Solubility of Nd:MVO$_4$ in different solvents

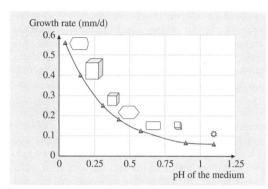

Fig. 18.39 Morphology variation with pH for Nd:YVO$_4$ crystals obtained from presintered nutrient in 1.5 M HCl + 3 M HNO$_3$ mineralizer (courtesy of K. Byrappa)

experimental conditions. Figure 18.39 shows the morphology of rare-earth vanadates and growth rate with pH of the medium. Both the chemical-reagent-grade nutrient and the presintered Nd:YVO$_4$ nutrient are used in the growth of rare-earth vanadates. In fact, the presintered nutrient gives better results. The most interesting part of this work is the need of an oxidizing agent such as hydrogen peroxide in the system, otherwise the crystals appear dark and opaque (Fig. 18.40). Often the crystals show twinning, depending upon the growth conditions. Some attempts at in situ morphology control of these vanadates have been made. The readers can refer to the works of the present author [18.174, 177, 178].

On the whole there is a slow decline in activity on bulk growth of other materials such as potassium titanyl phosphate, potassium titanyl arsenate, mixed-framework rare-earth silicates, germinates, rare-earth tungstates, borates, phosphates, oxides of various metal like tellurium, zirconium, hafnium, and a variety of sulfides of lead, copper, mercury, silver, cadmium, etc. The growth of these crystals was extremely popular from the 1960s to the 1980s. Readers can obtain more information on the growth of these crystals in earlier publications [18.1, 2, 4, 7]. The trend in hydrothermal research shifted towards the growth of fine crystals of various compounds, especially a large family of piezoelectric ceramic crystals such as lead zirconium titanates (PZT), ferrites, hydroxyapatites, etc., during the 1990s and towards the turn of the 20th century focus shifted to the growth of nanocrystals. Hence, the growth of selected fine and nanocrystals is now discussed. Accordingly the chapter refers to polyscale crystals, because it covers bulk crystals to nanocrystals. It should be noted that bulk growth of other important current technological materials such as ZnO, GaN, etc. has been discussed separately by Callaghan et al. in Chap. 19 of this handbook.

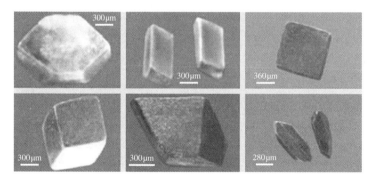

Fig. 18.40 Characteristic photographs of Nd:YVO$_4$ crystals (courtesy of K. Byrappa)

18.5 Hydrothermal Growth of Fine Crystals

The growth of fine crystals under hydrothermal conditions has been known since hydrothermal technology was born. The majority of early hydrothermal experiments carried out during the 1840s to early 1900s mainly dealt with fine to nanocrystalline products, which were discarded as failures due to the lack of sophisticated tools to examine the fine products except some chemical techniques [18.6]. During this period many experiments were carried out on the synthesis of fine crystals of zeolites, clays, some silicates, hydroxides, etc. [18.22]. When Barrer reported the hydrothermal synthesis of fine particles of zeolites during the 1940s, it opened a new branch of science, viz. molecular sieve technology. During the late 1960s and 1970s, attempts were made to synthesize fine crystals of metal oxides using the hydrothermal method. This was a very popular field of research at that time [18.4, 182, 183]. Hydrothermal research during the 1990s marked the beginning of work on processing of fine to ultrafine crystals with controlled size and morphology. Today, it has evolved to be one of the most efficient methods of soft chemistry for the preparation of advanced materials such as fine to nanocrystals with controlled size and shape. Currently, the annual market value of electronic ceramics is over US$ 1 billion, and the market for nanocrystals processing (US$ 120 billion in 2002) is increasing at 15% annually, to reach US$ 370 billion by 2010, and will jump to become a trillion-dollar industry by 2015, according to National Science Foundation (NSF) predictions.

Of all the ceramics, the PZT family has been studied most extensively using the hydrothermal technique. Since the early 1980s several thousands of reports have appeared on the preparation of these ceramics. Thermodynamic calculation and kinetics of these systems have been studied extensively (Figs. 18.4 and 18.5) [18.48, 50]. Several new variants/approaches to the processing of these electronic ceramic crystals have been reported to enhance the kinetics, shorten the processing time, control the size and shape, maintain the homogeneity of the phases, and achieve reproducibility. A great variety of precursors and solvents have been attempted in the processing of these fine crystals. Similarly fine film formation of these on an appropriate substrate has been accomplished by several workers [18.184, 185]. Here only some selected crystals such as PZT and HAP will be discussed.

The important step in the synthesis of fine crystals of advanced materials is the use of surfactants and chelates to control the nucleation of a desired phase, such that the desired phase homogeneity, size, shape, and dispersibility can be achieved during the crystallization of these fine crystals. This marked the beginning of the study of precursor preparation for different systems, surface interactions with capping agents or surfactants, and polymerized complexes. The surfaces of the crystals can be altered to become hydrophobic or hydrophilic, depending upon the applications [18.186, 187]. Today this approach is playing a key role in preparing highly dispersed, oriented, and self-assembled fine to nanocrystals. Figure 18.41 shows the new chemical approach for preparation of precursors. Using such an approach a wide range of advanced materials such as the PZT family of ceramics, ferrites, phosphates, sulfides, oxides, hydroxyapatites, etc., as well as composites have been prepared as fine crystals for technological ap-

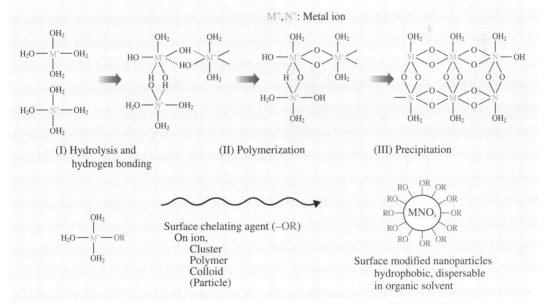

Fig. 18.41 A new preparative chemical approach for precursor preparation (courtesy of Prof. M. Yoshimura)

plications with preferred morphology such as whiskers, rods, needles, plates, spheres, etc., depending upon their application. This precursor-based chemical approach to hydrothermal synthesis has made tremendous progress in recent years and has also drastically reduced the temperature and pressure conditions required for crystal growth.

Riman and group have done extensive work on intelligent engineering of fine crystals of the PZT family and HA based on the thermodynamic modeling approach, and also calculated their crystallization kinetics in detail. Such studies not only helped in the hydrothermal synthesis of these fine crystals, but also in controlling their precise shape and size as per application requirements. Figures 18.42 and 18.43 show designer fine crystals of PZT, LiMn$_2$O$_4$, HA, etc., prepared under hydrothermal conditions.

The majority of the PZT systems incorporate intolerable amounts of alkaline metals, which are introduced in the form of mineralizers. In recent years organic mineralizers have become popular with a large number of workers. For example, *Riman* and group have found that tetramethylammonium hydroxide [N(CH$_3$)$_4$OH] is a favorable substitute for alkaline metal hydroxide mineralizer in producing phase-pure PZT [18.89]. Phase-pure MeTiO$_3$ (Me = Ca, Sr, Ba) can be obtained at input molalities of Ba, Sr, and Ca greater than 7×10^{-5}, 10^{-6}, and 5×10^{-5} M respectively. Otherwise, the relative location of the 99.995% yield regions for the three titanates will be similar to the pattern noted in the stability diagrams. In concentrated solutions, the consumption of OH$^-$ ions is caused by the predominant

Fig. 18.43 SEM photograph of HAp crystals (courtesy of Dr. W. Suchanek)

reaction

$$\text{Me}^{2+} + \text{TiO}_2 + 2\text{OH}^- \rightarrow \text{MeTiO}_3 + \text{H}_2\text{O} \; .$$

Such an approach to understand the crystallization mechanism of the PZT family of crystals has been made by several workers and there are many commercial producers of these ceramic crystals, especially in the USA, Japan, and Europe.

Gersten has extensively reviewed the processing parameters for the synthesis of fine ferroelectric perovskite crystals by the hydrothermal method [18.49]. Accordingly, the first step in the growth of these fine particles is thermodynamic verification of the correct processing conditions for the reaction of the desired product. The chemical purity of the precursors should be high, and the pH adjusters or other additives should be decomposable at the calcination temperature. The supersaturation is influenced by the initial reagent concentration, pH, experimental temperature, stirring rate, type of mineralizer, and time. An increase in the supersaturation will result in a decrease in crystal size.

In recent years interest in fine HA crystals with desired size and shape has resulted in a rapid increase in the number of publications on HAp. Starting chemicals such as H$_3$PO$_4$, Ca(OH)$_2$, and lactic acid or other solvents are taken in a Teflon beaker, inserted into an autoclave, and hydrothermally treated in the temperature range 150–200 °C for a few hours under autogenous pressure. The molar ratios of solvent/Ca and Ca/P are adjusted appropriately to get the desired phase of calcium phosphate. The morphology of the resultant HA can be controlled through the initial precursors and their ratios, besides the experimental temperature and duration. Even stirring influences the morphology of the crystals. For example, the diameter of the grains increases with increasing Ca/P molar ratio in the starting solution and is generally larger for high lactic acid/Ca molar ratios. The aspect ratio of HAp crystals is in the

Fig. 18.42a–d Hydrothermally synthesized fine crystals: (a) PZT; (b) LiMn$_2$O$_4$; (c) PbTiO$_3$; (d) BaTiO$_3$ (courtesy of Prof. Richard E. Riman)

range of 5–20. It decreases with increasing Ca/P ratio and is lower in the case of high lactic acid/Ca ratios. When the lactic acid/Ca and Ca/P starting ratios are low, crystals have the shape of whiskers, whereas in other cases large, elongated grains form [18.188].

There is a steady flow of publications relating to the growth of fine crystals of oxides, silicates, vanadates, phosphates, tungstates, titanates, etc., for various applications. The interesting aspect of their growth is the adaptability of the hydrothermal technique for all these compounds, which can be synthesized through spontaneous nucleation without many complications as in the case of bulk crystal growth. The reaction mixtures can be stirred at different rates and the crystals can also be produced continuously with the use of a flow reactor. There are many publications related to the use of flow reactors for continuous production of such fine crystals [18.97, 98].

18.6 Hydrothermal Growth of Nanocrystals

The hydrothermal technique is becoming one of the most important tools to synthesize nanocrystals for a wide variety of technological applications such as electronics, optoelectronics, catalysis, ceramics, magnetic data storage, biomedicine, biophotonics, etc. On the whole 21st century hydrothermal technology is more inclined towards nanotechnology owing to its advantages in obtaining high-quality, monodispersed, homogeneous nanocrystals with controlled size and shape, as well as the lower-PT conditions of the synthesis, simple apparatus, shorter duration, and lower cost of production. The most important advantage is that nanocrystals with desired physicochemical characteristics can be prepared, and desired surface charge can be introduced onto the nanocrystals in situ with the help of surface modifiers, capping agents, etc.

Hundreds of types of nanocrystals have been synthesized using hydrothermal technique, with over 10 000 publications in the last 8 years. The number of publications is increasing year by year and covers all groups of advanced materials such as metals, metal oxides, semiconductors including II–VI and III–V compounds, silicates, sulfides, hydroxides, tungstates, titanates, carbon, zeolites, etc. It is not possible to discuss the synthesis of all these nanocrystals using hydrothermal technology. Instead, the synthesis of some representative and technologically important nanocrystals will be discussed.

In recent years noble-metal particles (such as Au, Ag, Pt, etc.), magnetic metals (such as Co, Ni, and Fe), metal alloys (such as FePt, CoPt), multilayers (such as Cu/Co, Co/Pt), etc. have attracted the attention of researchers owing to their new interesting fundamental properties and potential applications as advanced materials with electronic, magnetic, optical, thermal, and catalytic properties [18.189–192]. The intrinsic properties of noble-metal nanoparticles strongly depend upon their morphology and structure. The synthesis and study of these metals have implications for the fundamental study of the crystal growth process and shape control. The majority of the nanostructures of these metal alloys and multilayers form under conditions far from equilibrium [18.193]. Among these metals, alloys, and multilayers, shape anisotropy exhibits interesting properties. Both the hydrothermal and hydrothermal supercritical water techniques have been extensively used in the preparation of these nanoparticles.

The synthesis of metal oxide nanocrystals under hydrothermal conditions is important because of its advantages in the preparation of highly monodispersed nanocrystals with control over size and morphology. There are thousands of reports in the literature, which also include a vast number of publications on supercritical water (SCW) technology for the preparation of metal oxides. The most popular among these metal oxides are TiO_2, ZnO, CeO_2, ZrO_2, CuO, Al_2O_3, Dy_2O_3, In_2O_3, Co_3O_4, NiO, etc. Metal oxide nanocrystals are of practical interest in a variety of applications, including high-density information storage, magnetic resonance imaging, targeted drug delivery, bio-imaging, cancer therapy, hyperthermia, neutron capture therapy, photocatalytic, luminescent, electronic, catalytic, optical, etc. The majority of these applications require nanocrystals of predetermined size and narrow size distribution with high dispersibility. Hence, a great variety of modifications are used in the hydrothermal technique.

Al'myasheva et al. and *Jiao* et al. have reviewed the hydrothermal synthesis of corundum nanocrystals under hydrothermal conditions [18.194, 195]. A high specific surface area corundum has been synthesized through the conversion of diaspore to corundum under hydrothermal conditions. This nanosized alumina has great application potential. The authors were able to develop a new transitional alumina reaction sequence that gave rise to an intermediate alpha structure, α'-Al_2O_3, with a very high surface area. Also they have investigated the thermodynamic basis and equilibrium relationships for the nanocrystalline phases. *Qian* and his group has studied extensively low temperature hydrothermal synthesis of a large variety of metal oxides under hydrothermal and solvothermal conditions [18.196]. Among the nanocrystals of the metal oxides, TiO_2 and ZnO occupy a unique place. Since the preparation of ZnO has been discussed in Chap. 19, only the synthesis of TiO_2 nanocrystals is discussed here. The synthesis of TiO_2 is usually carried out in small autoclaves of Morey type, provided with Teflon liners. The conditions selected for the synthesis of TiO_2 particles are: $T \leq 200\,°C$, $p < 100$ bar. Such pressure/temperature conditions facilitate the use of autoclaves of simple design provided with Teflon liners. The use of Teflon liners helps to obtain pure and homogeneous TiO_2 particles. Though the experimental temperature is low ($\approx 150\,°C$), TiO_2 particles with a high degree of crystallinity and desired size and shape could be achieved through a systematic understanding of the hydrothermal chemistry of the media [18.197]. A variety of surfactants are used to produce nanocrystals of the desired shape and size. Figure 18.44 shows TiO_2 nanocrystals obtained under supercritical hydrothermal conditions (400 °C and pressure 30 MPa) in the presence of hexaldehyde.

Adschiri and co-workers [18.198, 199] have worked out in detail a continuous synthesis of fine metal oxide particles using supercritical water as the reaction medium. They have shown that fine metal oxide particles are formed when a variety of metal nitrates are contacted with supercritical water in a flow system. They postulated that the fine particles were produced because supercritical water causes the metal hydroxides to rapidly dehydrate before significant growth takes place. The two overall reactions that lead from metal salts to metal oxides are hydrolysis and dehydration

$$M(NO_3)_2 + xH_2O \to M(OH)_x + xHNO_3 \,,$$
$$M(OH)_x \to MO_{x/2} + \tfrac{1}{2}xH_2O \,.$$

Processing in SCW increases the rate of dehydration such that this step occurs while the particle size is small and the reaction rate is less affected by diffusion through the particle. Furthermore, the gas-like viscosity and diffusivity of water in the critical region lead to a negligible mass-transfer limitation. The net effect is that the overall synthesis rate is very large. The high temperature also contributes to the high reaction rate. Several metal oxides, including α-Fe_2O_3, Fe_3O_4, Co_3O_4, NiO, ZrO_2, CeO_2, $LiCoO_2$, α-$NiFe_2O_4$, $Ce_{1-x}Zr_xO_2$, etc., have been prepared by this technique.

Hydrothermal synthesis of sulfides of various divalent, trivalent, and pentavalent metals constitutes an important group of materials for a variety of technological applications. They popularly form II–VI, III–VI, and V–VI group semiconductors, which are being studied extensively with respect to their different morphologies and particle size, which in turn greatly influence their properties. There are several hundred reports on these sulfides, such as CdS, PbS, ZnS, CuS, NiS, NiS_2, NiS_7, Bi_2S_3, $AgIn_5S_8$, MoS, FeS_2, InS, and Ag_2S, prepared through hydrothermal or solvothermal routes with or without capping agents/surfactants/additives to alter their morphologies and sizes as desired.

Among II–VI group semiconductor nanocrystals, AX (A = Cd, Pb, Zn, X = S, Se, Te), CdS is an important one. These AX nanocrystals have important applications in solar cells, light-emitting diodes, nonlinear optical materials, optoelectronic and electronic devices, biological labeling, thermoelectric coolers, thermoelectronic and optical recording materials, etc. Furthermore, these compounds can exhibit varying structures such as zincblende, wurtzite, halite, etc. Several papers have been published recently reporting the synthesis of chalcogenides by the hydrothermal method [18.200–203]. On the whole, for crystallization of sulfides, nonaqueous solvents are found to be more favorable, also in terms of decreasing the PT conditions

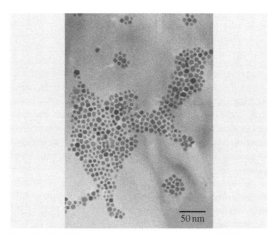

Fig. 18.44 TiO_2 nanocrystals prepared at 400 °C and 30 MPa pressure using hexaldehyde as the surface modifier (courtesy of Prof. T. Adschiri)

Fig. 18.45a–d TEM and SEM images of CdS products obtained at 180 °C for 5 h in mixed solvents with different volume ratios: (**a**) 5% ethylene, TEM image with SEM image as *inset*; (**b**) 15%, TEM image with SEM image as *inset*; (**c**) 65%, TEM image; (**d**) 100%, SEM image with the *upper right inset* showing a magnified picture of the hexagonal ends of the long rods, and the *lower left inset* showing the high-resolution TEM (HRTEM) image of a nanorod. The *scale bars* in the TEM and SEM images all represent 100 nm. The *scale bar* in the HRTEM image is 5 nm (courtesy of Prof. Yan Li)

required for crystallization. *Qian*'s group has reported the hydrothermal synthesis (using nonaqueous solvents) of nanocrystalline CdS in some coordinating solvents such as ethylenediamine and pyridine [18.204–206]. *Li* et al. have used thioacetamide as the sulfide source, as it easily releases sulfide ions, a process which is beneficial for lowering the reaction temperature and shortening the reaction period [18.207]. The hydrothermal route is more popular than all the other methods reported in the literature because of the lower temperature, shorter duration, and control over the size and morphology. The experiments are usually carried out in the temperature range 150–200 °C. Figure 18.45 shows transmission electron microscopy (TEM) and scanning electron microscopy (SEM) images of CdS nanocrystals.

The synthesis of various carbon polymorphs such as graphite, diamond or diamond-like carbon, fullerenes, etc., has attracted considerable interest for a long time because of their importance in science and technology. There are uncertainties about the phase stabilities of these polymorphs, as some of them do not find a place in the carbon pressure–temperature (PT) diagram and are also known for their contrasting physical properties. The exact physicochemical phenomena responsible for their formation are yet to be understood. Attempts to synthesize these forms under various conditions and with various techniques, sometimes even violating thermodynamic principles, have met with a fair amount of success. The stabilities of graphite and diamond in nature were mainly controlled by p–T–f_{O_2} in the C–O–H system [18.208–211]. The role of C–O–H fluids [18.212, 213], as well as the hydrothermal and organic origin of these polymorphs, especially with reference to diamond genesis, prompted material scientists to explore the possibility of synthesizing them at fairly low-pressure/temperature conditions. The hydrothermal technique is highly promising for reactions involving volatiles, as they attain the supercritical fluid state, and supercritical fluids are known for their greater ability to dissolve nonvolatile solids [18.214]. Silicon carbide powder has been used for the synthesis of carbon polymorphs [18.215, 216], and *Gogotsi* et al. [18.217] have reported decomposition of silicon carbide in supercritical water and discussed the formation of various carbon polymorphs. *Basavalingu* et al. have explored the possibilities of producing carbon polymorphs under hydrothermal conditions through decomposition of silicon carbide in the presence of organic compounds instead of pure water [18.218]. The organic compounds decompose into various C–O–H fluids; the main components are CO, OH, CO_2, and C_1H_x radicals. It is very well known that these fluids play a significant role in creating a highly reducing environment in the system and also assist in the dissociation of silicon carbide and precipitation of the carbon phase. The study of solid and gaseous inclusions in diamond also indicated the C–O–H fluids as the source for nucleation and growth of diamonds in nature [18.211].

Fig. 18.46 SEM images of diamond nanocrystals with well-developed octahedral facets adhered to the inner walls of the broken spherical particles (courtesy of K. Byrappa)

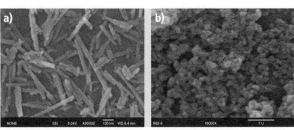

Fig. 18.47a,b SEM images of LaPO$_4$ and Nd:LaPO$_4$. (**a**) LaPO$_4$ synthesized at 120 °C for about 16 h; (**b**) LaPO$_4$ synthesized at 120 °C for 30 h

The experiments were carried out in the pressure temperature range of 200–300 MPa and 600–800 °C using externally heated Roy–Tuttle test-tube-type autoclaves. Figure 18.46 shows SEM images of diamond nanocrystals with well-developed octahedral facets adhered to the inner walls of the broken spherical particles.

Very recently, great efforts have been made to develop new phosphor systems for white-LED, FED, and programmed data processor (PDP) applications based on rare-earth phosphate systems. Several authors have reported the synthesis of rare-earth phosphate compounds via different methods. However, the hydrothermal method has distinct advantages over the other methods for the preparation of these rare-earth orthophosphates as good-quality nanocrystals. The present author's laboratory has prepared LaPO$_4$ and Nd^{3+}-doped LaPO$_4$ under hydrothermal conditions. The effect of experimental temperature, solvents, the ratio of the starting materials, stirring, and experimental duration has been studied in great detail with respect to crystal size, morphology, and the crystallinity of the resultant products in order to find the optimum experimental conditions for preparation of biophotonic materials. The experiments were carried out using Teflon-lined Morey autoclaves, Tuttle–Roy autoclaves, and stirred autoclaves. In a typical preparation, La$_2$O$_3$ and H$_3$PO$_4$ were taken in 1 : 1.2 molar ratio in a beaker containing water (14 ml) to prepare LaPO$_4$, and Nd$_2$O$_3$ (1–6 mol %) was added for the preparation of Nd^{3+}:LaPO$_4$. The pH of the solution was adjusted to 1.4–2 using ethanol. Usually the experimental duration was from 12 to 16 h. Figure 18.47 shows nanocrystals of rare-earth orthophosphates prepared under hydrothermal conditions.

There are hundreds of reports on the hydrothermal preparation of nanocrystals for various applications: to mention a few, the preparation of AlPO$_4$, ferrites, phosphors such as Eu:Y$_2$Sn$_2$O$_7$, Nd:YAlO$_3$, GaN, GaP, Ga$_3$P, vanadates, fluorides, carbonates, garnets, C$_3$N$_4$, hydroxides, etc. using both hydrothermal and solvothermal routes [18.219–225]. On the whole, solvothermal routes or the use of nonaqueous solutions is becoming more popular even for ultra-high-melting compounds such as GaN and diamond. Furthermore, the solvothermal route can minimize the entry of undesired carbonate or hydroxyl molecules into the final compounds. Similarly, the stoichiometry of the starting materials, and in some cases the change in valency of the metals, can be well controlled when using both the hydrothermal and solvothermal routes. However, the experimenter has to bear in mind that, though nonaqueous solvents show very high reactivity, one has to understand the exothermic and endothermic reactions taking place inside the autoclave, the pressure surge, and the release or formation of highly dangerous components with high volatility. If these things are checked in advance then the method can be well suited to advanced nanomaterial synthesis.

18.7 Concluding Remarks

The hydrothermal technique has evolved into one of the most efficient methods to grow crystals of different size with desired properties. The importance of the hydrothermal technique for the preparation of highly strategic materials such as GaN and ZnO has been realized in recent years. Numerical modeling has contributed greatly to understand the hydrodynamic conditions, which in turn assists in improving the quality of the crystals without any macro- or microdefects even for an industrial-scale autoclave. The recent progress in the thermodynamic modeling and also the solution chemistry of the hydrothermal media has greatly contributed to the drastic reduction in the PT conditions of the growth of crystals, even with ultralow solubility and high melting temperature. The generation of yield diagrams or stability field diagrams helps in the intelligent engineering of the crystal growth process, so that the yield is almost 100%, i.e., all the solute is converted into the product without any loss of the nutrient. The study of solubility in the successful growth of crystals under hydrothermal conditions is essential, as is the kinetics of crystallization, which helps greatly in enhancing the growth rate. The application of additional energy such as microwave, sonar, mechanochemical, electrochemistry, biomolecules, etc. takes us into a different field of chemistry at the speed of light, because of the increased growth kinetics. The stirring of the precursor during the crystal growth under hydrothermal conditions also contributes to the size and shape of the crystals, especially in the case of fine crystal growth. In recent years, organic molecules, capping agents, and surfactants have been extensively used to introduce a desired surface charge to the crystal surface, and also the properties of the crystal surface can be altered to either hydrophobic or hydrophilic. Similarly the crystal size and shape can be altered with the help of such organic agents. This area of in situ crystal growth process is fast growing and making the hydrothermal technique into one of the most efficient methods in crystal growth. However, our knowledge today on the growth of crystals such as diamond, gallium nitride, gallium berlinite, etc., is comparable to that of quartz some 60 years ago. Therefore, a collective multidisciplinary approach is essential to understand the hydrothermal technique in order to grow crystals of desired shape, size, and properties. Knowledge on the hydrothermal phase equilibrium is also essential and has to be supported by thermodynamic modeling and computation techniques. Although much of the recent work on the hydrothermal growth of bulk crystals such as quartz, gemstones, gallium nitride, etc., is treated as classified, there is still enormous scope for the application of this technique to grow crystals of technological significance. The number of publications on this important tool of materials processing is also increasing year by year.

18.A Appendix

Table 18.10 List of the polycrystals prepared under hydrothermal conditions. Auts. = autogenous pressure

Compound	Mineralizer(s)	Temperature (°C)	Pressure (kbar)	References
Native elements				
Au, Ag, Pt, Co, Ni,	NaOH, KOH, HCl,	400–600	> 0.60	[18.189–192]
Te, As, Bi, C	HI, HBR			[18.218, 226]
Se	NaOH, KOH	–	–	[18.227]
Cd	NaOH, KOH	–	–	[18.228]
Pb, Cu	NaOH, KOH	450–550	> 0.05	[18.229]
Oxides				
CuO, Cu_2O	NaOH, KOH	350–450	0.65–0.8	[18.230]
BeO	NaOH, KOH	450–550	0.8	[18.231]
ZnO	NaOH, KOH, NH_4Cl	350–600	0.78–1.0	[18.232]
CdO	NaOH, Na_2CO_3, $(NH_4)_2SO_4$	250–500	0.3–2.0	[18.233]
NiO	NH_4Cl	300	0.3	[18.234]
HgO	NaOH	200–300	0.1	[18.235]
PbO	LiOH	450	0.6	[18.236]
Al_2O_3, emerald	NaOH, KOH, Na_2CO_3, K_2CO_3	300–620	0.5–0.7	[18.27, 194, 195]
RE_2O_3	H_2O	> 500	0.2–0.9	[18.237]
In_2O_3	NaOH, NH_4Cl	350–600	0.4–3.0	[18.238]
V_2O_3	NaOH	550–700	1.3	[18.239]
Sb_2O_3	KF, Na_2CO_3	460–600	1.0–1.5	[18.240]
Bi_2O_3	NaOH, KOH	360–600	0.4–1.2	[18.241]
Au_2O_3	Na_2CO_3	300	3.0	[18.242]

Table 18.10 (continued)

Compound	Mineralizer(s)	Temperature (°C)	Pressure (kbar)	References
Fe_3O_4	NaOH	350–405	0.75–0.8	[18.243]
MnO_2	NaOH	450	0.75–0.8	[18.244]
TiO_2	KF, NaF, NH_4F, NaOH, KOH	100–700	0.5–3.0	[18.197]
ZrO_2, HfO_2	NaF, KF, NH_4F	520–690	1.0	[18.2, 34]
SnO_2	NaOH, KOH	450–600	0.7	[18.245]
TeO_2	HCl, HF, HI, HBr, HNO_3	250–385	0.05–0.8	[18.246]
PbO_2	NH_4OH	230–290	0.5	[18.247]
CrO_2	$HCLO_4$, HIO_4, NH_4ClO_4	350–400	2.0	[18.248]
SiO_2, quartz	$NaOH + Na_2CO_3$, $NaCl + KCl$	350–370	1.5	[18.35]
Amethyst	NH_4F	210–450	–	[18.167]
GeO_2	NaOH, KOH	350	1.0	[18.249]
$K_2Te_4O_9 \cdot 3H_2O$	$K_3PO_4 + H_2O$	380	–	[18.250]
$Y_3Fe_5O_{12}$	KOH	420	2.0	[18.251]
$Y_3Al_5O_{12}$	K_2CO_3, Na_2CO_3, NaOH	350	2.0	[18.252]
$ABi_{12}O_{20}$ (A = Ti, Si, Ge)	NaOH	500–570	1.5–2.0	[18.253]
$NiFe_2O_3$	NH_4Cl	475	–	[18.109]
Hydroxide				
$M(OH)_2$ (M = Be, Mg, Ca, Cd, Mn, Ni, Co	NaOH, H_2O	100–550	0.6–4.0	[18.254]
$RE(OH)_3$ (RE = La – Gd, Dy, Er, Yb)	NaOH	350–450	0.6	[18.254]
$M(OH)_3$ (M = Cr, In)	NaOH	300–450	0.4–1.5	[18.254]
$Al(OH)_3$	H_2O	< 150	0.04	[18.255]
MOOH, MOOD, $In(OD)_3$ M = In, Cr, Gd, Sn	H_2O, D_2O	180–600	0.02–1.4	[18.256]
α-ScOOH, β-ScOOH	H_2O	162–350	0.006–0.17	[18.257]
$Sr_3Cr_2(OH)_{12}$	H_2O	150–200	–	[18.258]
Malachite, azurite	–	–	–	[18.259]
Garnets				
$A_3B_2Si_3O_{12}$ (A = Mn, Ca, Fe, Cd; B = Al, Fe, In, V, Cr)	$CaCl_2$, $CdCl_2$, NH_4Cl	400–900	0.4–30.0	[18.260, 261]
Silicates				
Li_2SiO_3	LiOH	450	0.15	[18.262]
$KHSiO_2O_5$	KOH	300	–	[18.263]
Al_2SiO_5	H_2O	900	20.0	[18.264]
$CaBSiO_4(OOH)$	$CaCl_2 + HCl$	300–400	0.3–0.35	[18.264]
Zn_2SiO_4	HCl	400–500	0.4	[18.265]
$Zn_2SiO_4:Mn^{2+}$	Ethanol	220	Auts.	[18.266]
$R_2Si_2O_7$ (R = La – Lu, Sc, Y, Cd)	H_2O	700	2.5	[18.267]
$Na_2Fe_5TiSiO_{20}$	H_2O	500–700	0.5	[18.268]
$Na_2R_6Si_6O_{24}(OH)_2$ (R = La, Y)	H_2O	500–700	2.0	[18.269]

Table 18.10 (continued)

Compound	Mineralizer(s)	Temperature (°C)	Pressure (kbar)	References
$M_4R_6Si_6O_{24}(OH)_2$ (M = Ba, Ca, Sr, Pb, Mn; R = La, Nd, Sm, Gd, Dy, Y)	H_2O	500–700	2.0	[18.269]
$MPbSiO_4$ (M = Mn, Fe, Co, Ni, Zn, Mg, Cd, Be)	H_2O	300–500	2.0	[18.270]
$MBe_2Si_2O_7$ (M = Ba, Sr, Pb)	H_2O	450–750	1.5–3.0	[18.271]
$Na_2Be_2Si_6O_{14}$	NaOH	400–600	1.5–3.0	[18.272]
$Na_2BeSi_2O_6$	NaOH, H_2O	400–600	–	[18.272]
$Ca_2MgSi_2O_7$	H_2O	798	4.06	[18.273]
$Na_2TiZnSiO_7$	NaOH	350–550	> 3.0	[18.274]
$Ba_2TiSi_2O_8$	$Ba(OH)_2$	200–500	> 2.0	[18.275]
$Li_4Sn_2Si_{12}O_{30}$	LaOH	500–600	> 2.0	[18.276]
$K_2SnSi_3O_9$	KOH	400–650	> 1.5	[18.277]
$RbFeSi_3O_8$	RbOH	480	1.6	[18.278]
$K_2CeSi_6O_{15}$	KOH	450	2.0	[18.279]
$K_3NdSi_6O_{15}$	KOH + K_2CO_3	400–600	2.0	[18.280]
$K(Mn,Fe)_2(Zn,Mn)_3Si_{12}O_{30}$	KOH + KCl	580–600	0.8	[18.281]
$K_2Mn_2Zn_2Si_4O_{15}$	KOH + KCl + $MnCl_2$	580–600	0.8	[18.282]
$FeMgAl_4Si_2O_{10}(OH)_4$	H_2O	650	9.2	[18.283]
$Na_2MnZn(SiO_3OH)_2(OH)_2$	NaOH	450	> 0.5	[18.284]
$Na_4Fe_{10}Ti_2Si_{12}O_{40}$	H_2O	700	1.0	[18.285]
$Na_3CaMg_2Si_8O_{22}(OH)_2$	Na_2CO_3	400–970	0.1–0.8	[18.286]
$LiAlSi_2O_6 \cdot H_2O$	H_2O	300–350	2.0	[18.287]
$Na_8SnSi_6O_{18}$	NaOH	600	1.0	[18.288]
$K_2ZrSi_2O_7$	KOH	310	0.5–1.0	[18.289]
$RERE'SiO_5$ (RE = La, Ce, Pr, Sm)	NaOH + HCl	135–700	1.0	[18.290]
$K_2ZrSi_3O_8$	KOH	350–450	0.5	[18.291]
$(Na_2Ca)ZrSi_3O_9 \cdot H_2O$	Na_2CO_3	350–530	–	[18.292]
$PbCa_2Si_3O_9$	H_2O	350	2.0	[18.293]
$KLi_2AlSi_4O_{10}F_2$	H_2O	400–670	2.0–5.0	[18.294]
$K_3Li_3Al_3Si_6O_{20}(F,OH)_4$	H_2O	400–670	2.0–5.0	[18.295]
Sodalite				
Mica	KOH, K_2CO_3, $KHCO_3$	–	–	
$Na_8Al_6Si_6O_{24}(OH)_2 \cdot nH_2O$	NaOH	200–450	0.5	[18.296]
Na_2MgSiO_4	NaOH	700	3.0	[18.297]
$Na_4Zr_2Si_3O_{12}$	NaOH	600	3.0	[18.298]
$NaAlSiO_4$, $Na_2Zn_2Si_2O_7$, $Na_2ZnSi_3O_8$, Na_2ZnSiO_4, $Na_2Mn_2Si_2O_7$	NaOH, NaOH + Na_2CO_3	375–450	1.0–2.0	[18.299]
Germanates				
$MGeO_4$ (M = Zr, Hf, U)	H_2O	150–300	1.0	[18.300]
Al_2GeO_5	H_2O	400–700	0.5	[18.301]
Pb_3GeO_5	KOH	450–500	0.6–1.2	[18.302]
$Sb_2Ge_2O_7$	KF	400–550	0.6–1.2	[18.303]
$Sb_4Ge_3O_{12}$	K_2CO_3	450–550	0.6–1.2	[18.304]

Table 18.10 (continued)

Compound	Mineralizer(s)	Temperature (°C)	Pressure (kbar)	References
$Bi_4Ge_3O_{12}$	NaOH	350–450	0.6–0.8	[18.305]
$CaGeO_3$	H_2O	150–450	0.8–1.8	[18.306]
$SrGeO_3$	NaOH	300–600	> 1.0	[18.307]
$Mg_5GeO_6(OH)_2$	H_2O	470–570	> 0.3	[18.308]
$Y_2GeO_6(OH,F)_4$	KF	450	1.5	[18.309]
$Yb_4Ge_3O_9(OH,F)_4$	KF	450	1.0	[18.310]
$Er_4Ge_3O_9(OH,F)_4$	CsF	450	0.6	[18.311]
$M_6Ge_4O_{10}(OH)_4$ (M = Na, Li, Rb, K, Cs)	H_2O	180–500	0.1	[18.312]
$Na_2LaGe_4(OH)$	NaOH	450	1.0	[18.313]
Na_2TiGeO_5, $Na_2Zn_2TiGeO_7$	NaOH	400–550		[18.314]
$Na_3ZrHGe_2O_8$	NaOH	500	1.2–2.0	[18.315]
Na_2ZrGeO_5, $Na_2ZrGe_2O_7$, $Na_4Zr_2Ge_3O_{12}$, Na_2BeSiO_2	NaOH	450–500	1.0	[18.316]
$KAlGe_2O_6$, $NaFeGe_2O_6$	KOH, NaOH	450	1.0	[18.317]
$NaAlGe_3O_9$	H_2O	800	10.0	[18.318]
$AgAlGe_3O_8$	$H_2O + O_2$	600	1.5	[18.319]
$TiAlGe_3O_8$	H_2O	1000	2.5	[18.319]
$MAlGe_3O_8$ (M = Li, Na, K, Rb, Cs)	H_2O	800–900	13.0–25.0	[18.319]
$Na_6Sn_4Ge_5O_{20}(OH)_2$		450	1.0	[18.320]
$Na_2SnGe_2O_6(OH)_2$		450	0.5	[18.321]
$Na_3REGe_2O_7$ (RE = La, Nd, Eu)		800	1.5	[18.322]
$Y_7Ge_2O_{12}(F,OH)_5$, $Sm_4Ge_3O_9(OH)_6$, $Y_4GeO_6(OH,F)_4$, Nd_4GeO_8, $Yb_2Ge_2O_7$, $RE_3GeO_5(OH)_3$ (RE = Gd, Sm, Dy, Er)	KF, CsOH, RbOH, NaOH	450–500	1.5–2.0	[18.323]
Phosphates				
$AlPO_4$	H_3PO_4, H_2SO_4, HCl, HCOOH	150–300	0.03–0.2	[18.324]
ABO_4 (A = Fe, Bi, Mn, Cr, Al, B; B = P, V, As)	NaOH, H_3PO_4	200–900	0.03–50.0	[18.325, 326]
$Ce(HPO_4)_2 \cdot 33H_2O$	H_3PO_4	160	–	[18.327]
$CuTi_2(PO_4)_3$	H_3PO_4	500	3.0	[18.328]
$RbNbOPO_4$	H_3PO_4	600	1.8	[18.329]
$Cu_3(PO_4)_2 \cdot H_2O$	$H_3PO_4 + H_2O$	220	–	[18.330]
$NH_4Zr_2(PO_4)_3$	NH_4Cl	250–300	0.1	[18.331]
$(H_3O)Zr_2(PO_4)_3$	H_2O	200	< 0.1	[18.332]
$KMnO_2 \cdot O(PO_4)(H_3PO_4)$	H_2O	400	3.0	[18.333]
NdP_5O_{14}, TmP_5O_{14}	H_3PO_4, $HCl + H_3PO_4$	240–300	0.05–0.1	[18.334]
$MREP_4O_{12}$ (M = Le, Na, K, Rb, Cs; RE = La–Nd)	H_3PO_4	300–500	0.1–0.5	[18.335]
$(Na_{2/3}Zr_{1/3})_2P_2O_7$	$H_3PO_4 + HCl$	250	< 0.01	[18.336]
$Na_2H_3Al(P_2O_7)_2$	H_3PO_4	200–250	0.01	[18.337]

Table 18.10 (continued)

Compound	Mineralizer(s)	Temperature (°C)	Pressure (kbar)	References
$NaHMP_2O_7$ (M = Ni, Co)	H_3PO_4	200–250	0.01	[18.338]
KH_2MoPO_7	H_3PO_4	180	< 0.01	[18.339]
$NaH_2(MoO)_2(PO_4)_3 \cdot \frac{1}{2}H_2O$	H_3PO_4	600	1.8	[18.340]
KTP	$K_2HPO_4 + KPO_3$	425–600	10 kpsi	[18.53]
Vanadates				
$TMAV_3O_7$, $TMAV_4O_{10}$	TMAOH, LiOH	200	Auts.	[18.174, 341–343]
$TMAV_8O_{10}$		–	–	
$Li_xV_{2-\delta}O_{4-\delta}$		–	–	
$R:MVO_4$ (R = Nd, Er, Eu; M = Y, Gd)		–	–	
γ-LiV_2O_5		–	–	
Borates				
$Zn_4B_6O_{12}$	NaOH	500	2.4	[18.344]
$Li_2B_4O_7$	HCl	250	0.12	[18.345]
$Li_3B_5O_8(OH)_2$	–	–	–	[18.346]
$REBO_3$ (RE = Sm, Er, Gd, Eu, Tm, Yb)	1.4 Butanediol	315	Auts.	[18.347]
$LiBO_2$, $LiZnBO_3$	NaOH, KOH	> 300	1.0	[18.344]
$LiH_2B_5O_9$	HCOOH	> 240	0.01	[18.348]
Carbonates				
MCO_3 (M = Ca, Mg, Sr, Ba)	NH_4Cl, LiCl, H_2CO_3, HNO_3	200–450	0.6–0.8	[18.139]
Chalcogenides				
Cu_9S_5, Ag_2S		180–343	0.01–0.15	[18.349]
Cu_2S		312	0.01	[18.350]
α-MnS	Thiourea, benzene	100–200	–	[18.351]
PbS	Toluene	220	Auts.	[18.352]
MoS_2	$C_2H_4NS + NH_3$	160–195	Auts.	[18.353]
Ag_2Se		180–343	0.01–0.5	[18.354]
MSe (M = Zn, Cd, Hg, Co, Pb)		180–500	0.01–1.5	[18.355]
MTe (M = Zn, Cd, Pb, Hg, Co)		180–400	0.05–2.0	[18.356]
MS (M = Cu, Zn, Hg, Pb, Fe, Cd, Mn, Co, Ni)		190–640	0.05–2.0	[18.357]
Bi_2Se, Sb_2Se		180–343	0.01–1.4	[18.355]
Sb_2Si_3, In_2Si_3, Bi_2S_3		180–343	0.01–0.15	[18.358]
$Bi_2Fe_3Sb_2Te_3$		180–343	0.01–0.15	[18.359]
$AgMS_2$ (M = Bi, Ga, In, Sb, Be)		180–343	0.01–0.15	[18.358]
$CuMS_2$ (M = Ga, In, Fe)		–	–	
Cu_8MS_6, M = Ge, Sb		312	0.1	[18.350]
$CuInTe_2$		265–343	0.05–0.15	[18.360]
$Cu_6ZnAs_2S_7$, $Cu_6FeAs_2S_7$		180–343	0.01–0.05	[18.361]
AX (A = Cd, Pb, Zn, In; X = S, Se, Te)	Ethylenediamine	150–200	Auts.	[18.204–206]
Titanates				
$PbTiO_3$, $PbTi_3O_7$, $PbZrO_3$	KF	450–700	0.8–3.0	[18.362]
$MTiO_3$ (M = Ca, Co, Mn, Ba, Sr)	NaOH, KOH, KF	200–700	0.3–2.0	

Table 18.10 (continued)

Compound	Mineralizer(s)	Temperature (°C)	Pressure (kbar)	References
$Na_2Ti_2O_5$, $Na_2Ti_6O_{13}$, $Na_2Ti_3O_7$, $Na_2Ti_3O_7 \cdot H_2O$	NaOH, HCl, HCOOH	200–500	0.08–3.0	[18.363, 364]
Molybdates				
$MMoO_4$ (M = Cd, Sr, Pb, Ba)	NH_4Cl, NaOH, KOH	350–500	–	[18.365]
$KLn(MoO_4)_2$ (Ln = La – Yb)	K_2MoO_4	525–600	–	[18.366]
$K_5Ln(MoO_4)$ (Ln = La – Yb)		–	–	
Ln_2MoO_6 (Ln = Pr – Er)		–	–	
Tunstates				
MWO_4 (M = Ba, Sr, Cd)	NH_4Cl	450	–	[18.365]
$Li_2Fe(WO_4)_2$	$LiCl + Na_2WO_4$	575–600	1.0–1.5	[18.367]
$MLn(WO_4)_2$	LiCl, KCl, NH_4Cl	400–720	1.2–1.7	[18.368, 369]
$LnWO_4 \cdot OH$ (M = Li, Na, K, NH_4; Ln = Ce, Pr, Nd)	$LiCl + NH_4Cl$	–	–	[18.366]
Niobates, tantalates, arsenates, gallates, etc.				
ABO_4 (A = Al, Fe, Ga, B, Bi; B = P, V, As)	NaOH, NaF, KF, HBr, HI	> 500	2.0	[18.370]
$LiNbO_3$	LiOH	650	2.0	[18.371]
$LiGaO_2$	NaOH	360–420	–	[18.372]
$Gd_3Ga_5O_{12}$	NaOH, Na_2CO_3, K_2CO_3	350–605	1.5	[18.252]
Potassium titanyl arsenate (KTA)	KOH	590	1.8	[18.373]
Chlorides, bromides, fluorides				
$Fe_6O_{11}Cl_2$, $TeOCl_2$, CuBr, PbBr	HCl, HBr, HI	100–200	0.1	[18.370]
$AREF_4$, $AREF_6$, A_2REF_5 (A = Li, Na, K, Rb, Cs; RE = Nd, Sm, Eu, Gd, Tb, Yb, Ho, Y)	KF, H_2O	450–500	Auts.	[18.374]
Nitrides				
AlN, GaN, InN	Toluene, THF, trioctylamine	265–290	Auts.	
	$NH_3 + NH_4Cl$, $NH_3 + NH_4Br$, $NH_3 + NH_4I$, (or admixture of all)	360–550	1.4	[18.375, 376]
Miscellaneous				
α-ZrP	H_3PO_4, H_2O	120–280	Auts.	[18.377]
β-TiP	H_3PO_4	300	Auts.	[18.378]

References

18.1 K. Byrappa, M. Yoshimura: *Handbook of Hydrothermal Technology* (Noyes, New York 2001)

18.2 M. Yoshimura, K. Byrappa: J. Mater. Sci. **43**, 2085–2103 (2007)

18.3 K. Byrappa, S. Ohara, T. Adschiri: Adv. Drug Del. Rev. **60**, 299–327 (2007)

18.4 K. Byrappa (Ed.): Hydrothermal growth of crystals, Prog. Cryst. Growth Charact. Mater. **21**, 1–365 (1990)

18.5 K. Byrappa: Hydrothermal processing of advanced materials. In: *Kirk-Othmer Encyclopedia of Chemical Technology* (Wiley, London 2005)
18.6 K. Byrappa, T. Adschiri: Prog. Cryst. Growth. Charact. Mater. **53**, 117–166 (2007)
18.7 K. Byrappa: Hydrothermal growth of crystals. In: *Handbook of Crystal Growth*, Vol. 2, ed. by D.T.J. Hurle (Elsevier, Amsterdam 1994) p. 465
18.8 K.F.E. Schafthaul: Gelehrte Anz. Bayer. Akad. **20**, 557 (1845)
18.9 R. Bunsen: Bemerkungen zu einigen Einwürfen gegen mehrere Ansichten über die chemisch-geologischen Erscheinungen in Island, Annalen **65**, 70 (1848)
18.10 F. Wöhler: Annalen **65**, 80 (1848), cited by R. Bunsen
18.11 H. De Senarmount: Ann. Chim. Phys. **32**, 142 (1851)
18.12 M. Daurree: Ann. Mines **12**, 289 (1857)
18.13 H.E. Saint-Claire Deville: Ann. Chim. Phys. **61**, 309–314 (1857)
18.14 K. von Chrustschoff: Am. Chem. **3**, 281 (1873)
18.15 J.B. Hannay: Proc. R. Soc. Lond. **30**, 178–189 (1880)
18.16 R. Moissan: *Experimental Petrology, Basic Principles and Techniques* (Clarendon, Oxford 1973) p. 5, cited by A.D. Edgar
18.17 C. Friedel, E. Sarasin: C. R. **92**, 1374–1378 (1881)
18.18 C. Friedel: Bull. Soc. Min. **14**, 7–10 (1891)
18.19 G. Spezia: Atti. Accad. Sci. Torino **35**, 95–107 (1900)
18.20 G. Spezia: Atti. Accad. Sci. Torino **40**, 254–262 (1905)
18.21 G.W. Morey, P. Niggli: J. Am. Chem. Soc. **35**, 1086–1130 (1913)
18.22 G.W. Morey: J. Am. Ceram. Soc. **36**, 279–285 (1953)
18.23 C. Barus: Am. J. Sci. **6**, 270 (1898)
18.24 R. Nacken: Chem. Z. **74**, 745–749 (1950)
18.25 R. Nacken: Captured German Reports RDRC/13/18 (1946)
18.26 R. Nacken: US Office of Technical Services Reports PB-6948 (1946)
18.27 K. Nassau: J. Cryst. Growth **35**, 211–222 (1976)
18.28 A.A. Shternberg: private communication (Moscow 1981, 1982)
18.29 J. Asahara, K. Nagai, S. Harada: Synthetic crystals by large autoclaves – The reality and characterization, Proc. 1st Int. Symp. Hydrotherm. React., ed. by S. Somiya (Gakujutsu Bunken Fukyu-Kai 1982) pp. 430–441
18.30 E.U. Franck: Survey of selected non-thermodynamic properties and chemical phenomena of fluids and fluids mixtures. In: *Chemistry and Geochemistry of Solutions at High Temperatures and Pressure*, Proc. Nobel Symp., Vol. 13/14, ed. by D.T. Rickard, F.E. Wickman (Pergamon, New York 1979) pp. 65–88
18.31 T.M. Seward: Metal complex formation in aqueous solutions at elevated temperatures and pressures. In: *Chemistry and Geochemistry of Solutions at High Temperatures and Pressure*, Proc. Nobel Symp., Vol. 13/14, ed. by D.T. Rickard, F.E. Wickman (Pergamon, New York 1979) pp. 113–132
18.32 H.C. Helgeson: Prediction of the thermodynamic properties of electrolytes at high pressures and temperatures. In: *Chemistry and Geochemistry of Solutions at High Temperatures and Pressure*, Proc. Nobel Symp., Vol. 13/14, ed. by D.T. Rickard, F.E. Wickman (Pergamon, New York 1979), pp. 133–178
18.33 K.S. Pitzer: Characteristics of very concentrated aqueous solutions. In: *Chemistry and Geochemistry of Solutions at High Temperatures and Pressure*, Proc. Nobel Symp., Vol. 13/14, ed. by D.T. Rickard, F.E. Wickman (Pergamon, New York 1979) pp. 249–272
18.34 S. Somiya (Ed.): Hydrothermal reactions, Proc. 1st Int. Symp. Hydrotherm. React. (Gakujutsu Bunken Fukyu-Kai, 1982) pp. 430–441
18.35 R.A. Laudise, J.W. Nielsen: Solid State Phys. **12**, 149 (1961)
18.36 V.A. Klipov, N.N. Shmakov: Influence of convective flow on the growth of quartz crystals, Proc. 45th Annu. Symp. Freq. Control, IEEE (New York 1991) p. 29
18.37 A.B. Ezersky, A. Garcimartin, J. Burguete, H.L. Mancini, C. Pérez-García: Phys. Rev. E. **47**, 1126–1131 (1993)
18.38 A.B. Ezersky, A. Garcimartin, H.L. Mancini, C. Pérez-García: Phys. Rev. E. **48**, 4414–4422 (1993)
18.39 B. Roux, O. Louchart, O. Terhmina: J. Phys. (France) **4**(C), 2–3 (1994)
18.40 Q.S. Chen, V. Prasad, A. Chatterjee: Modelling of fluid flow and heat transfer in a hydrothermal crystal growth system, Proc. Am. Soc. Mech. Eng. Heat Transf. Div. (1998) p. 119, HTD 361-4
18.41 Q.S. Chen, V. Prasad, A. Chatterjee, J. Larkin: J. Cryst. Growth **198/199**, 710 (1999)
18.42 E.A. Evans, H. Li, G.-X. Wang: Bulk flow of solution in an autoclave for crystal growth, Proc. CD IMECE 2001 (New York 2001), Paper No. HTD 24342
18.43 H. Li, E.A. Evans, G.-X. Wang: J. Cryst. Growth **256**, 146–155 (2003)
18.44 H. Li, G.-X. Wang, E.A. Evans: J. Cryst. Growth **271**, 257–267 (2004)
18.45 H. Li, E.A. Evans, G.-X. Wang: J. Cryst. Growth **275**, 561–571 (2005)
18.46 R.-U. Barz, M. Grassl: J. Cryst. Growth **249**, 345–353 (2003)
18.47 V.N. Popov, Y.S. Tsivinskaya, T.B. Bekker, K.A. Kokh, A.E. Kokh: J. Cryst. Growth **289**, 652–658 (2006)
18.48 M.M. Lencka, R.E. Riman: Thermodynamics of multicomponent perovskite synthesis in hydrothermal solution. In: *Handbook of Crystal Growth Technology*, ed. by K. Byrappa, T. Ohashi (Springer, Berlin Heidelberg 2003) p. 271
18.49 B. Gersten: Growth of multicomponent perovskite oxide crystals: Synthesis conditions of the hydrothermal growth of ferroelectric powders. In:

18.50 *Handbook of Crystal Growth Technology*, ed. by K. Byrappa, T. Ohashi (Springer, Berlin Heidelberg 2003) p. 299

18.50 R.E. Riman, W.L. Suchanek, K. Byrappa, C.W. Chen, P. Shuk, C.S. Oakes: Solid State Ion. **151**, 393 (2002)

18.51 O.V. Dimitrova: Investigations of the phase formations in the system $Na_2O-RE_2O_3-SiO_2-H_2O$ under hydrothermal conditions. Ph.D. Thesis (Moscow State University, Moscow 1975)

18.52 K. Byrappa, J.R. Paramesha: Mater. Sci. Forum **315-317**, 514–518 (1999)

18.53 R.A. Laudise, W.A. Sunder, R.F. Belt, G. Gashurov: J. Cryst. Growth **102**, 427–433 (1990)

18.54 R.A. Laudise, A.A. Ballman, J.C. King: J. Phys. Chem. Solids **26**, 1305 (1965)

18.55 E.U. Franck: Pure Appl. Chem. **24**, 13–30 (1970)

18.56 R.C. Weast (Ed.): *Handbook of Physics and Chemistry*, 64th edn. (CRC, Boca Raton 1983)

18.57 H.C. Helgeson: Phys. Chem. Earth **13/14**, 133 (1981)

18.58 J.B. Hasted, D.M. Ritson, C.H. Collie: J. Chem. Phys. **16**, 1 (1948)

18.59 M. Uematsu: Phase equilibria and static properties. In: *Supercritical Fluids – Molecular Interaction, Physical Properties, and New Applications*, Springer Series in Materials Processing, ed. by Y. Arai, T. Sako, Y. Takebayaschi (Springer, Berlin Heidelberg 2002) p. 71

18.60 E.U. Franck: Int. Corros. Conf. Ser., 109 (1973)

18.61 G.C. Kennedy: Am. J. Sci. **248**, 540–543 (1950)

18.62 S.D. Haman: Phys. Chem. Earth **13/14**, 89 (1981)

18.63 K. Todheide: Ber. Bunsenges. Phys. Chem. **86**, 1005 (1982)

18.64 E.D. Kolb, P.L. Key, R.A. Laudise, E.E. Simpson: Bell Syst. Tech. J. **61**, 639 (1983)

18.65 W.L. Marshall, E.V. Jones, J. Jones: Inorg. Nucl. Chem. **36**, 2313 (1974)

18.66 A.A. Shternberg: Controlling the growth of crystals in autoclaves. In: *Crystallization Processes Under Hydrothermal Conditions*, ed. by A.N. Lobachev (Consultants Bureau, New York 1973) pp. 225–240

18.67 L.N. Demianets, E.N. Emelyanova, O.K. Melnikov: Solubility of sodalite in aqueous solutions of NaOH under hydrothermal conditions. In: *Crystallization Processes Under Hydrothermal Conditions*, ed. by A.N. Lobachev (Consultants Bureau, New York 1973) pp. 125–150

18.68 G.W. Morey, J.M. Hesselgesser: Am. J. Sci., 367 (1952)

18.69 R.A. Laudise: J. Am. Chem. Soc. **81**, 562 (1959)

18.70 R.A. Laudise: Hydrothermal growth of crystals. In: *Progress in Inorganic Chemistry*, Vol. 3, ed. by F.A. Cotton (Wiley, New York 1962)

18.71 D.J. Marshall, R.A. Laudise: Crystal growth by hydrothermal technique. In: *Crystal Growth*, ed. by H.S. Peiser (Pergamon, New York 1966) p. 557

18.72 N.Y. Ikornikova, A.N. Lobachev, A.R. Vasenin, V.M. Egrov, A.V. Autoshin: Apparatus for precision research in hydrothermal experiments. In: *Crystallization Processes Under Hydrothermal Conditions*, ed. by A.N. Lobachev (Consultants Bureau, New York 1973) p. 241

18.73 J.B. Hannay, J. Hogarth: Proc. R. Soc. Lond. **30**, 178 (1880)

18.74 M. Hosaka, S. Taki: J. Cryst. Growth **100**, 343 (1990)

18.75 R.A. Laudise, A.A. Ballman: J. Phys. Chem. **65**, 1396–1400 (1961)

18.76 V.M. John, S. Kordes: Chem. Earth **70**, 75–89 (1953)

18.77 J.M. Stanley: Ind. Eng. Chem. **468**, 1684–1689 (1954)

18.78 K. Byrappa, V. Venkatachalapathy, B. Puttaraju: J. Mater. Sci. **19**, 2855–2862 (1984)

18.79 M. Cochez, A. Ibanez, A. Goiffon, E. Philippot: Eur. J. Solid State Ionorg. Chem. **30**(55), 509–519 (1993)

18.80 O.V. Zvereva, M.Y. Mininzon, L.N. Demianets: J. Phys. (France) **4**(C2), 19–24 (1994)

18.81 P. Yot, O. Cambon, D. Balitsky, A. Goiffon, E. Philippot, B. Capelle, J. Detaint: J. Cryst. Growth **224**, 294–302 (2001)

18.82 R.-U. Barz, M. Grassl, P. Gille: J. Cryst. Growth **245**, 273–277 (2002)

18.83 O.Y. Samoilov: *The Structure of Aqueous Solutions and the Hydration of Ions* (Akademy Nauk, Moscow 1957)

18.84 V.A. Kuznetsov: Sov. Phys. Crystallogr. **12**, 608–611 (1968)

18.85 S. Kaneko, F. Imoto: Nippon Kagaku Kaishi **6**, 985–990 (1975)

18.86 N.A. Ovramenko, L.I. Shvets, F.D. Ovcharenko, B.Y. Kornilovich: Izv. Akad. Nauk USSR, Inorg. Mater. **15**, 1982–1985 (1979)

18.87 W. Hertl: J. Am. Ceram. Soc. **71**, 879–883 (1988)

18.88 M.M. Lencka, R.E. Riman: Chem. Mater. **53**, 31–41 (1981), .

18.89 J.O. Eckert Jr., C.C. Hung-Houston, B.L. Gersten, M.M. Lencka, R.E. Riman: J. Am. Ceram. Soc. **79**, 2929–2939 (1996)

18.90 S. Komarneni: Enhanced reaction kinetics under microwave-hydrothermal conditions, Proc. 2nd Int. Conf. Solvotherm. React. (Takamatsu 1996) pp. 97–100

18.91 S.H. Zhung, J.H. Lee, P.M. Forster, G. Ferey, A.K. Cheetham, J.S. Chang: Chem. Eur. J. **12**, 7899–7905 (2006)

18.92 V.I. Popolitov: Hydrotherm. Growth Crystals Prog. Cryst. Growth Charact. Mater. **21**, 255–297 (1990)

18.93 B.N. Litvin, D.A. Tules: Apparatus for hydrothermal synthesis and growth of monocrystals. In: *Crystallization Process under Hydrothermal Conditions*, Studies in Soviet Science, ed. by A.N. Lobachev (Consultant Bureau, New York 1973), p.139

18.94 R.A. Laudise, E. Kaldis (Ed.): *Crystal Growth of Electronic Materials* (Elsevier, Amsterdam 1985) p. 159

18.95 Leco Catalogue, Tem-Press Research Division, (Bellefonte, Pennsylvania 2005)

18.96 R. Roy, E.F. Osborn: Econ. Geol. **47**, 717–721 (1952)

18.97 T. Adschiri, K. Arai: Hydrothermal synthesis of metal oxide nanoparticles under supercritical con-

18.97 ditions. In: *Supercritical Fluid Technology in Materials Science and Engineering*, ed. by Y.-P. Sun (Marcel Dekker, New York 2002) pp. 311–325
18.98 E. Reverchon, R. Adam: J. Supercrit. Fluids **37**, 1–22 (2006)
18.99 M. Goto (Ed.): Proc. 8th Int. Symp. Supercrit. Fluids (Kyoto 2006)
18.100 Catalogue of M/s Berghof, Germany; M/s Nittokoatsu, Japan
18.101 A.C. Walker: Ind. Eng. Chem. **36**, 250–256 (1953)
18.102 M. Hosaka, S. Taki: J Cryst. Growth **52**, 837 (1981)
18.103 F. Lafon, G. Demazean: J. Phys. (France) **4**(C2), 177–182 (1994)
18.104 R.A. Laudise, A.A. Ballman, J.C. King: J. Phys. Chem. Solids. **26**, 1305–1308 (1965)
18.105 R.A. Laudise: Hydrothermal crystal growth – some recent results. In: *Advanced Crystal Growth*, ed. by P.M. Dryburgh, B. Cockayne, K.G. Barraclough (Prentice Hall, NewYork 1987) pp. 267–288
18.106 R.A. Laudise, R.A. Sullivan: Chem. Eng. Prog. **55**, 55–59 (1959)
18.107 R.L. Barns, E.D. Kolb, R.A. Laudise, E.E. Simpson, K.M. Kroupa: J. Cryst. Growth **34**, 189–197 (1976)
18.108 X. Buisson, R. Arnaud: J. Phys.(France) **4**(C2), 25–32 (1994)
18.109 A.A. Ballman, R.A. Laudise: Solution growth. In: *Art and Science of Growing Crystals*, ed. by J.J. Gilman (Wiley, New York 1963) p. 231
18.110 J.J. Martin, A.F. Armington: J. Cryst. Growth **62**, 203–206 (1983)
18.111 A.F. Armington, J.J. Larkin: J. Cryst. Growth **71**, 799 (1985)
18.112 M. Hosaka, T. Miyata: Mater. Res. Bull. **28**, 1201–1208 (1993)
18.113 G.R. Johnson, R.A. Irvine, J.W. Foise: A parametric study of the variables involved in quartz growth, IEEE Proc. 44th Annu. Symp. Freq. Control (1990) pp. 216–221
18.114 M. Deleuze, O. Cambon, A. Goiffon, A. Ibanez, E. Philippot: J. Phys. (France) **4**(C2), 79–84 (1994)
18.115 K. Bräuer, E. Müller: Cryst. Res. Technol. **19**, 101–109 (1984)
18.116 O. Cambon, M. Deleuze, J.P. Michel, J.P. Aubry, A. Goiffon, E. Philippot: J. Phys. (France) **4**(C2), 85–91 (1994)
18.117 H. Iwasaki, F. Iwasaki, E.A. Marina, L.V. Balitskaya: Process for producing unsintered cristobalite silica, US Patent 4853198 (1989)
18.118 H. Iwasaki, F. Iwasaki, V.S. Balitsky, L.V. Balitskaya: J. Cryst. Growth **187**, 481 (1998)
18.119 H. Iwasaki, F. Iwasaki, M. Kurashige, K. Oba: J. Cryst. Growth **234**, 711 (2002)
18.120 Z.P. Chang, G.R. Barsch: IEEE Trans. Sonics Ultrasonics **23**, 127 (1976)
18.121 R.M. O'Connell, P.H. Corr: Temperature compensated cuts of berlinite and β-eucryptite: for saw devices, Proc. 31st Annal. Freq. Control Symp. (1977) pp. 182–186
18.122 E.D. Kolb, A.M. Glass, R.L. Rosenberg, J.C. Grenier, R.A. Laudise: Frequency dependence in quartz seed orientation, Proc. 35th Ultrasonic Symp. (New York 1981)
18.123 E. Philippot, D. Palmier, M. Pintard, A.A. Goiffon: J. Solid State Chem. **123**, 1–13 (1996)
18.124 K. Byrappa, S. Srikantaswamy: Prog. Cryst. Growth. Charact. Mater. **21**, 199 (1990)
18.125 M. Beaurain, P. Armand, P. Papet: J. Cryst. Growth **294**, 396–400 (2006)
18.126 B.H.T. Chai, M.L. Shand, I. Bucher, M.A. Gillee: Berlinite synthesis and SAW characteristics, Proc. IEEE Ultrasonic Symp. (1979) p. 557
18.127 V.I. Popolitov, I.M. Yaroslavskii: Izv. Akad. Nauk USSR, Neorg. Mater. **26**, 892 (1990)
18.128 I.M. Yaroshavskii, V.I. Popolitov: Neorg. Mater. Izv. Akad. Nank SSSR **26**, 1055–1059 (1990)
18.129 N. Prud'homme, D. Cachau-Herreillat, P. Papet, O. Cambon: J. Cryst. Growth **286**, 102–107 (2006)
18.130 D.V. Balitsky, E. Philippot, P. Papet, V.S. Balitsky, F. Pey: J. Cryst. Growth **275**, 887–894 (2005)
18.131 M. Haouas, F. Taulelle, N. Prud'homme, O. Cambon: J. Cryst. Growth **296**, 197–206 (2006)
18.132 W. Wallnofer, P.W. Krempl, F. Krispel, V. Willfurth: J. Cryst. Growth **198/199**, 487 (1999)
18.133 R.-U. Barz, M. Grassl, P. Gille: J. Cryst. Growth **237–239**, 843–847 (2002)
18.134 P. Hofmann, U. Juda, K. Jacobs: J. Cryst. Growth **275**, 1883–1888 (2005)
18.135 M. Grassl, R.-U. Barz, P. Gille: Cryst. Res. Technol. **37**, 531–539 (2002)
18.136 K. Jacobs, P. Hofmann, D. Klimm: J. Cryst. Growth **237–239**, 837–842 (2002)
18.137 D.R. Kinloch, R.F. Belt, R.C. Puttbach: J. Cryst. Growth **24/25**, 610–613 (1974)
18.138 S. Hirano, K. Kikuta: J. Cryst. Growth **79**, 223–226 (1986)
18.139 N.Y. Ikornikova: *Hydrothermal Synthesis of Crystals in Chloride Systems* (Nauka, Moscow 1975) pp. 1–222, in Russian
18.140 S. Hirano, K. Kikuta: J. Cryst. Growth **94**, 351–356 (1989)
18.141 S. Hirano, T. Yogo, K. Kikuta, Y. Yoneta: J. Ceram. Soc. Jpn. **101**, 113–117 (1993)
18.142 K. Yanagisawa, Q. Feng, K. Ioku, N. Yamasaki: J. Cryst. Growth **163**, 285–294 (1996)
18.143 K. Yanagisawa, K. Kageyama, Q. Feng, I. Matsushita: J. Cryst. Growth **229**, 440–444 (2001)
18.144 V.N. Rumyantsev, I.G. Ganeev, I.S. Rez: Solubility of SiO_2 in alkali and carbonate solutions. In: *Crystal Growth*, Vol. 9 (Nauka, Moscow 1972) pp. 51–54
18.145 V.A. Kuznetsov, A.A. Shternberg: Sov. Phys. Crystallogr. **12**, 280–285 (1967)
18.146 R.L. Barus, R.A. Laudise, R.M. Shields: J. Phys. Chem. **67**, 835–840 (1963)
18.147 G. Yamaguchi, H. Yanagida, S. Sojma: Bull. Soc. Chem. Jpn. **35**, 1789–1791 (1962)

18.148 K.F. Kashkurov, P.I. Nikitichev, V.V. Osipov, L.D. Sizova, A.V. Simonov: Sov. Phys. Crystallogr. **12**, 837–839 (1968)

18.149 R.R. Monchamp, R.C. Puttbach, J.W. Nielson: J. Cryst. Growth **2**, 178 (1968)

18.150 R.A. Laudise, A.A. Ballman: J. Am. Chem. Soc. **80**, 2655–2657 (1958)

18.151 V.S. Balitsky, E.E. Lisinstina: *Synthetic Analogues and Imitation of Natural Gemstones* (Nedra, Moscow 1981)

18.152 P. Hautefeuille, A. Perrey: C. R. Acad. Sci. **106**, 1800–1810 (1888)

18.153 R. Webster: *Gems*, 3rd edn. (Newnes–Butterworths, London 1975)

18.154 M. Hosak: Hydrothermal Growth of Crystals, Prog. Crystal Growth Charact. **21**, 71–96 (1990)

18.155 B.W. Anderson: *Gem Testing*, 8th edn. (Butterworths, London 1971)

18.156 R.T. Liddicoat Jr.: *Handbook of Gem Identification*, 10th edn. (Gemmological Institute of America, Los Angeles 1975)

18.157 P.J. Yancey: Hydrothermal process for growing crystals having the structure of beryl in highly acid chloride medium, US Patent 3723337 (1973)

18.158 E.M. Flanigen: Hydrothermal process for growing crystals having the structure of beryl in an alkaline halide medium, US Patent 3567642 (1971)

18.159 E.M. Flanigen, N.R. Mumbach: Hydrothermal process for growing crystals having the structure of beryl in an acid halide medium, US Patent 3567643 (1971)

18.160 R.E. Kane, R.T. Liddicoat Jr.: Gems Gemol. **21**(3), 156 (1985)

18.161 K. Scarratt: J. Gemmol. **21**(5), 294 (1989)

18.162 K. Schmetzer: J. Gemmol. **21**(3), 145 (1988)

18.163 Z. Chen, G. Zhang, H. Shen, C. Huang: J. Cryst. Growth **244**, 339–341 (2002)

18.164 V.G. Thomas, S.P. Demin, D.A. Foursenko, T.B. Bekker: J. Cryst. Growth **206**, 203–214 (1999)

18.165 L.I. Tsinobar, I.G. Chentsova: Kristallografiya **4**, 633 (1959)

18.166 V.S. Balitsky, I.B. Machina, A.A. Marin, J.E. Shigley, G.R. Rossman, T. Lu: J. Cryst. Growth **212**, 255–260 (2000)

18.167 V.S. Balitsky: J. Cryst. Growth **41**, 100 (1977)

18.168 M. Hosaka, S. Taki: J. Cryst. Growth **64**, 572 (1983)

18.169 K. Nassau: *Gemstone Enhancement*, 2nd edn. (Butterworths, Boston 1994)

18.170 K. Byrappa, M.K. Devaraju, P. Madhusudan, A.S. Dayananda, B.V. Suresh Kumar, H.N. Girish, S. Ananda, K.M.L. Rai, P. Javeri: J. Mater. Sci. **41**, 1395 (2006)

18.171 B.C. Chakoumakos, M.M. Abraham, L.A. Batner: J. Solid State Chem. **109**, 197 (1994)

18.172 M. Prasad, A.K. Pandit, T.H. Ansari, R.A. Singh, B.M. Wanklyn: Phys. Lett. A **138**, 61 (1989)

18.173 B.H.T. Chai, G. Loutts, X.X. Chang, P. Hong, M. Bass, I.A. Shcherbakov, A.I. Zagumennyi: *Advanced Solid State Lasers, Technical Digest*, Vol. 20 (Optical Society of America, Washington 1994) p. 41

18.174 K. Byrappa, K.M.L. Rai, B. Nirmala, M. Yoshimura: Mater. Sci. Forum **506**, 315 (1999)

18.175 F.M. Levin: J. Am. Ceram. Soc. **50**, 381 (1967)

18.176 L.G. Vanuitret, R.C. Linares, R.R. Soden, A.A. Ballman: J. Chem. Phys. **36**, 702 (1962)

18.177 K. Byrappa, C. Ramaningaiah, K. Chandrashekar, K.M.L. Rai, B. Basavalingu: J. Mater. Sci. **506**, 315 (1999)

18.178 K. Byrappa, B. Nirmala, K.M. Lokanatha Rai, M. Yoshimura: Crystal growth, size, and morphology control of Nd:RVO$_4$ under hydrothermal conditions. In: *Crystal Growth Technology*, ed. by K. Byrappa, T. Ohachi (Springer, Berlin, Heidelberg 2003) p. 335

18.179 H. Wu, H. Xu, Q. Su, T. Chen, M. Wu: J. Mater. Chem. **13**, 1223 (2003)

18.180 S. Erdei, B.M. Jin, F.W. Ainger: J. Cryst. Growth **174**, 328 (1997)

18.181 S. Erdei, M. Kilmkiewicz, F.W. Ainger, B. Keszei, J. Vandlik, A. Suveges: Mater. Lett. **24**, 301 (1995)

18.182 T. Mitsuda: Ceram. Jpn. **15**, 184 (1980)

18.183 S. Somiya: J. Mater. Sci. **41**, 1307 (2006)

18.184 K. Kajiyoshi, K. Tomono, Y. Hamaji, T. Kasanami, M. Yoshimura: J. Am. Ceram. Soc. **78**, 1521 (1995)

18.185 W.S. Cho, M. Yashima, M. Kakihana, A. Kudo, T. Sakata, M. Yoshimura: J. Am. Ceram. Soc. **80**, 765 (1997)

18.186 S. Ohara, T. Mousavand, T. Sasaki, M. Umetsu, T. Naka, T. Adschiri: J. Mater. Sci. **41**, 1445 (2006)

18.187 T. Mousavand: Synthesis of organic–inorganic hybrid nanoparticles by in-situ surface modification under supercritical hydrothermal conditions. Ph.D. Thesis (Tohoku University, Sendai 2007)

18.188 M. Yoshimura, H. Suda, K. Okamoto, K. Ioku: J. Mater. Sci. **29**, 3399–3402 (1994)

18.189 S. Forster, M. Antonietti: Adv. Mater. **10**, 195 (1998)

18.190 Y. Zhu, H. Zheng, Y. Li, L. Gao, Z. Yang, Y.T. Qian: Mater. Res. Bull. **38**, 1829 (2003)

18.191 V.F. Puntes, K.M. Drishnan, A.P. Alivisatos: Science **291**, 2115 (2001)

18.192 Q. Xie, Z. Dai, W. Huang, J. Liang, C. Jiang, Y.T. Qian: Nanotechnology **16**, 2958 (2005)

18.193 Z. Tian, J. Liu, J.A. Voigt, H. Xu, M.J. Mcdermott: Nano Lett. **3**, 89 (2003)

18.194 O.V. Al'myasheva, E.N. Korytkova, A.V. Maslov, V.V. Gusarov: Inorg. Mater. **41**, 460 (2005)

18.195 X. Jiao, D. Chen, L. Xiao: J. Cryst. Growth **258**, 158 (2003)

18.196 M. Wu, Y. Xiong, Y. Jia, J. Ye, K. Zhang, Q. Chen: Appl. Phys. A **81**, 1355 (2005)

18.197 K. Byrappa, K.M.L. Rai, M. Yoshimura: Environ. Technol. **21**, 1085 (2000)

18.198 T. Adschiri, K. Kanaszawa, K. Arai: J. Am. Ceram. Soc. **75**, 1019 (1992)

18.199 T. Adschiri, Y. Hakuta, K. Arai: Ind. Eng. Chem. Res. **39**, 4901 (2000)

18.200 M. Schur, H. Rijnberk, C. Nather: Polyhedron **18**, 101 (1998)

18.201 C.L. Cahill, B. Gugliotta, J.B. Parise: Chem Commun. **16**, 1715 (1998)

18.202 G.C. Guo, R.M.W. Kwok, T.C.W. Mak: Inorg. Chem. **36**, 2475 (1997)

18.203 W. Wang, Y. Geng, P. Yan, F. Liu, Y. Xie, Y. Qian: J. Am. Chem. Soc. **121**, 4062 (1999)

18.204 Y. Li, H. Liao, Y. Fan, Y. Zhang, Y. Qian: Inorg. Chem. **38**, 1382 (1999)

18.205 S. Yu, J. Yang, Z. Han, Y. Zhou, R. Yang, Y. Qian, Y. Zhang: J. Mater. Chem. **9**, 1283 (1999)

18.206 Y. Li, H. Liao, Y. Fan, L. Li, Y. Qian: Mater. Chem. Phys. **58**, 87 (1999)

18.207 Y. Li, F. Huang, Q. Zhang, Z. Gu: J. Mater. Sci. **35**, 5933 (2000)

18.208 C.E. Melto, A.A. Giardini: Am. Mineral. **59**, 775 (1974)

18.209 M. Schrauder, O. Navon: Nature **365**, 42 (1999)

18.210 O. Navon: Nature **353**, 746 (1991)

18.211 G.D.J. Guthrie, D.R. Veblen, O. Navon, G.R. Rossman: Earth Planet Sci. Lett. **105**, 1 (1991)

18.212 R.C. De Vries: Synthesis of diamond under metastable conditions, Ann. Rev. Mater. Sci. **17**, 161–187 (1987)

18.213 R.C. DeVries, R. Roy, S. Somiya, S. Yamada: Trans. Mater. Res. Soc. Jpn. B **14**, 641 (1994)

18.214 C.A. Eckert, B.L. Knutsan, P.G. Debenedetti: Nature **383**, 313 (1996)

18.215 Y.G. Gogotsi, K.G. Nickel, P.J. Kofstad: Mater. Chem. **5**, 2313 (1995)

18.216 R. Roy, D. Ravichandran, P. Ravindranathan, A. Badzian: J. Mater. Res. **11**, 1164 (1996)

18.217 Y.G. Gogotsi, P. Kofstad, M. Yoshmura, K.G. Nickel: Diam. Rel. Mater. **5**, 151 (1996)

18.218 B. Basavalingu, J.M. Calderon Moreno, K. Byrappa, Y.G. Gogotsi: Carbon **39**, 1763 (2001)

18.219 X. Zhang, H. Liu, W. He, J. Wang, X. Zi, R.I. Boughton: J. Cryst. Growth **275**, 1913 (2005)

18.220 S. Krupanidhi, C.N.R. Rao: Adv. Mater. **16**, 425 (2004)

18.221 B. Huang, J.M. Hong, X.T. Chen, Z. Yu, X.Z. You: Mater. Lett. **59**, 430 (2005)

18.222 Y. Zhang, H. Guan: Mater. Res. Bull. **40**, 1536 (2005)

18.223 U.K. Gautam, K. Sardar, F.L. Deepak, C.N.R. Rao: Pramana **65**, 549 (2005)

18.224 H. Xu, W. He, H. Wang, H. Yan: J. Cryst. Growth **260**, 447 (2004)

18.225 M. Cao, X. He, X. Wu, C. Hu: Nanotechnology **16**, 2129 (2005)

18.226 A. Rabenau, H. Rau: Naturwissenschaften **55**, 336 (1968)

18.227 J.F. Balasico, R.B. White, R. Roy: Mater. Res. Bull. **2**, 913 (1967)

18.228 L.N. Demianets: *Hydrothermal Synthesis of Crystals* (Nauka, Moscow 1968), in Russian

18.229 L.N. Demianets, L.S. Garashina, B.N. Litvin: Kristallografiya **8**, 800 (1963)

18.230 I.P. Kuzmina: Geol. Ore Depos. **6**, 101 (1963)

18.231 P. Hartman: Phys. Status Solidi **2**, 585 (1962)

18.232 D. Ehrentraut, H. Sato, Y. Kagamitani, H. Sato, A. Yoshikawa, T. Fukuda: Prog. Cryst. Growth Charact. Mater. **52**, 280 (2006)

18.233 J. Bauer, P. Kaczerovsky: Vysok. Skola Chem.-Tech. Praze G. **12**, 153 (1970)

18.234 E. Prochazkova, D. Rykl, V. Seidl: Czechoslovakian Patent 170391 (1973), No. 5939–73

18.235 I.P. Kuzmina, B.N. Litvin: Crystal Growth **4**, 160 (1964)

18.236 C.J.M. Rooymans, W.F. Langenhoff: J. Cryst. Growth **3–4**, 411 (1968)

18.237 V.B. Glushkova: *Polymorphism of Rare Earth Oxides* (Nauka, Leningrad 1967), in Russian

18.238 I.P. Kuzmina, N.M. Khaidukov: *Crystal Growth from High Temperature Aqueous Solutions* (Nauka, Moscow 1977), in Russian

18.239 H. Guggenheim: Solid State Phys. **12**, 780 (1961)

18.240 V.I. Popolitov, A.N. Lobachev, V.F. Perkin: Kristallografiya **17**, 436 (1972)

18.241 B.N. Litvin, V.I. Popolitov: *Hydrothermal Synthesis of Inorganic Compounds* (Nauka, Moscow 1984), in Russian

18.242 E.D. Kolb, A.J. Caporaso, R.A. Laudise: J. Cryst. Growth **19**, 242 (1973)

18.243 K. Byrappa, S. Srikantaswamy, G.S. Gopalakrishna, V. Venkatachalapathy: J. Mater. Sci. **21**, 2202 (1986)

18.244 S.P. Fedoseeva: J. Cryst. Growth **8**, 59 (1972)

18.245 I.P. Kuzmina, B.N. Litvin: Kristallografiya **7**, 478 (1963)

18.246 V.I. Popolitov, A.N. Lobachev: Izv. Akad. Nauk USSR: Inorg. Mater. **8**, 960 (1972)

18.247 T. Eiiti, K. Iodji, K. Kendzi, S. Tatsuro: Patent 5248959 (1975), Japan No. 50-102484

18.248 G. Demazeau, P. Maestro: Hydrothermal growth of Cr_2O_3, Proc. 7th Int. AIRAPT Conf. (Le Creusot), Vol. 1 (1979) p. 572

18.249 I.P. Kuzmina: *Author's Abstract IKAN* (SSSR, Moscow 1968), in Russian

18.250 B.N. Litvin, V.I. Popolitov: *Hydrothermal Synthesis of Inorganic Compounds* (Nauka, Moscow 1984) p. 165, in Russian

18.251 R.A. Laudise, E.E. Kolb: J. Am. Ceram. Soc. **45**, 51 (1962)

18.252 G.I. Distler, S.A. Kobzareva, A.N. Lobachev, O.K. Melnikov, N.S. Triodina: Krist. Tech. **13**, 1025 (1978)

18.253 S.C. Abraham, P.B. Jameson, J.L. Bernstein: J. Chem. Phys. **47**, 4034 (1967)

18.254 B.N. Litvin, V.I. Popolitov: *Hydrothermal Synthesis of Inorganic Compounds* (Nauka, Moscow 1984) p. 216, in Russian

18.255 G. Ervin, E.F. Osborn: J. Geol. **59**, 381 (1951)

18.256 A.N. Christensen: Acta Chem. Scand. **20**, 896 (1966)

18.257 A.N. Christensen: Acta Chem. Scand. **30**, 133 (1976)

18.258 H. Schwarz: Z. Naturforsch. **226**, 554 (1967)

18.259 V.S. Balitsky, T. Bublikova: Prog. Cryst. Growth. Charact. Mater. **21**, 139 (1990)

18.260 B.V. Mill: Dokl. Akad. Nauk SSSR **156**, 814 (1964)
18.261 J. Ito, C. Frondel: Am. Mineral. **53**, 1276 (1968)
18.262 B.A. Maksimov, Y.A. Kharitonov, V.V. Ilyukhin, N.V. Belov: Dokl. Akad. Nauk SSSR **178**, 980 (1968)
18.263 M. Bihan, T.A. Katt, R. Wey: Bull. Soc. France Mineral. Crystallogr. **941**, 15 (1971)
18.264 D.E. Appleman, J.R. Clark: Am. Mineral. **50**, 679 (1965)
18.265 K. Kodaira, S. Ito, T. Matsushita: J. Cryst. Growth **29**, 123 (1975)
18.266 J. Wan, Z. Wang, X. Chen, L. Mu, W. Yu, Y. Qian: J. Lumin. **121**, 32 (2006)
18.267 Z. Ito, I. Harold: Am. Mineral. **53**, 778 (1968)
18.268 D.H. Lindsley: *Carnegie Inst. Annu. Rep.* (Direct. Geophys. Lab., Washington 1971) p. 188
18.269 J. Ito: Am. Mineral. **53**, 782 (1968)
18.270 J. Ito, C. Frondel: Am. Mineral. **52**, 1077 (1967)
18.271 J. Ito, C. Frondel: Arkiv. Min. Geol. **58**, 391 (1968)
18.272 G.V. Bukin: Mineralogical Museum, Akad. Nauk SSSR **19**, 131 (1969)
18.273 R.J. Harker, O.G. Tuttle: Am. J. Sci. **254**, 468 (1956)
18.274 P.A. Sandomirskii, M.A. Smirnov, A.V. Arakcheeva, N.V. Belov: Dokl. Akad. Nauk SSSR **227**, 856 (1976)
18.275 B.Y. Kornilovich, N.A. Ovramenko, F.D. Ovcharenko: Dokl. Akad. Nauk SSSR **261**, 245 (1981)
18.276 I.Y. Nekrasov, T.P. Dadze: J. Mineral. **3**, 287 (1980)
18.277 I.Y. Nekrasov, T.P. Dadze, N.V. Zayakina: Dokl. Akad. Nauk SSSR **261**, 479 (1981)
18.278 G.D. Brunton, L.A. Harris, O.C. Kopp: Am. Mineral. **57**, 1720 (1972)
18.279 E.E. Strelkova, O.G. Karpov, B.N. Litvin: Kristallografiya **22**, 174 (1977)
18.280 D.Y. Pushcharovskii, O.G. Karpov, E.A. Pobedimskaya, N.V. Belov: Dokl. Akad. Nauk SSSR **234**, 1323 (1977)
18.281 D.Y. Pushcharovskii, T. Baatarin, E.A. Pobedimskaya, N.V. Belov: Kristallografiya **16**, 899 (1971)
18.282 D.Y. Pushcharovskii, E.A. Pobedimskaya, N.V. Belov: Dokl. Akad.Nauk SSSR **185**, 395 (1969)
18.283 L.B. Halferdahl: J. Petrol. **2**, 49 (1961)
18.284 D.Y. Pushcharovskii, E.A. Pobedimskaya, B.N. Litvin, N.V. Belov: Dokl. Akad. Nauk SSSR **214**, 91 (1974)
18.285 R.N. Thomson, J.E. Chisholm: Mineral. Mag. **37**, 253 (1969)
18.286 C.F. Warren: Am. Mineral. **56**, 997 (1971)
18.287 D.J. Drysdale: Am. Mineral. **56**, 187 (1971)
18.288 A.N. Safronov, N.N. Nevskii, V.I. Ilyukhin, N.V. Belov: Dokl. Akad. Nauk SSSR **255**, 278 (1980)
18.289 R. Caruba, A. Baumer, G. Turco: C. R. Acad. Sci. Paris **270**, 1 (1970)
18.290 V.G. Chykhlanstev, K.V. Alyamovskaya: Izv. Akad. Nauk SSSR: Inorg. Mater. **6**, 1639 (1970)
18.291 N.G. Shymyaskaya, V.A. Blinov, A.A. Voronkov: Dokl. Akad. Nauk SSSR **208**, 1876 (1973)
18.292 M.C. Michel-Levy: Bull. Soc. France Mineral. Crystallogr. **84**, 2989 (1967)
18.293 J. Ito: Am. Mineral. **53**, 998 (1968)
18.294 J.L. Munoz: Am. Mineral. **53**, 1490 (1968)
18.295 A. Baronnet, M. Amouric, B. Chabot, F. Corny: J. Cryst. Growth **43**, 255 (1978)
18.296 L.N. Demianets, E.N. Emelyanova, O.K. Melnikov: Solubility of sodalite under hydrothermal conditions. In: *Crystallization Process Under Hydrothermal Conditions*, ed. by A.N. Lobachev (Consultant Bureau, New York 1973) p. 151
18.297 R.D. Shannon: Phys. Chem. Mineral. **4**, 139 (1979)
18.298 D. Tranqui, J.J. Capponi, J.C. Joubert, R.D. Shannon: J. Solid State Chem. **73**, 325 (1988)
18.299 J.W. Cobble: Rapp. Tech. Cent. Belge Etude Corros. **142**, 119 (1982)
18.300 R. Caruba, A. Baumer, G. Turco: Geochim. Cosmochim. Acta **39**, 11 (1975)
18.301 A. Baumer, G. Turco: C.R. Acad. Sci. D **270**, 1197 (1970)
18.302 A.A. Bush, S.A. Ivanov, S.Y. Stefanovich: Izv. Akad. Nauk SSSR: Inorg. Mater. **13**, 1656 (1977)
18.303 M.N. Tseitlin, G.F. Plakhov, A.N. Lobachev: Kristallografiya **18**, 836 (1973)
18.304 V.I. Popolitov, G.F. Plakhov, S.Y. Stefanovich: Izv. Akad. Nauk SSSR: Inorg. Mater. **17**, 1841 (1981)
18.305 B.N. Litvin: Hydrothermal chemistry of silicates and germinates. Ph.D. Thesis (Academy of Sciences Russia, Moscow 1978)
18.306 A.N. Lazarev, A.K. Shirivinskaya: Izv. Akad. Nauk SSSR: Inorg. Mater. **12**, 771 (1976)
18.307 T.N. Nadezhdina, E.A. Pobedimskaya, V.V. Ilyukhin, N.V. Belov: Dokl. Akad. Nauk SSSR **223**, 1086 (1977)
18.308 S.R. Lyon, E.S. Ehlers: Am. Mineral. **56**, 118 (1970)
18.309 B.A. Maksimov, Y.V. Nikolskii, Y.A. Kharitonov: Dokl. Akad. Nauk SSSR **239**, 87 (1978)
18.310 A.M. Dago, D.Y. Pushcharovskii, E.A. Pobedimskaya, N.V. Belov: Dokl. Akad. Nauk SSSR **250**, 857 (1980)
18.311 D.Y. Pushcharovskii, A.M. Dago, E.A. Pobedimskaya, N.V. Belov: Dokl. Akad. Nauk SSSR **251**, 354 (1980)
18.312 D.M. Roy, R. Roy: Am. Mineral. **39**, 957 (1954)
18.313 D.Y. Pushcharovskii, E.A. Pobedimskaya, O.V. Kudryasteva, B. Hettash: Kristallografiya **1**, 1126 (1976)
18.314 I.P. Kuzmina, O.K. Melnikov, B.N. Litvin: *Hydrothermal Synthesis of Crystals* (Nauka, Moscow 1968) p. 41
18.315 N.A. Nosirev: Investigations of germanate system in hydrothermal condition. Ph.D. Thesis (Institute of Crystallography, Moscow 1975)
18.316 Y.K. Egorov-Tismenko, M.A. Simonov, N.V. Belov: Dokl. Akad. Nauk SSSR **227**, 2 (1976)
18.317 I.I. Soloveva, V.V. Bakanin: Kristallografiya **12**, 591 (1967)
18.318 S. Kume, S. Ueda, M. Koizumi: J. Geophys. Res. **74**, 2145 (1969)
18.319 N. Kinomura: J. Am. Ceram. Soc. **56**, 344 (1973)
18.320 N.V. Belov: Dokl. Akad. Nauk SSSR **268**, 360 (1983)
18.321 A.N. Christensen: Acta Chem. Scand. **24**, 1287 (1970)
18.322 O. Jarchow, K.H. Kalska, H. Schenk: Naturwissenschaften **68**, 475 (1981)

18.323 L.N. Demianets, A.N. Lobachev, G.A. Emelchenko: *Rare Earth Germanates* (Nauka, Moscow 1980)
18.324 E.D. Kolb, R.A. Laudise: J. Cryst. Growth **43**, 313 (1978)
18.325 F. Dachille, R. Roy: Z. Kristallogr. **111**, 451 (1959)
18.326 S. Srikantaswamy: Phases and crystallization in the system aluminum orthophosphate. Ph.D. Thesis (University of Mysore, India 1988)
18.327 R.G. Hermann: Ph.D. Thesis (Ohio University, Ohio 1972)
18.328 E.M. McCarron, J.L. Calabrene, M.A. Subramanian: Mater. Res. Bull. **22**, 1421 (1987)
18.329 J.M. Congo, P. Kierkegaard: Acta Chem. Scand. **20**, 72 (1966)
18.330 H. Effenberger: J. Solid State Chem. **57**, 1240 (1985)
18.331 A. Clearfield, B.D. Roberts, A. Clearfield: Mater. Res. Bull. **19**, 219 (1984)
18.332 M.A. Subramanian, B.D. Roberts, A. Clearfield: Mater. Res. Bull. **19**, 1417 (1984)
18.333 P. Lightfoot, A.K. Cheetam, A.W. Sleight: J. Solid State Chem. **73**, 325 (1988)
18.334 K. Byrappa, S. Srikantaswamy, S. Gali: J. Mater. Sci. Lett. **9**, 235 (1990)
18.335 K. Byrappa, B.N. Litvin: J. Mater. Sci. **18**, 703 (1983)
18.336 S. Gali, K. Byrappa: Acta Crystallogr. C **46**, 2011 (1990)
18.337 S. Gali, A. Cardenas, K. Byrappa, G.S. Gopalakrishna: Acta Cryst. C **48**, 1650 (1992)
18.338 K. Byrappa, B.V. Umesh Dutt, D. Poojary, A. Clearfield: J. Mater. Res. **9**, 1519 (1994)
18.339 A. Clearfield: Prog. Cryst. Growth. Charact. Mater. **21**, 1 (1990)
18.340 R. Peascoe: Ph.D. Thesis (Texas University, Austin 1989)
18.341 T. Chirayil, P.Y. Zavalij, M.S. Whittingham: Chem. Mater. **10**, 2629 (1998)
18.342 K. Byrappa, C.K. Chandrashekar, B. Basavalingu, K.M.L. Rai, S. Ananda, M. Yoshimura: J. Cryst. Growth **306**, 94 (2007)
18.343 Y.W. Wang, H.Y. Xu, H. Wang, Y.C. Zhang, Z.Q. Song, H. Yan, C.R. Wan: Solid State Ion. **167**, 419 (2004)
18.344 L.N. Demianets: Prog. Cryst. Growth. Charact. Mater. **21**, 299 (1990)
18.345 K. Byrappa, K.V.K. Shekar: J. Mater. Chem. **2**, 13 (1992)
18.346 K. Byrappa, K.V.K. Shekar, S. Gali: Cryst. Res. Technol. **6**, 768 (1992)
18.347 S. Hosokawa, Y. Tanaka, S. Iwamoto, M. Inoue: J. Mater. Sci. **43**, 3079 (2008)
18.348 A. Cardenas, J. Solans, K. Byrappa, K.V.K. Shekar: Acta Cryst. C **49**, 645 (1992)
18.349 M. Elli, L. Cambi: Atti. Accad. Naz. Lincei, Rend. Cl. Sci. Fis. Mat. Nat. **39**, 87 (1965)
18.350 L. Cambi, M. Elli: Chim. Ind. Ital. **47**, 136 (1968)
18.351 S. Biswas, S. Kar, S. Chaudhuri: J. Cryst. Growth **284**, 129 (2005)
18.352 U.K. Gautam, R. Seshadri: Mater. Res. Bull. **39**, 669 (2004)
18.353 J.R. Ota, S.K. Srivastava: J. Nanosci. Nanotechnol. **6**, 168 (2006)
18.354 L. Cambi, M. Elli: Chim. Ind. Ital. **51**, 3 (1969)
18.355 L. Cambi, M. Elli: Atti. Accad. Naz. Lincei. Rend. Cl. Sci. Fis. Mat. Nat. **40**, 553 (1966)
18.356 E.D. Kolb, A.J. Caporaso, R.A. Laudise: J. Cryst. Growth **3/4**, 422 (1968)
18.357 N.K. Abrikosov, V.F. Bankina, L.V. Porestkaya: J. Cryst. Growth **7**, 176 (1967)
18.358 L. Cambi, M. Elli: Chim. Ind. Ital. **48**, 944 (1966)
18.359 L. Cambi, M. Elli: Atti. Accad. Naz. Lincei. Rend. Cl. Sci. Fis. Mat. Nat. **41**, 241 (1966)
18.360 L. Cambi, M. Elli: Chim. Ind. Ital. **50**, 94 (1968)
18.361 V.S. Balitsky, V.V. Komova, N.A. Ozerova: Izv. Akad. Nauk SSSR Ser. Geol. **12**, 93 (1971)
18.362 V.A. Kuznetsov, A.N. Lobachev (Ed.): *Crystal Processes Under Hydrothermal Conditions* (Consultant Bureau, New York 1973)
18.363 I. Keeman: Z. Anorg. Allg. Chem. **346**, 30 (1966)
18.364 K. Byrappa, B.V. Umesh Dutt, R.R. Clemente, S. Gali, A.B. Kulkarni: In: *Current Trends in Crystal Growth and Characterization*, ed. by K. Byrappa (MIT, Bangalore 1991) p. 272
18.365 L.N. Demianets: *Hydrothermal Synthesis of Crystals* (Nauka, Moscow 1968) p. 93, in Russian
18.366 L.Y. Kharchenko, V.I. Protasova, P.V. Klevstov: J. Inorg. Chem. **22**, 986 (1977)
18.367 P.V. Klevstov, N.A. Novgorodsteva, L.Y. Kharchenko: Dokl. Akad. Nauk SSSR **183**, 1313 (1968)
18.368 K. Byrappa, A. Jain: J. Mater. Res. **11**, 2869 (1996)
18.369 R.F. Klevstova, L.Y. Kharchenko, S.V. Borisov, V.A. Efremov, P.V. Klevstov: Kristallografiya **24**, 446 (1979)
18.370 V.I. Popolitov, B.N. Litvin: *Growth of Crystals Under Hydrothermal Conditions* (Nauka, Moscow 1986) p. 38
18.371 V.G. Hill, K. Zimmerman: J. Electrochem. Soc. Solid State **115**, 978 (1968)
18.372 E.D. Kolb, A.J. Caporoso, R.A. Laudise: J. Cryst. Growth **8**, 354 (1971)
18.373 R.F. Belt, J.B. Ings: J. Cryst. Growth **128**, 956 (1993)
18.374 A.V. Novoselova: Izv. Akad. Nauk SSSR, Inorg. Mater. **20**, 967 (1984)
18.375 J. Choi, E.G. Gillan: J. Mater. Chem. **16**, 3774 (2006)
18.376 D. Ehrentraut, N. Hoshino, Y. Kagamitani, A. Yoshikawa, T. Fukuda, H. Itoh, S. Kawabata: J. Mater. Chem. **17**, 886 (2007)
18.377 A. Clearfield, G.D. Smith: Inorg. Chem. **7**, 431 (1969)
18.378 S. Alluli, C. Ferraggiva, A. Laginestra, M.A. Massucci, N. Tomassini: J. Inorg. Nucl. Chem. **399**, 1043 (1977)

19. Hydrothermal and Ammonothermal Growth of ZnO and GaN

Michael J. Callahan, Qi-Sheng Chen

Zinc oxide (ZnO) and gallium nitride (GaN) are wide-bandgap semiconductors with a wide array of applications in optoelectronic and electronics. The lack of low-cost, low-defect ZnO and GaN substrates has slowed development and hampered performance of devices based on these two materials. Their anisotropic crystal structure allows the polar solvents, water and ammonia, to dissolve and crystallize ZnO and GaN at high pressure. Applying the techniques used for hydrothermal production of industrial single-crystal quartz to ZnO and GaN opens a pathway for the inexpensive growth of relatively larger crystals that can be processed into semiconductor wafers. This chapter will focus on the specifics of the hydrothermal growth of ZnO and the ammonothermal growth of GaN, emphasizing requirements for industrial scale growth of large crystals. Phase stability and solubility of hydrothermal ZnO and ammonothermal GaN is covered. Modeling of thermal and fluid flow gradients is discussed and simulations of thermal and temperature profiles in research-grade pressure systems are shown. Growth kinetics for ZnO and GaN respectively are reviewed with special interest in the effects of crystalline anisotropy on thermodynamics and kinetics. Finally, the incorporation of dopants and impurities in ZnO and GaN and how their incorporation modifies electrical and optical properties are discussed.

- 19.1 Overview of Hydrothermal and Ammonothermal Growth of Large Crystals 657
  - 19.1.1 Comparison of Ammonia and Water as Solvents 657
  - 19.1.2 Growth of Large Crystals by the Transport Growth Model 659
- 19.2 Requirements for Growth of Large, Low-Defect Crystals 661
  - 19.2.1 Thermodynamics: Solubility and Phase Stability 661
  - 19.2.2 Environmental Effects on Growth Kinetics and Structure Perfection (Extended and Point Defects) 664
  - 19.2.3 Doping and Alloying 665
- 19.3 Physical and Mathematical Models 666
  - 19.3.1 Flow and Heat Transfer 666
  - 19.3.2 Porous-Media-Based Transport Model 666
  - 19.3.3 Numerical Scheme 667
- 19.4 Process Simulations 669
  - 19.4.1 Typical Flow Pattern and Growth Mechanism 669
  - 19.4.2 Effect of Permeability on the Porous Bed..................... 670
  - 19.4.3 Baffle Design Effect on Flow and Temperature Patterns 670
  - 19.4.4 Effect of Porous Bed Height on the Flow Pattern.................... 672
  - 19.4.5 Simulation of Reverse-Grade Soluble Systems 672
- 19.5 Hydrothermal Growth of ZnO Crystals 674
  - 19.5.1 Growth Kinetics and Morphology . 674
  - 19.5.2 Structural Perfection – Extended Imperfections (Dislocations, Voids, etc.) 676
  - 19.5.3 Impurities, Doping, and Electrical Properties 678
  - 19.5.4 Optical Properties 679
- 19.6 Ammonothermal GaN.......................... 681
  - 19.6.1 Alkaline Seeded Growth 681
  - 19.6.2 Acidic Seeded Growth 683
  - 19.6.3 Doping, Alloying, and Challenges . 684
- 19.7 Conclusion 685
- References 685

Gallium nitride (GaN) and zinc oxide (ZnO) are emerging semiconductor materials that will have an enormous impact in electronics and optoelectronics. GaN and ZnO both have the anisotropic wurtzite crystal structure (space group $P6_3mc$) as their thermodynamically stable phase. This anisotropy, along with a large ionic component in their chemical bonds, accounts for the high spontaneous and piezoelectric polarization found in both GaN and ZnO [19.1]. GaN and ZnO also have large saturation velocities, high mobilities, large radiation resistance, nonlinear optical properties, and are chemically and thermally robust. They both have direct bandgaps in the ultraviolet (UV) region (GaN, $E_g = 3.42$ eV at 300 K; ZnO, $E_g = 3.37$ eV at 300 K) and have the ability to form direct-bandgap alloys for the fabrication of quantum wells (AlN and InN with GaN and MgO and CdO with ZnO) [19.2–5].

Devices based on the GaN [19.2, 3] and ZnO [19.4, 5] material systems are currently or will be produced in the future for a myriad of applications including solid-state lighting, power electronics, high-power radio-frequency (RF) monolithic microwave integrated circuit (MMIC) arrays, terahertz detection, high-density optical storage, and UV–infrared (IR) detection. GaN-devices have seen rapid development in the last 10 years due to the overcoming of major technological hurdles in the early 1990s [19.6]. A multibillion-dollar market now exists for GaN green–violet light-emitting diodes (LEDs) and the overall GaN-device market is forecasted to increase by the tens of billions of dollars in the next several decades. GaN-based blue and violet laser diodes are in high demand for high-definition digital versatile disk (DVD) drives. ZnO-based devices have not seen penetration into the semiconductor marketplace because of its own set of technical hurdles yet to be overcome (P-doping, contacts, low-defect active layers, etc.) [19.7], but an intense research and development effort for commercial ZnO-based devices is currently progressing.

Semiconductor devices have typically been manufactured on *native* substrates that have nearly identical crystalline structure to the active device layers manufactured on them. Semiconductor substrates are normally cut from large single boules that have been grown by melt techniques, with variations of the Czochralski and Bridgman growth methods being the most common [19.8]. The melt techniques are generally preferred over vapor or solution methods for the growth of large single crystals, owing to the higher growth rates, which allow for lower costs and rapid scaling of boule diameters.

There is an important class of semiconductor materials where growth of large single crystals is problematic. These materials have different and in many cases superior electrical, optical, and structural properties compared with traditional semiconductors such as Si. ZnO and the group III nitrides: AlN, GaN, and InN are in this class of semiconductors, which also include SiC and diamond. The reactivity of molten ZnO and the relatively high oxygen overpressure (≈ 50 atm) required to melt ZnO makes the melt growth of large single-crystal ZnO boules difficult [19.9]. GaN also cannot be grown by traditional melt techniques due to the extreme pressures and temperatures required to prevent the disassociation of GaN to form molten GaN [19.10]. In fact, all of the group III nitrides decompose into their corresponding group III metal and N_2 well before their melting points when heated at atmospheric pressure.

Techniques for growing nitride thin films on Si, sapphire (Al_2O_3), and silicon carbide (SiC) substrates were developed in the 1980s and 1990s due to the lack of GaN substrates [19.6]. GaN thin-film growth on nonnative substrates such as sapphire produces a large number of threading dislocations and other defects caused by the large lattice and thermal expansion mismatch between GaN and the nonnative substrate. These defects are deleterious to device reliability and performance. Low-cost commercial GaN substrates would enable enhanced performance, increased yields, and allow rapid market penetration of nitride-based devices. Therefore, there has been a considerable amount of research in producing GaN wafers grown by non-molten techniques. The most widely developed is the hydride vapor-phase epitaxy (HVPE) technique. Several research institutes have demonstrated GaN-based devices, processed on HVPE GaN wafers with improved device metrics compared with nitride devices fabricated on SiC and sapphire wafers. Although an improvement over heteroepitaxy, GaN HVPE wafers are inferior in structural quality compared with Si or GaAs wafers. GaN-based laser diodes manufactured on HVPE GaN substrates have extremely low yields and reduced performance compared with the red diodes grown on low-defect GaAs substrates. HVPE GaN substrates are available only in limited quantities and are extremely expensive. Several other techniques such as high-pressure Ga flux and alkali fluxes have been investigated for the growth of large GaN crystals but have not produced the scale and quality that is required for high-volume low-cost GaN semiconductor wafers.

Large zinc oxide single crystals were first produced in the 1960s due to the interest in using single-crystal

ZnO for piezoelectric transducers and surface acoustical wave (SAW) devices. Renewed interest arose because of ZnO's potential as an isostructural, nearly lattice-matched substrate for group III nitride semiconductor device structures. Advances in fabricating ZnO and ZnMgO quantum wells on sapphire that exhibit strong optically stimulated UV emissions have further driven demand for development of ZnO substrates [19.11]. Zinc oxide boules up to 2 inch in diameter and 1 cm thick have been grown by vapor-phase transport [19.12], and melt-grown wafers of 2 in diameter are now also available [19.9]. Both processes typically yield wafers with dislocation densities of 10^4–10^5 cm^{-2}. Recently, the most promising method for the growth of inexpensive, large, low-defect ZnO boules is the hydrothermal method [19.13].

The similarity of water and ammonia as polar solvents allows GaN crystals to be grown in ammonia solvents (ammonothermal growth) similar to the hydrothermal growth of oxides crystals in high-pressure water solutions. Large quantities of low-cost ZnO and GaN wafers could theoretically be manufactured by applying the same scaling techniques used for quartz growth to the hydrothermal growth of ZnO and ammonothermal growth of GaN.

This chapter will focus on the specifics of hydrothermal growth of large single crystals of zinc oxide (ZnO) and ammonothermal growth of gallium nitride (GaN). Ammonia and water will be compared as solvents and a brief overview of the temperature gradient method, which is the predominant technique employed for large-scale hydrothermal growth of single crystals, will be presented. Phase stability, adequate solubility, and optimization of thermal and fluid flow gradients in hydrothermal systems through modeling and simulation will be reviewed. Finally, growth kinetics for hydrothermal ZnO and ammonothermal GaN and how the incorporation of dopants and impurities influence the electrical and optical properties on these two important semiconductor materials will be discussed.

19.1 Overview of Hydrothermal and Ammonothermal Growth of Large Crystals

19.1.1 Comparison of Ammonia and Water as Solvents

Water's abundance, low toxicity, high purity, liquid phase at room temperature and atmospheric pressure, high dielectric constant, and its amphoteric properties make it the most widely used solvent. Several large crystals have been grown in water at ambient conditions such as aluminum potassium sulfate (ALUM) and potassium dihydrogen phosphate (KDP), but many compounds need higher temperatures to obtain the high solubility and kinetics for the growth of large crystals. Hundreds of different crystalline compounds have been grown by the hydrothermal technique, but only quartz crystals have been produced in the size and quantities that are required for semiconductor substrates. Ammonia is the most common anhydrous solvent because of the many similarities with water, as shown in Table 19.1. Ammonia is readily available, with 109 000 metric tons produced worldwide in 2004 [19.14]. Costs for ultrahigh-purity anhydrous ammonia have been driven down due to its use in the synthesis of Si$_2$N$_3$ during complementary metal–oxide–semiconductor (CMOS) processing, and in the production of GaN LEDs.

Like water, ammonia is a polar molecule. Polar molecules have a nonuniform or anisotropic structure that causes positive and negative charges to form in opposite parts of the molecule. The old alchemist's

Table 19.1 Physical properties of ammonia and water

Property	Ammonia	Water
Boiling point (°C)	−33.4	100
Freezing point (°C)	−77.7	0
Critical temperature (°C)	132.5	374.2
Critical pressure (bar)	113	221
Density (g cm^{-3})	0.68 (−33 °C)	0.96 (100 °C)
Ionic product	≈ 10^{-29}	10^{-14}
Heat of vaporization (kcal/mol)	5.58	9.72
Heat of fusion (kcal/mol)	1.35	2.0
Viscosity of liquid at 25 °C (cP)	0.135	0.891
Dielectric constant	22 (−33 °C)	80 (0 °C)
Dipole moment (D)	1.46	1.84
Polarizability (cm^3) (×10^{24})	2.25	1.49
Specific conductance (Ω^{-1}cm^{-1})	4 × 10^{-10} (−15 °C)	4 × 10^{-8}

adage: *similia similibus solvuntur* (i.e., like dissolves like), a basic rule of solution growth, holds for polar solvents, which tend to dissolve ionic and polar solids. Ammonia has a lower, but still relatively high, dielectric constant and lower dipole moment than water (the dielectric constant and dipole moment are measures of the degree of polarization of a molecule), and therefore has less ability to dissolve highly ionic compounds than water, but a greater ability to dissolve organic molecules. Thus, ammonia-based solutions have been used predominately for chemical synthesis of fertilizers, pharmaceutical products, and plastics. Water and ammonia are also both protic solvents, which means they can donate a hydrogen bond to a solvated compound. Ammonia has a slightly higher proton affinity than that of water, and thus is a more basic solvent than water and enhances the acidity of many compounds. Table 19.1 lists some of the physical properties of ammonia and water.

Byrappa and *Yoshimura* define hydrothermal growth as [19.15]

> *any heterogeneous chemical reaction in the presence of a solvent (whether aqueous or nonaqueous) above room temperature at a pressure greater than 1 atm in a closed system.*

The term *solvothermal growth* has also been used generically when discussing both aqueous and nonaqueous solvents at above ambient conditions, but recently has been used more specifically when discussing the use of organic solvents at above ambient temperatures and pressures. The term *hydrothermal* is more commonly used when discussing aqueous (water-based) solvents, and the term *ammonothermal* has been recently adopted for discussing ammoniated (ammonia-based) solvents. Here we will define hydrothermal and ammonothermal as subclasses of the generic term of solvothermal growth as shown in Fig. 19.1 to avoid confusion while discussing differences and similarities of aqueous and ammoniated solvents.

Solvothermal crystal growth offers several advantages over better known methods such as melt growth. Solvothermal growth is a low-temperature process, which often makes possible the growth of materials that are difficult or impossible to melt, or materials which, on solidifying from a melt and cooling down, would undergo phase changes (because of such changes, α-quartz cannot be grown from the melt). Low-temperature solvothermal growth can also minimize or eliminate the incidence of temperature-induced point defects, as illustrated by the hydrothermal growth of $Bi_{12}SiO_{20}$ [19.16], and can produce large amounts of material by simultaneous growth on multiple seeds (over 4000 kg of single-crystal quartz has been produced in a single run [19.15]). Only a small amount of user intervention and monitoring is required during growth because of the extremely uniform temperature gradients that can be maintained and the absence of moving parts. Some disadvantages of solvothermal growth are the low growth rates and initial capital equipment costs, but these are offset by the ability to grow multiple crystals in a single run and the extended lifetimes of the autoclaves.

The autoclaves are made out of high-strength steels or special alloys. The vessel must be corrosion resistant and able to withstand the temperature and pressure requirements for long periods of time. Corrosive solutions employing concentrated acids and bases are sometimes required to increase the solubility to obtain acceptable growth rates. Therefore, to protect the autoclave, a noble-metal liner (e.g., silver, gold, platinum or Teflon depending upon the pressure and temperature (PT) conditions and the solvent medium) is used in this case. Superstrong high-content-nickel alloy autoclaves are used for ammonothermal growth of GaN due to the higher temperatures and pressures required for the growth of large crystals of GaN. These autoclaves currently have small volumes, but should be easily scalable to larger sizes, albeit at greater expense than current industrial-scale autoclaves. Four centimeter-diameter autoclaves have recently been used to grow GaN crystals [19.17].

Additional tasks must be preformed when ammonia is being used as a high-pressure solvent. Ammonia, which is a vapor at ambient temperatures and pressures, can be condensed at room temperature and atmospheric pressure into an autoclave immersed in a chilled alcohol bath, pumped into an autoclave using high-pressure pumps, or chilled and poured into an autoclave. Precautions have to be taken, such as the use of glove boxes and vacuum apparatus, for any reactants or products that are air or moisture sensitive, e.g., alkali metals and their amides and azides used as mineralizers. Ammonia is

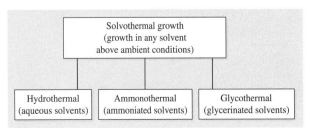

Fig. 19.1 Several subclasses of solvothermal growth

also toxic, so autoclaves with internal volumes greater than 50 cm^3 should be run inside chemical exhaust cabinets in case of a seal leak or rupture. There are several publications that cover specific details of the apparatus and autoclaves used in both hydrothermal [19.15, 18] and ammonothermal reactions [19.19, 20].

19.1.2 Growth of Large Crystals by the Transport Growth Model

Solvothermal growth is typically performed for two reasons:

1. The desired material is not thermodynamically favorable at ambient conditions;
2. The kinetics is such that the growth rate of the desired material is extremely slow at ambient conditions.

The *transport growth* model (also called the *temperature-differential* or *dissolution–crystallization* model) is predominately used for solvothermal growth of large single crystals, due to the low growth rate of quartz and many other technologically important materials grown in high-pressure solvents. Table 19.2 shows the dissolution–crystallization mechanism which is kinetically enhanced by the addition of *mineralizers*. Ammonia, like water, is amphoteric and can be made either acidic or alkaline depending on which compounds or *complexing agents* are combined with it. These complexing agents, or *mineralizers*, ionize in the solvent; it is these cations (acidic ions) or anions (alkaline ions) that promote the high solubility of many solid-phase compounds at low temperature. Source material is dissolved to form an intermediate species (left to right flow in the reversible chemical reactions in Table 19.2). Subsequently, the intermediate species becomes supersaturated by an external force and crystallizes out as single-crystal deposits (right to left flow in reactions in Table 19.2).

The *transport growth model* and all solvothermal growth mechanisms are based on supersaturation. Fig-

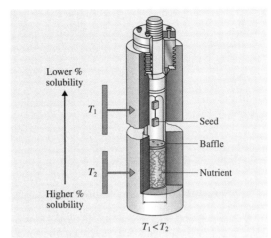

Fig. 19.2 Schematic of solvothermal crystal growth by the transport growth model (after [19.18])

ure 19.2 shows a schematic of the model. A temperature zone is established where the lower half of the autoclave, called the dissolution zone (T_2 in Fig. 19.2), is at a higher temperature than the upper half of the autoclave, called the crystallization zone (T_1 in Fig. 19.2). A baffle is placed between the two zones to help establish near-isothermal conditions in each zone. The solvent at the bottom of the vessel dissolves the nutrient until it reaches saturation. The solvated species is transported by the lower density fluid in the warmer dissolution zone, through natural fluid convection, to the cooler crystallization zone. Because of the lower temperature at the seed the solvated species becomes supersaturated, comes out of solution, and deposits on the seed (normal saturation conditions). Fluid convection returns the higher-density depleted solution to the lower-density fluid in the dissolution zone, where additional nutrient is dissolved. The cycle repeats as long as there is nutrient in the lower zone. Quartz [19.15], zinc oxide [19.21–27], GaN [19.20], and many other inorganic crystals have

Table 19.2 Comparison of dissolution–crystallization cycle of the hydrothermal and ammonothermal techniques

Hydrothermal: oxides	Ammonothermal: nitrides
Water: $2H_2O \rightarrow H_3O^+ + OH^-$	Ammonia: $2NH_3 \rightarrow NH_4^+ + NH_2^-$
ZnO in alkali water solution	GaN in ammonobasic solution
$ZnO + 2H_2O + 4OH^- \leftrightarrow Zn(H_2O)_2(OH)_4^-$	$GaN + (x-3)NH_2^- + 2NH_3 \leftrightarrow Ga(NH_2)_x^{3-x}$
Mineralizers	Mineralizers
Acids: HNO_3, HCl, HI	Acids: NH_4Cl, NH_4I, HI
Bases: KOH, NaOH, LiOH	Bases: KNH_2, $NaNH_2$, $LiNH_2$

Table 19.3 Status of technology, ca. 2006, for solvothermal growth of SiO_2, ZnO, and GaN crystals (after [19.28])

Parameter	SiO_2	ZnO	GaN
Autoclave inner diameter	0.65 m	≤ 0.2 m	≤ 0.03 m
Autoclave inner length	14 m	≤ 3 m	0.2–0.7 m
Volume	4.6 m^3	≤ 0.2 m^3	$\leq 5 \times 10^{-4}$ m^3
Seed size	$70 \times 45 \times 230$ mm^3	≈ 50 mm diameter	≤ 25 mm diameter
Seeds per batch	1400	112	1–4
Weight of crystal	1700 g	320 g; 20 mm thick	Few gram
Total yield per batch	2300 kg	36 kg	Few gram
Growth rate (c-axis)	25 µm/h	10 µm/h	≤ 2 µm/h
3-Run-per-year yield	6900 kg	108 kg	0.12 kg

been grown solvothermally by the transport growth model.

A recent publication by *Fukuda* and *Ehrentraut* [19.28] gives an outlook of industrial growth of hydrothermal ZnO crystals and ammonothermal GaN by projecting their current and future development path with that of quartz growth. Table 19.3 shows the current status of hydrothermal quartz, hydrothermal zinc oxide, and ammonothermal gallium nitride. Even with the slow growth rates of zinc oxide and gallium nitride (20–40% and 5–20% of the growth rate of quartz, respectively) under current solvothermal conditions it is theoretically possible to grow hundreds of kilograms of zinc oxide or gallium nitride in a single autoclave.

Hydrothermal zinc oxide crystals of several hundred grams have been grown with batch sizes in the tens of kilograms in small production autoclaves as shown in Table 19.3. An industrial-scale 500 l autoclave was recently used to grow 200 ZnO crystals at a growth temperature of 330 °C. The run lasted

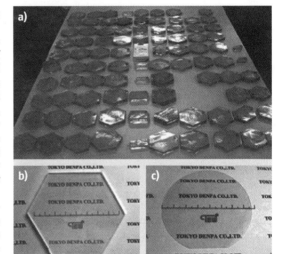

Fig. 19.3 (a) Two inch-size ZnO crystals produced during single growth run; (b) 3 inch ZnO crystal viewed down c-axis; (c) 3 inch (0001) ZnO wafer (after [19.13])

for 100 days and the weight of the crystals varied in the range 100–250 g [19.29]. If there is a large demand for ZnO wafers, economies of scale will allow for price reductions (200–400 US$/cm^2 in 2006), as there is no reason ZnO crystals could not be grown in large industrial-scale autoclaves currently used to grown single-crystal quartz. Figure 19.3a shows 2 inch-diameter ZnO crystals grown in an autoclave with the equivalent dimensions as the *ZnO autoclave* listed in Table 19.3 [19.13]. Figures 19.3b and 19.3c show the largest current ZnO crystal and corresponding 3 inch-diameter wafer cut from the crystal.

Fukuda and Ehrentraut go on to apply SiO_2 quartz development to future ZnO and GaN development in order to project the cost of ZnO and GaN wafers when

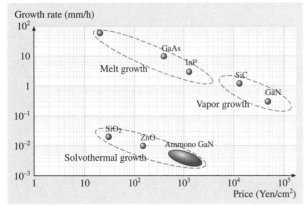

Fig. 19.4 Growth rates and cost of large-area single crystals (projected cost for ZnO and GaN) (after [19.28])

their development cycle fully matures. The authors claim that hydrothermal ZnO and ammonothermal GaN wafers will be cost-competitive with tradition compound semiconductor wafers such as GaAs and InP in the future. This comparison is shown in Fig. 19.4, which shows the potential of applying solvothermal growth techniques for industrial production of semiconductor wafers.

19.2 Requirements for Growth of Large, Low-Defect Crystals

19.2.1 Thermodynamics: Solubility and Phase Stability

In solvothermal growth of large crystals a *nutrient*, often the compound one intends to crystallize out as a single crystal, is dissolved in the solvent. The dissolved compound may form intermediate complexes or species in solution. In solvothermal growth, convective circulation and/or diffusion of these species throughout the solvent provides the primary mechanism for synthesizing crystalline compounds. A solvothermal growth system is designed to bring the soluble species into a region of the solvent medium where a change in conditions – e.g., in temperature, solvent composition, pH, or pressure – promotes crystal growth. Ideally, this change of conditions puts the soluble species in supersaturation, so the species must come out of solution – hopefully as the desired compound (e.g., ZnO, GaN) – until an equilibrium state (saturation) is achieved.

Thus, there are four fundamental thermodynamic requirements for the growth of high numbers of low-defect crystals by the hydrothermal/solvothermal method:

1. The desired material must have an adequate solubility at a given set of conditions (solvent, dissolution temperature, and pressure);
2. The solubility of the desired material must have adequate temperature dependence between the dissolution zone and crystallization zone;
3. The desired material must be the thermodynamically preferred material in the crystallization zone for a given set of conditions (solvent, crystallization temperature, and pressure);
4. The solvent must be thermodynamically stable at the temperatures and pressure needed to fulfill requirements 1–3 over long periods of time.

Low solubility will result in low growth rates; excessively high solubility will result in polycrystalline growth or spontaneous nucleation, which may be desirable for solvothermal powder synthesis, but prohibits growth of low-defect-density single crystals. Typically, the nutrient should be 1–10 wt.% soluble in the solvent. Solubility can be increased by adding a proper complexing agent (mineralizer). Most mineralizers used in solvothermal growth change the pH of the solvent, making the solvent either acidic or alkaline, i.e., increasing the number of cations (H_3O^+ in water, NH_4^+ in ammonia) or anions (OH^- in water, NH^- or NH_2^{2-} in ammonia) that attack the nutrient material in ionic solutions. For zinc oxide and gallium nitride, a solubility of approximately 5% by weight in alkaline solution yields high-quality single crystals at reasonable growth rates.

Because temperature is used to create supersaturation for large crystals, solubility of the compound must have temperature dependence. Natural convection allows solute to transport to the seed interface, where varying the temperature gradient between the nutrient and seed controls crystal growth kinetics.

The third requirement above can be difficult to achieve because of the multicomponent nature of solvothermal growth. If an oxide such as zinc oxide is desired, the hydride, hydroxide or hydrate should not be thermodynamically favored for the specific mineralizer, temperature, and pressure conditions employed. Also, chemical elements that are components in the mineralizer must not contribute to the formation of undesirable solid compounds at the growth interface.

The final requirement is not readily apparent in hydrothermal growth. Water is stable as a liquid or supercritical fluid to temperatures above the maximum operating conditions of even small super-high-pressure research autoclaves. Ammonia however starts decomposing well below 500 °C at atmospheric pressure, and even under 3–4 kbar of pressure a significant percentage of ammonia will decompose above 500 °C [19.19]. This has two deleterious effects, first the change in solvent composition by ammonia decomposition over long periods of time will change the kinetic and possibly the thermodynamics of the growing GaN crystals, and secondly, hydrogen could have adverse effects on the autoclave vessels through hydrogen embrittlement. The decomposition of ammonia is therefore one of the main issues that must be addressed for the growth of large crystals by the ammonothermal method.

The solubility and phase stability of ZnO in hydrothermal solvents and GaN in ammonothermal solvents are presented here.

Solubility of Hydrothermal ZnO

The solubility of hydrothermal zinc oxide in an OH^- alkaline medium is shown in Fig. 19.5 [19.23]. Note that solubility increases with temperature. This is called *normal or forward-grade solubility*.

Zinc oxide is an amphoteric oxide, meaning it acts as an acid in alkaline solutions and as a base in acidic solutions. It is possible to grow hydrothermal zinc oxide in acidic solutions as well as the alkaline solutions shown in Fig. 19.6. *McCandlish* and *Uhrin* recently studied the solubility of ZnO in an acidic medium and grew ZnO at 100–250 °C, with growth rates up to 0.25 mm/day [19.30]. Figure 19.6 illustrates the solubility in acidic regimes. The squares signify 2 molal aqueous nitric acid and the circles signify a proprietary acidic solution. Note that the nitric acid solution exhibits normal solubility, whereas the proprietary solution exhibits decreasing solubility with increasing temperature (*retrograde or reverse-grade solubility*). To grow ZnO crystals under conditions of normal solubility the seed is placed in a colder region than the source material (Fig. 19.2); to grow under conditions of retrograde solubility, the seed is placed in the hotter region.

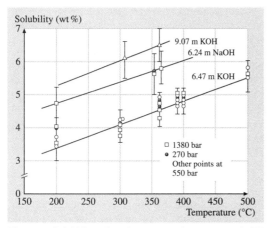

Fig. 19.5 Solubility of ZnO versus T in aqueous NaOH and KOH solutions (after [19.23])

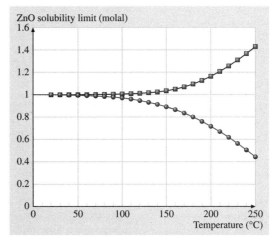

Fig. 19.6 Normal solubility of ZnO in 2 molal nitric acid (*squares*), and retrograde solubility of ZnO in a proprietary acid solution (*circles*) (after [19.30])

Phase Equilibrium of ZnO in Hydrothermal Solvents

Hüttig and *Möldner* studied the phase equilibrium of the $ZnO-H_2O$ system to 40 °C and found zinc oxide to be the stable solid phase at pressures above 50 torr and temperatures above 35 °C [19.31]. *Lu* and *Yeh* experimentally showed that zinc oxide is the stable product up to pH = 12.5 in an aqueous ammonia solution at 100 °C [19.32]. *Laudise* and *Ballman* grew large zinc oxide crystals in alkaline media and found zinc oxide to be the stable product at 200–400 °C in 1.0 M NaOH [19.33]; subsequently it was found that zinc oxide can be grown in hydroxide solutions up to 10 M at temperatures exceeding 300 °C [19.27].

Recent advances in thermodynamic modeling of aqueous solution chemistry can aid in choosing conditions that achieve crystal growth of zinc oxide and other materials. A thermodynamic model of aqueous-based chemistry has been developed that computes the stability of zinc oxide in different aqueous regimes. The model uses commercial software (OLI Systems Inc., Morris Plains) and is detailed in several publications [19.34–36]. McCandlish and Uhrin initially modeled zinc oxide in the hydroxide system to validate the model against experimental data. Subsequently, a thermodynamic model was created for the growth of zinc oxide in acidic environments. Figure 19.7 shows the computed stability of ZnO at 150 °C as a function of pH with HNO_3 as the mineralizer [19.30].

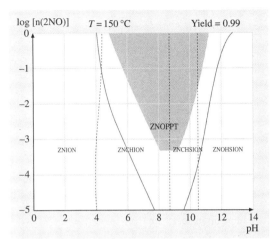

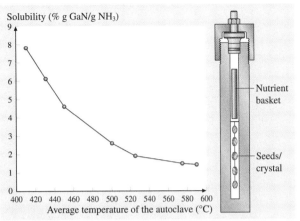

Fig. 19.7 Yield diagram for the precipitation of ZnO in 2 molal nitric acid at 150 °C as a function of pH; log [n(2NO)] signifies the log of the molar concentration of nitric acid in solution; ZNION, ZNCHION, ZNCHSION and ZNOHSION signifies the basic atomic chemical composition of the ionic compounds in solution for a given pH: Zn ion, ZnCH ion, ZnCHS ion, ZnOHS ion; ZnOPPT signifies ZnO crystallites that participated out of solution for a given pH (after [19.30])

Solubility of Ammonothermal GaN

Ammonothermal growth of GaN has retrograde solubility in alkaline solutions and normal solubility in acidic solutions. Figure 19.8 shows the solubility of polycrystalline GaN after a series of runs in the KNH_2–NH_3 system ($\pm 10\%$ variability in filling and molarity) [19.42]. The KNH_2 concentration is about 3.5 ± 0.5 M, the temperature gradient is around 10 °C/cm, and the pressure is 1.3–2.4 kbar. The solubility of GaN in solutions of greater than 1 molal NaN_3–NH_3, KN_3–NH_3, KNH_2–NH_3, and $NaNH_2$–NH_3

Fig. 19.8 Solubility of GaN polycrystalline in KNH_2–NH_3 system, 3.5 ± 0.5 M KNH_2 concentration (after [19.42])

has similar tendencies: a negative solubility coefficient and rather high solubility, in the range of 1–10% between 400 °C and 600 °C.

Using acidic mineralizers such as 0.4 M NH_4Cl, GaN has a normal solubility in ammonia [19.43]. The acidic conditions required the use of a Pt inner liner to protect the autoclave from corrosion. A growth rate of 0.02–0.03 mm/day can be achieved at a temperature of 550 °C and a pressure of less than 1.5 kbar. Normal solubility has also been claimed using other ammonium halide mineralizers [19.41, 43].

Phase Equilibrium of GaN in Ammonothermal Solvents

The early work on ammonothermal synthesis of nitride powders was performed by *Jacobs* and *Schmidt* [19.19]. They synthesized several novel nitrogen-based compounds and designed apparatus specifically for ammonothermal powder synthesis. Later *Peters* [19.44]

Table 19.4 Growth conditions of ammonothermal synthesis of GaN microcrystals

Investigators	Mineralizers	Temperature (°C)	Pressure (kbar)	Growth rate and habit
Dwilinski et al. [19.37]	KNH_2	550	1–5	Hexagonal microcrystals
Purdy [19.38]	NH_4Cl, NH_4Br or NH_4I	250–500	0.7	Mixed cubic and hexagonal microcrystals
Ketchum and *Kolis* [19.39]	KNH_2/KI	400	2.4	Hexagonal submillimeter-size plates
Yoshikawa et al. [19.40]	NH_4Cl	500	1.2	Hexagonal needle shape 0.02–0.03 mm/day
Lan et al. [19.41]	NH_4F, NH_4Cl, NH_4Br or NH_4I	450–600	1–2	Hexagonal microcrystals

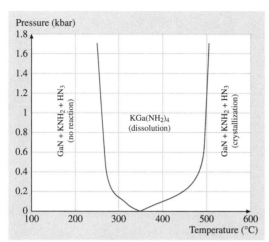

Fig. 19.9 Phase diagram of GaN in the GaN–KNH$_2$–NH$_3$ system (after [19.45]) (*curves* based on experiments between 0.3–2.4 kbar (5–35 kpsi); below 0.3 kbar the tendency is estimated).

and *Dwilinski* et al. [19.37] were the first to synthesize AlN and GaN microcrystals ammonothermally. The work on ammonothermal microcrystalline synthesis is summarized in Table 19.4.

Dwilinski et al. [19.37] obtained microcrystals of BN, AlN, and GaN by the ammonothermal method using lithium or potassium amide as mineralizer at pressures in the range 1–5 kbar and temperatures up to 550 °C. *Ketchum* and *Kolis* [19.39] grew ammonothermal single crystals of gallium nitride in supercritical ammonia at 400 °C and 2.4 kbar by using potassium amide (KNH$_2$) and potassium iodide (KI) as mineral-

izers. Hexagonal GaN crystals of $0.5 \times 0.2 \times 0.1$ mm^3 were obtained. They also used potassium azide (KN$_3$) or sodium azide (NaN$_3$) to increase the solubility of GaN in ammonia [19.48].

The phase diagram of GaN in the GaN–KNH$_2$–NH$_3$ system is shown in Fig. 19.9 [19.45]. Unlike hydrothermal ZnO, high temperatures are needed to precipitate GaN out of an ammonothermal solution. The reaction KGa(NH$_2$)$_4$ ↔ KNH$_2$ + GaN + 2NH$_3$ did not produce GaN at temperatures below 400 °C. Pressure variations had little effect on the thermodynamics of GaN formation.

GaN has a metastable cubic phase that has been formed ammonothermally. *Purdy* et al. synthesized both cubic and hexagonal GaN by ammonothermal reactions of gallium metal or GaI$_3$ under acidic (NH$_4$Cl, NH$_4$Br or NH$_4$I) conditions [19.38, 49]. The reaction temperatures were 250–500 °C and pressures were up to 10 000 psi (0.6895 kbar).

Figure 19.10a shows well-defined cubic crystals grown in ammonia with the addition of lithium chloride to an acidic solution [19.46]. *Hashimoto* et al. [19.50] have shown that the ammonium halides (acidic mineralizer) and the alkali halide (neutral mineralizer) can form mixed cubic and hexagonal phases of GaN. *Ehrentraut* et al. [19.47] showed that pure hexagonal phases can be obtained in acidic ammonia solutions. Figure 19.10b shows that hexagonal formation is favored at lower temperatures in solutions for successively smaller halide cations. Pure phase hexagonal GaN was obtained at ≥ 470 °C for NH$_4$Cl mineralizer, ≥ 500 °C for NH$_4$Br mineralizer, and ≥ 550 °C for NH$_4$I mineralizer. If one takes into account the metastable nature of cubic GaN (*Purdy* [19.51] showed a correlation between cubic GaN formation and short-duration experiments), a phase conversion of the metastable cubic GaN to the stable hexagonal GaN could be possible in long-duration experiments as performed by *Ehrentraut* et al. [19.47].

19.2.2 Environmental Effects on Growth Kinetics and Structure Perfection (Extended and Point Defects)

The stable phase of ZnO and GaN has the wurtzite crystal structure, which is hexagonal with a space group of $P6_3mc$. The noncentrosymmetric structure of the wurtzite structure produces an anisotropy in which the opposite sides of a basal plane wafer have different atomic arrangements at their surfaces. This anisotropy in hexagonal GaN and ZnO also causes a nonsymmet-

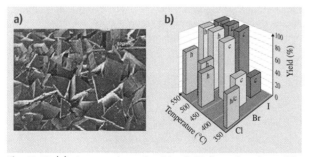

Fig. 19.10 (a) Ammonothermal cubic GaN formed under acidic conditions with the addition of lithium (after [19.46]); (b) temperature and mineralizer effect on phase stability of gallium nitride synthesized under acidic ammonothermal conditions (after [19.47]) (h – hexagonal, c – cubic)

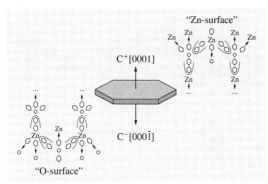

Fig. 19.11 Electronic charge distribution of zinc oxide basal faces *(GaN has the same crystal structure as ZnO)* (after [19.21, 52])

rical charge state due to its anisotropic crystal structure. The C^+ side of the basal plane is comprised of a Zn-rich layer for ZnO or a Ga-rich layer for GaN, and the C^- is comprised of an O-rich layer for ZnO or a N-rich layer for GaN, as illustrated in Fig. 19.11 (a zinc oxide crystal is shown; gallium nitride has the exact same configuration).

The C^+ plane has a net positive surface charge because of the greater number of positive dangling bonds (Zn or Ga) on its surface, in contrast to the C^- plane, which has a net negative surface charge due to the greater number of negative dangling bonds (O or N). The resulting distribution of electric charge causes disparities, among the various growth planes, in growth rates, as well as in impurity incorporation, chemical etching, and optical and electrical properties.

Many anisotropic or polar crystals including inorganic and organic materials have been grown from solutions. When growing polar crystals from solutions, especially a highly polarized solution such as water or ammonia, investigations into the surface chemistry of the crystals and the composition of the growth medium need to be taken into consideration. Typically the solvent, the intermediate species in solution, and the crystal itself all have surface charge states. The intermediate species can be broken down into fundamental growth units that react with the crystal's growth facets. The structure of these growth units determines growth kinetics on the various polar faces of the crystal due in large part to the charge state of these faces in relation to the charge states of the fundamental growth units.

Impurities in solutions cause changes in molecular diffusion and atomic absorption on advancing crystal surfaces, which in turn, influences the growth kinetics of the crystal. Because ZnO and GaN have anisotropic crystal structures, impurities, even at small levels, can effect a change in growth rates along specific crystallographic axes which induce point and line defects. In addition, the concentrations of impurities incorporated in the various growth sectors of a polar crystal can vary. ZnO and GaN grown on their positive polar surfaces have different impurity concentrations, growth rates, and morphologies than material grown on the negative polar surfaces. In short, the growth characteristics (growth morphology, impurities distribution, and crystal quality) are influenced by the anisotropic nature of polar crystals. The influence of impurities on hydrothermal ZnO and ammonothermal GaN will be discussed in Sects. 19.4 and 19.5, respectively.

19.2.3 Doping and Alloying

Semiconductor wafers ideally should be conducting or insulating. The previous section discusses how impurities can dramatically influence the growth kinetics. Therefore, great care must be taken while applying doping in a solvothermal medium. The majority of dopants slow down growth rates because they block the diffusion and/or absorption of the matrix atoms to the lattice sites of the various surfaces of the crystal. Dopants can have different segregation coefficients in solution growth, similar to impurities in molten solidification. It is difficult to control growth morphology in anisotropic crystals grown from solution with high levels of impurities or dopants. As in any semiconductor crystal growth process, impurities must be reduced to the lowest possible levels. This will allow for the smallest levels of dopants introduced for the synthesis of semiconductor boules with the desired conductivity levels. Finally solvothermal has several advantages over molten techniques for the growth of alloy wafers, such as solid sources and low temperature gradients. Preliminary experiments of hydrothermal ZnMgO [19.53] and solvothermal AlGaN [19.54] have been conducted.

Fluid flow and optimization of temperature zones are extremely important for the formation of low-defect crystals. The next two sections will discuss how to model fluid velocity and temperature gradient in a solvothermal system.

19.3 Physical and Mathematical Models

19.3.1 Flow and Heat Transfer

The hydrothermal and ammonothermal growth processes employ aqueous and ammoniated solvents under high temperatures and high pressures to dissolve and recrystallize materials that are relatively insoluble under ordinary conditions. After the system is pressurized, the solvent occupies most of the volume. The convection system for hydrothermal/ammonothermal growth consists of a porous bed whose height changes with the growth, a fluid layer overlying this porous bed, a metal baffle with holes (which lies above the porous bed), and solid seed plates whose size increase with the growth. Figure 19.12 shows the schematic of growth system that has been used experimentally to synthesize GaN [19.39]. The autoclave has an internal diameter of 0.932 cm, external diameter of 3.5 cm, internal height of 18.4 cm, and external height of 20.3 cm (Tem-Press MRA 138R with a volume of 12.5 ml). The thicknesses of the sidewall and bottom of the autoclave are 1.28 cm. The charge height is 1 cm, and the gap between the baffle and charge bed is 2 cm. A baffle made from 0.28 mm Ag foil is used to divide the autoclave into two parts: the upper and lower portions.

Hence the upper portion can be considered as a fluid layer with the assumption of incompressible flow and the Boussinesq approximation [19.56, 57], and the Navier–Stokes equations can be used in the fluid layer. Suppose that the density has a linear temperature dependence of the form

$$\rho = \rho_0[1 - \beta(T - T_0)], \qquad (19.1)$$

where ρ, β, and T are density, isobaric coefficient of expansion, and temperature, and ρ_0 and T_0 are constant reference values for the density and temperature, respectively.

In the solid region which comprises the autoclave walls, the baffle, and the seeds, only conductive heat transfer is considered. In the fluid region, convective heat transfer is considered.

19.3.2 Porous-Media-Based Transport Model

The nutrient particles (e.g., ZnO or GaN) in the bottom of the autoclave can be considered as a porous medium. In this case, the Darcy–Brinkman–Forchheimer model can be employed in the porous layer [19.58, 59]. The dimensionless parameters of the system are listed as follows

$$A = H/R, \quad \mathrm{Gr} = g\beta R^3 \Delta T/\nu^2, \quad \mathrm{Pr} = \nu/\alpha,$$
$$\mathrm{Da} = K/R^2, \quad \mathrm{Fs} = b/R,$$

where A, Gr, Pr, Da, and Fs denote the aspect ratio, Grashof number, Prandtl number, Darcy number, and Forchheimer number, respectively. H is the internal height of the autoclave, R is the internal radius of the autoclave, g is acceleration due to gravity, ΔT is the maximum temperature difference on the sidewall of the autoclave, ν is kinematic viscosity, α is thermal diffusivity, the permeability of porous matrix $K = d_p^2 \varepsilon^3/[150(1-\varepsilon)^2]$ with d_p as the average diameter of the nutrient particles, and the Forchheimer coefficient $b = 1.75 K^{0.5}/(\sqrt{150}\varepsilon^{1.5})$.

The governing equations in the porous and fluid layers can be combined by defining a binary parameter B as: $B = 0$ in the fluid layer and $B = 1$ in the porous layer, respectively. The porosity is $\varepsilon = 0$ in solid, $0 < \varepsilon < 1$ in porous layer, and $\varepsilon = 1$ in fluid layer, respectively. The combined governing equations in a cylindrical coordinate system are

$$\frac{\partial(\varepsilon \rho_f)}{\partial t} + \nabla \cdot (\rho_f \boldsymbol{u}) = 0, \qquad (19.2a)$$

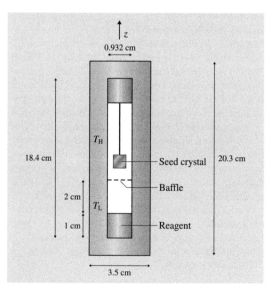

Fig. 19.12 Schematic of an ammonothermal growth system (after [19.55])

$$\frac{\rho_f}{\varepsilon}\frac{\partial \boldsymbol{u}}{\partial t} + \frac{\rho_f}{\varepsilon}(\boldsymbol{u}\cdot\nabla)\frac{\boldsymbol{u}}{\varepsilon}$$
$$= -\nabla p - \rho_f \beta (T-T_0)\boldsymbol{g}$$
$$+ \nabla\cdot(\mu_e \nabla \boldsymbol{u}) - B\left[\left(\frac{\mu_f}{K} + \frac{\rho_f b}{K}|\boldsymbol{u}|\right)\boldsymbol{u}\right], \quad (19.2b)$$

$$(\rho c_p)_e \frac{\partial T}{\partial t} + (\rho c_p)_f [(\boldsymbol{u}\cdot\nabla)T] = \nabla\cdot(k_e \nabla T), \quad (19.2c)$$

where μ, k, and c_p denote dynamic viscosity, thermal conductivity, and specific heat, respectively, and $\boldsymbol{g}$ is the gravity vector. Subscripts f and e denote fluid and effective, respectively.

The following scales are used to nondimensionalize the governing equations: length, R; velocity, $u_0 = \nu/R$; time, $t_0 = R^2/\nu$; pressure, $\rho\nu^2/R^2$; temperature, $T_H - T_L$. T_H and T_L are the high and low temperature applied on the sidewall of the autoclave, respectively. The resulting nondimensionalized equations are

$$\frac{\partial \varepsilon\bar{\rho}}{\partial t} + \frac{1}{r}\frac{\partial}{\partial r}(r\bar{\rho}u) + \frac{\partial}{\partial z}(\bar{\rho}w) = 0, \quad (19.3a)$$

$$\frac{\partial}{\partial t}\left(\frac{1}{\varepsilon}\bar{\rho}u\right) + \frac{1}{r}\frac{\partial}{\partial r}\left(\frac{1}{\varepsilon^2}r\bar{\rho}uu\right) + \frac{\partial}{\partial z}\left(\frac{1}{\varepsilon^2}\bar{\rho}wu\right)$$
$$= \bar{\mu}\left[\frac{1}{r}\frac{\partial}{\partial r}\left(r\frac{\partial u}{\partial r}\right) + \frac{\partial^2 u}{\partial z^2} - \frac{u}{r^2}\right] - \frac{\partial p}{\partial r}$$
$$- B\left(\frac{1}{\text{ReDa}} + \frac{\text{Fs}}{\text{Da}}|\boldsymbol{u}|\right)u, \quad (19.3b)$$

$$\frac{\partial}{\partial t}\left(\frac{1}{\varepsilon}\bar{\rho}w\right) + \frac{1}{r}\frac{\partial}{\partial r}\left(\frac{1}{\varepsilon^2}r\bar{\rho}uw\right) + \frac{\partial}{\partial z}\left(\frac{1}{\varepsilon^2}\bar{\rho}ww\right)$$
$$= \bar{\mu}\left[\frac{1}{r}\frac{\partial}{\partial r}\left(r\frac{\partial w}{\partial r}\right) + \frac{\partial^2 w}{\partial z^2}\right] - \frac{\partial p}{\partial z} + \text{Gr}\Theta$$
$$- B\left(\frac{1}{\text{ReDa}} + \frac{\text{Fs}}{\text{Da}}|\boldsymbol{u}|\right)w, \quad (19.3c)$$

$$\bar{\rho}\bar{c}_p\frac{\partial \Theta}{\partial t} + \frac{1}{r}\frac{\partial}{\partial r}(r\bar{\rho}u\Theta) + \frac{\partial}{\partial z}(\bar{\rho}w\Theta)$$
$$= \frac{1}{\text{Pr}}\bar{k}\left[\frac{1}{r}\frac{\partial}{\partial r}\left(r\frac{\partial \Theta}{\partial r}\right) + \frac{\partial}{\partial z}\left(\frac{\partial \Theta}{\partial z}\right)\right], \quad (19.3d)$$

where $\bar{\rho} = \frac{\rho_e}{\rho_f}$, $\bar{\mu} = \frac{\mu_e}{\mu_f}$, $\bar{c}_p = \frac{c_{p_e}}{c_{p_f}}$, $\bar{k} = \frac{k_e}{k_f}$.

19.3.3 Numerical Scheme

The momentum equations (19.3b,c) and energy equation (19.3d) are solved using an in-house-developed finite-volume algorithm [19.60, 61]. The above conservation equations (19.3b–d) can be written in the following general form

$$\frac{\partial}{\partial t}(rc\bar{\rho}\phi) + \frac{1}{r}\frac{\partial}{\partial r}(dr\bar{\rho}u\phi) + \frac{\partial}{\partial z}(d\bar{\rho}w\phi)$$
$$= \frac{\partial}{\partial r}\left(r\Gamma\frac{\partial \phi}{\partial r}\right) + \frac{\partial}{\partial z}\left(r\Gamma\frac{\partial \phi}{\partial z}\right) + r(S_C + S_p\phi), \quad (19.4)$$

where ϕ is the generalized variable, Γ is the diffusion coefficient, and S_C is the volumetric source. The coefficients are defined as

$$c = \frac{1}{\varepsilon}, \quad d = \frac{1}{\varepsilon^2}, \quad \Gamma = \bar{\mu}, \quad S_C = -\frac{\partial p}{\partial r},$$

and

$$S_p = -B\left(\frac{1}{\text{ReDa}} + \frac{\text{Fs}}{\text{Da}}|\boldsymbol{u}|\right) - \frac{1}{r^2} \text{ for (19.3b)}$$
$$(\phi = u);$$

$$c = \frac{1}{\varepsilon}, \quad d = \frac{1}{\varepsilon^2}, \quad \Gamma = \bar{\mu}, \quad S_C = -\frac{\partial p}{\partial z} + \text{Gr}\Theta,$$

and

$$S_p = -B\left(\frac{1}{\text{ReDa}} + \frac{\text{Fs}}{\text{Da}}|\boldsymbol{u}|\right) \text{ for (19.3c) } (\phi = w);$$

$$c = \bar{\rho}\bar{c}_p, \quad d = 1, \quad \Gamma = \frac{\bar{k}}{\text{Pr}}, \quad S_C = 0,$$

and

$$S_p = 0 \text{ for (19.3d) } (\phi = \Theta).$$

The grid used for this task is a structured trapezoidal mesh. For a typical primary point, the discretized form of the continuity (19.3a) in a generalized coordinate system (ξ, η) is then written as

$$\frac{\varepsilon\bar{\rho} - \varepsilon^0\bar{\rho}^0}{\Delta t}rJa\Delta\xi\Delta\eta$$
$$+ [(r\bar{\rho}\alpha_\xi u_\xi)_e - (r\bar{\rho}\alpha_\xi u_\xi)_w]\Delta\eta$$
$$+ [(r\bar{\rho}\alpha_\eta u_\eta)_n - (r\bar{\rho}\alpha_\eta u_\eta)_s]\Delta\xi$$
$$= S_{NO}\Delta\xi\Delta\eta, \quad (19.5)$$

where Δt is the time step, the curvature source term arising from the nonorthogonal grid $S_{NO} = [(r\bar{\rho}\beta_\xi u_\eta)_e - (r\bar{\rho}\beta_\xi u_\eta)_w] \times \Delta\eta + [(r\bar{\rho}\beta_\eta u_\xi)_n - (r\bar{\rho}\beta_\eta u_\xi)_s]\Delta\xi$, α_ξ and β_ξ are the primary area and the secondary area over the control-volume face, which is represented by $\xi = $ const, e.g., $\boldsymbol{u}\cdot\boldsymbol{e}^\xi h_\eta = \alpha_\xi u_\xi - \beta_\xi u_\eta$, where $\boldsymbol{e}^\xi$ is the contravariant base vector, and h_η is the scale factor. So, $\alpha_\xi = h_\xi h_\eta^2/(Ja)$ and $\beta_\xi = h_\xi h_\eta^2(\boldsymbol{e}_\xi\cdot\boldsymbol{e}_\eta)/(Ja)$ [19.60, 61].

The momentum equations (19.3b,c) can be written as

$$\frac{(rc\bar{\rho}\phi Ja - rc\bar{\rho}^0\phi^0 Ja^0)_P \Delta\xi\Delta\eta}{\Delta t}$$
$$+ [(r\alpha_\xi J_\xi)_e - (r\alpha_\xi J_\xi)_w]\Delta\eta$$
$$+ [(r\alpha_\eta J_\eta)_n - (r\alpha_\eta J_\eta)_s]\Delta\xi$$
$$= [r(S_C + S_P\phi)Ja + S_\phi]\Delta\xi\Delta\eta, \quad (19.6)$$

where the curvature source term S_ϕ arises from the nonorthogonal grid, and is given by $S_\phi = [(r\beta_\xi J_\eta)_e - (r\beta_\xi J_\eta)_w]\Delta\eta + [(r\beta_\eta J_\xi)_n - (r\beta_\eta J_\xi)_s]\Delta\xi$. The flux components in the ξ and η directions are $J_\xi = d\bar{\rho}u_\xi\phi - \frac{1}{h_\xi}\Gamma\frac{\partial\phi}{\partial\xi}$ and $J_\eta = d\bar{\rho}u_\eta\phi - \frac{1}{h_\eta}\Gamma\frac{\partial\phi}{\partial\eta}$, respectively.

Multiplying (19.5) by $d\phi$, subtracting it from (19.6), and multiplying the resulting equation by $(1+i)$ yields the discretized equation for u and v in the control volume [19.55, 61]

$$a_P\phi_P = a_E\phi_E + a_W\phi_W + a_N\phi_N + a_S\phi_S$$
$$- (a_E + a_W + a_N + a_S)_{\text{conv}}\phi_P + S_\phi$$
$$- dS_{\text{NO}}\phi + b, \quad (19.7)$$

where

$$a_P = (1+i)rc\bar{\rho}Ja\frac{1}{\Delta t} - (1+i)rJaS_P$$
$$+ (a_E + a_W + a_N + a_S)_{\text{cond}},$$
$$b = (1+i)rc\bar{\rho}^0 Ja^0 \frac{1}{\Delta t}\phi_P^0 + (1+i)rJaS_C + iS_{\text{conv}},$$

and

$$S_{\text{conv}} = a_E\phi_E^0 + a_W\phi_W^0 + a_N\phi_N^0 + a_S\phi_S^0$$
$$- (a_E + a_W + a_N + a_S)\phi_P^0 + S_\phi^0 - dS_{\text{NO}}\phi^0.$$

The subscripts *conv* and *cond* indicate convective and conductive terms, respectively. The discretization in time is first order when $i = 0$ and second order when $i = 1$.

The momentum equations can be written as

$$A_P u_P = (H_u)_P - (1+i)\frac{\partial p}{\partial r}, \quad (19.8a)$$

$$A_P w_P = (H_w)_P - (1+i)\frac{\partial p}{\partial z}, \quad (19.8b)$$

where the subscript P represents the central point of a finite volume

$$A = (1+i)c\bar{\rho}\frac{1}{\Delta t} - (1+i)S_P$$
$$+ [(a_E + a_W + a_N + a_S)_{\text{cond}}]/(rJa),$$
$$(H_u)_P = [a_E u_E + a_W u_W + a_N u_N + a_S u_S$$
$$- (a_E + a_W + a_N + a_S)_{\text{conv}}\phi_P + S_\phi$$
$$- dS_{\text{NO}}\phi + b']/(rJa),$$

and b' represents b without the pressure term. The velocity component and pressure gradient in ξ direction are, respectively,

$$u_\xi = \frac{r'_\xi u + z'_\xi w}{h_\xi}, \quad (19.9a)$$

$$\frac{\partial p}{\partial \xi} = \frac{\partial p}{\partial r}r'_\xi + \frac{\partial p}{\partial z}z'_\xi, \quad (19.9b)$$

where the prime denotes the differential. By setting $(1+i)p$ as p, combining (19.8a,b), and using the procedures for pressure treatment as in [19.55, 62], we obtain

$$(u_\xi)_P = \frac{(H_u - A_P u_P)r'_\xi + (H_w - A_P w_P)z'_\xi - \left(\frac{\partial p}{\partial \xi}\right)_P}{h_\xi A_P}$$
$$+ (u_\xi)_P^0, \quad (19.10)$$

and

$$(\bar{\rho}u_\xi\alpha_\xi)_P = \left[(H_u - A_P u_P)r'_\xi + (H_w - A_P w_P)z'_\xi \right.$$
$$\left. - \left(\frac{\partial p}{\partial \xi}\right)_P\right]\bar{\rho}\alpha_\xi/(h_\xi A_P) + (\bar{\rho}u_\xi\alpha_\xi)_P^0. \quad (19.11)$$

Substituting the above into the continuity equation, we obtain the pressure equation as

$$a_P p_P = a_E p_E + a_W p_W + a_N p_N + a_S p_S + b, \quad (19.12)$$

where

$$a_E = D_e \Delta\eta,$$
$$a_P = a_E + a_W + a_N + a_S,$$
$$D_e = [\bar{\rho}\alpha_\xi/(h_\xi A)]_e,$$
$$b = -\{[(H_\xi - Au_\xi)h_\xi]_e D_e + (\rho u_\xi\alpha_\xi)_e^0\}\Delta\eta$$
$$+ \{[(H_\xi - Au_\xi)h_\xi]_w D_w + (\rho u_\xi\alpha_\xi)_w^0\}\Delta\eta$$
$$- \{[(H_\eta - Au_\eta)h_\eta]_n D_n + (\rho u_\eta\alpha_\eta)_n^0\}\Delta\xi$$
$$+ \{[(H_\eta - Au_\eta)h_\eta]_s D_s + (\rho u_\eta\alpha_\eta)_s^0\}\Delta\xi$$
$$+ S_{\text{NO}},$$

and

$$H_\xi = \frac{H_u r'_\xi + H_w z'_\xi}{h_\xi}.$$

For the temperature equation, profiled temperature boundary conditions are applied on the outer surfaces of the autoclave. The temperature profile set on the sidewall of the autoclave is $T = T_H$, $z < H_B - 0.5\delta_T$; $T = T_H - (T_H - T_L)(z - H_B + 0.5\delta_T)/\delta_T$; $H_B - 0.5\delta_T \le$

$z < H_B + 0.5\delta_T$; $T = T_L$, $z > H_B + 0.5\delta_T$, where H_B is the height of the baffle and δ_T is the length of the portion of the wall where the temperature changes from T_H to T_L. The top and bottom of the autoclave are considered adiabatic. The temperature distribution is considered axisymmetric, $\partial T/\partial r = 0$, at $r = 0$.

For solving the momentum equations and the pressure equation inside the autoclave, the fluid boundaries were searched inside the autoclave in the r and z directions, respectively [19.55]. For example, when solving the equations using the tridiagonal matrix algorithm (TDMA) method, the fluid boundaries were searched in the r or z direction separately, and the equations were solved in different intervals of fluid space in this direction. In this way, the fluid field was obtained inside the autoclave that contains different shapes of baffles and seeds. A mesh size of 302×77 was used in the simulation, and the nondimensional time step was $\Delta t = 10^{-6}$.

19.4 Process Simulations

19.4.1 Typical Flow Pattern and Growth Mechanism

For the solubility curve with a positive coefficient of temperature, the growth zone is maintained at a lower temperature than that in the dissolving zone, thus the nutrient becomes supersaturated in the growth zone. The critical properties of ammonia are $T_c = 405.5$ K and $P_c = 112.8$ bar. The reduced pressure and reduced temperature at 2 kbar and 250 °C for the growth condition in [19.39] are $P_r = 2000/112.8 = 17.7$ and $T_r = 523/405.5 = 1.3$. For $P_r = 10$ and $T_r = 1.3$, the viscosity and conductivity of ammonia are $\mu/\mu_1 = 4.3$ and $k/k_1 = 5.0$, where μ_1 and k_1 are the dynamic viscosity and the thermal conductivity at 250 °C and atmospheric pressure [19.63].

Solubility data of GaN for mineralizers of KN_3, KNH_2/KI were obtained in [19.39, 55]. The solubility of GaN for mineralizer of 1.6 M KN_3 is high in the temperature range of 300–450 °C. It seems that, by using azide as mineralizer, a high solubility of GaN can be obtained at low growth temperatures. For 0.8 M KNH_2/KI, the solubility is low in the range of 350–550 °C. Mineralizer of 2–6 M KNH_2 or $NaNH_2$ was used to increase the solubility of GaN [19.55].

For charge particle size of 0.6 mm, $T_H - T_L$ applied on the sidewall of 50 K, and $\delta_T = 1$ cm, the aspect ratio, Grashof number, Prandtl number, Darcy number, and Forchheimer number for the system in [19.39] are $A = 40$, $Gr = 4.46 \times 10^6$, $Pr = 0.73$, $Da = 2.2 \times 10^{-5}$, and $Fs = 2.6 \times 10^{-3}$, respectively. The baffle has an opening of 30% in the cross-sectional area, including the central opening of 20% and a gap between the baffle and autoclave of 10% in the cross-sectional area. The reference velocity and time scale are $u_0 = 3.5 \times 10^{-5}$ m/s and $t_0 = 131$ s, respectively.

The flow pattern is shown in Fig. 19.13a. There are two flow cells rotating in different directions under the baffle. The flow goes up along the sidewall driven by the buoyancy, which is caused by the high temperature applied on the lower part of the sidewall. Some flow penetrates through the gap between the baffle and the sidewall of the autoclave, and some flow goes inward along the baffle and then downward near the central opening of the baffle. The flow in the porous layer is much weaker than that in the fluid layer. The modified Grashof number can be used to measure the flow strength in the porous charge, $Gr^* = Gr \cdot Da$. In this case, the modified Grashof number is $Gr^* = 98.2$. The fluid flow cannot penetrate the porous layer, and heat and mass transfer in the porous layer are mainly by conduction and diffusion, respectively.

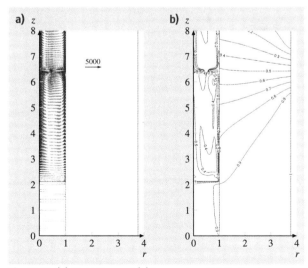

Fig. 19.13 (a) Fluid flow and (b) temperature distribution in an autoclave with internal diameter of 0.932 cm, internal height of 18.4 cm, particle size of 0.6 mm, and $\Delta T = 50$ K (after [19.55])

The temperature distribution is shown in Fig. 19.13b. The temperatures in the porous charge are almost the same as the high temperature T_H applied on the lower part of the sidewall of the autoclave. A large temperature gradient occurs near the fluid–charge interface. The aspect ratio in this case is 40 and the temperatures at $z > 20$ within the autoclave are almost the same as the low temperature T_L applied on the upper part of the sidewall of the autoclave.

19.4.2 Effect of Permeability on the Porous Bed

The optimum precursor sizes found in the ZnO growth experiments have been confirmed by numerical simulations [19.55]. When the charge particle size is increased from 0.6 mm to 3 mm, the Darcy number in the charge increases by 25 times and the modified Grashof number is $Gr^* = 2455$. Figures 19.14a and 19.14b show the flow pattern and temperature distribution in the autoclave, respectively. Significant convective effects are seen occurring in the charge (Fig. 19.14a), and the velocity is large in the porous layer. There are again two flow cells below the baffle. The flow moves upward through the gap between the baffle and the sidewall, and fluid flow is oscillating in the central hole. The flow structure above the baffle is complex and oscillating.

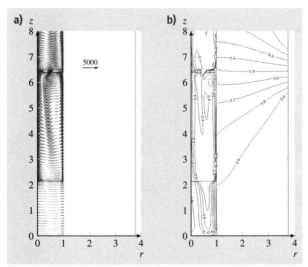

Fig. 19.14 (a) Fluid flow and (b) temperature distribution in a growth system with particle size of 3 mm and $\Delta T = 50$ K. The autoclave has an internal diameter of 0.932 cm and internal height of 18.4 cm (after [19.55])

As can be seen from Fig. 19.14b, the temperature distribution in the charge is in the convection mode, and large temperature gradients appear near the interface between the charge and the sidewall of the autoclave. It is obvious that the particle size is an important factor to consider for successful growth of GaN by the ammonothermal method.

The constraints for ammonothermal growth include dissolving of the charge, nucleation on the sidewall, transfer of nutrient from charge to seed, and growth kinetics. Mass transfer between the charge and the fluid layer is important for successful growth. The flow strength in the fluid layer depends on the Grashof number, which is proportional to the temperature difference on the sidewall and the cube of the internal radius of the autoclave. Flow in the charge layer depends on the product of the Grashof and Darcy numbers, which is proportional to the square of the average diameter of particles. The flow strength in the porous layer is increased by increasing the size of the particles, or by putting particles in bundles as in the hydrothermal growth.

19.4.3 Baffle Design Effect on Flow and Temperature Patterns

The optimization of the baffle design has been performed numerically in [19.62] for the growth system used in [19.64] which has an internal diameter of 0.875 inch (2.22 cm), external diameter of 3 inch (7.62 cm), internal height of 14 inch (35.56 cm), and external height of 15 inch (38.10 cm) (Tem-Press MRA 378R with a volume of 134 ml) [19.64]. The thickness of the sidewall of the autoclave is 1 inch (2.54 cm). The baffle is located at a distance of 6 inch (15.24 cm) from the bottom of the autoclave. The charge particle size is 0.6 mm. $\Delta T = 50$ K is applied on the sidewall and the baffle thickness is 0.28 mm. The aspect ratio, Grashof number, Prandtl number, Darcy number, and Forchheimer number are, $A = 16$, $Gr = 6.0 \times 10^7$, $Pr = 0.73$, $Da = 3.8 \times 10^{-6}$, and $Fs = 1.1 \times 10^{-3}$, respectively. The reference velocity and time scale are $u_0 = 1.4 \times 10^{-5}$ m/s and $t_0 = 748$ s, respectively.

Figure 19.15a shows the flow pattern for a system with a baffle opening of 15% in the cross-sectional area, e.g., 10% in the central hole and 5% in the ring opening between the baffle and the sidewall of the autoclave. The flow is very weak in the porous layer, and the flow in the fluid layer is much stronger. The modified Grashof number, which is used to measure the flow strength in the porous charge, is $Gr^* = 228$. Thus, heat

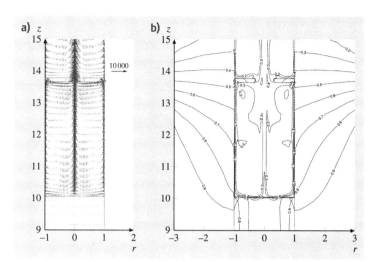

Fig. 19.15 (a) Fluid flow and (b) temperature field in a system with a baffle opening of 15% in the cross-sectional area (central opening of 10% and ring opening of 5%) (after [19.62])

and mass transfer in the porous layer is mainly by conduction and diffusion. This will constrain the nutrient transport between the charge and fluid layer, and cause nutrient deposition on the sidewall of the autoclave near the fluid–charge interface, as observed in experiments.

The temperature distribution is shown in Fig. 19.15b. The charge has a temperature of T_H as applied on the lower part of the autoclave. A large temperature gradient exists at the fluid–charge interface and the fluid–autoclave interface. Supersaturation in the fluid is related to the temperature difference between the charge and the fluid layer. A large temperature gradient at the fluid–charge interface may cause a large supersaturation, and subsequently nucleation near the fluid–charge interface.

The mixing of flow across the baffle has been investigated. Figure 19.16 shows the changes of the vertical velocity at the center of the central hole opening in certain time period. The patterns of oscillations of velocity are repeatable for a longer time period than that

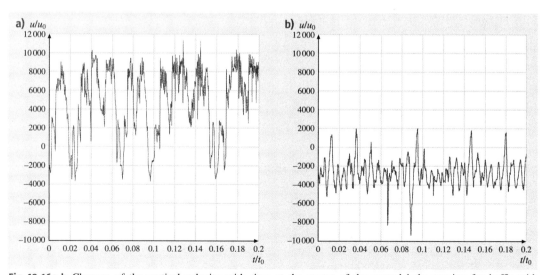

Fig. 19.16a,b Changes of the vertical velocity with time at the center of the central hole opening for baffle with (a) opening of 15% in the cross-sectional area, and (b) opening of 20% (after [19.62])

shown in Fig. 19.16. The heating on the bottom and cooling on the top promote Bénard-type convection in the fluid layer, which interacts with the vertical temperature boundary layer near the sidewall of the autoclave, causing the unsteady and oscillatory flow. For a long time period, it is observed that the amplitude of velocity oscillation in the center of the central hole is larger in the case of the 15% opening (Fig. 19.16a) than in the case of the 20% opening (Fig. 19.16b). The vertical velocity in the center of the central hole changes direction over time in the case of the 15% opening, while it is negative most of the time in the case of the 20% opening. In the case of the 20% opening, the fluid can go up through the ring opening of 10% in the cross-sectional area and return back through the central opening of the same size as the ring opening, so the flow is mixed more thoroughly across the baffle. Oscillation of the vertical velocity in the central hole can be decreased by reducing the difference between the sizes of the central opening and ring opening.

19.4.4 Effect of Porous Bed Height on the Flow Pattern

The influence of the height of the porous bed on transport phenomena in a hydrothermal system was investigated in [19.59]. Since the hydrothermal growth is a very slow process, it can be considered as quasisteady, and the flow and temperature fields for given porous bed height under steady-state condition were obtained in [19.59]. Note that the height of the porous bed decreases as the polycrystalline charge dissolves and the solute moves up for deposition on the seed. Evidently the flow and temperature patterns will change with the porous bed height. Chen et al. [19.59] examined the effect of decreasing height in a case with $A = 3$, $\eta = 0.4$, $\Pr = 1$, $\mathrm{Da} = 10^{-4}$, and $\mathrm{Gr} = 10^5$, where η denotes the ratio of the porous bed height and the overall height.

As shown in Fig. 19.17a, a small portion of the hot surface lies in the fluid region, which promotes a local recirculation on top of the porous layer because of the increased buoyancy effect in this region. The small cell now acts as a buffer restricting the fluid moving from the porous region to the fluid region. It gains fluid from both of the stronger cells and feeds back to them. When the Grashof number is low, the flow may show an axisymmetric pattern with two strong convective rolls and one weak cell. However, the axisymmetric nature of the flow field is completely destroyed at higher Grashof number. A convective roll may then appear in the central region directly above the porous bed and, depending on its location, the large convective rolls in the fluid region may shift. As can be expected, this does not influence the flow and temperature fields in the porous region in any appreciable manner as long as the Darcy number or permeability is low. From growth considerations, a change in flow pattern in the porous bed has no special meaning. However, a major shift in flow characteristics in the fluid region can significantly change the growth dynamics and quality of the crystal. The isotherm pattern is shown in Fig. 19.17b, which is distorted accordingly even though in the central region of the system.

19.4.5 Simulation of Reverse-Grade Soluble Systems

For the ammonothermal growth of GaN with a retrograde solubility, the predetermined amount of GaN particles is loaded above the baffle inside an autoclave (Fig. 19.18). GaN seeds are hung on a wire below the baffle. In this case, the solubility of GaN has a negative coefficient with respect to temperature, so that the growth zone in the lower part is maintained at a higher temperature than that in the upper part. The baffle opening is used to control the mixing of nutrients in two

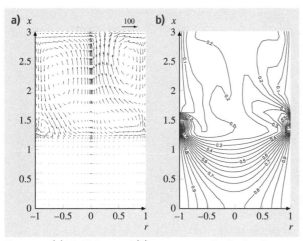

Fig. 19.17 (a) Fluid flow and (b) temperature distribution in a vertical cross-section of the autoclave; $A = 3$, $\eta = 0.4$, $\Pr = 1$, $\mathrm{Da} = 10^{-4}$, and $\mathrm{Gr} = 10^5$ ($\mathrm{Gr}^* = 10$) (after [19.59])

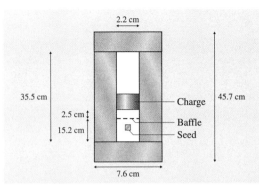

Fig. 19.18 Schematic of an ammonothermal growth system with retrograde solubility (after [19.65])

zones, and cause the transfer of nutrient from the upper part to the lower part.

By using a mineralizer of 2–3 M KNH_2, GaN has a retrograde solubility in ammonia as in [19.64]. The solubility of GaN in ammonia changes from 10% by weight to 2% if temperature increases from 350 °C to 600 °C. With a fill of 60–85% and temperature of 600 °C, pressure is in the range of 25–45 kpsi. A typical run takes 14–21 days and deposition can be observed both on the seeds and on walls of the autoclave.

Numerical studies were performed for an autoclave used in [19.64]. The baffle is located at a distance of 6 inch (15.24 cm) from the bottom of the autoclave. The charge of 4 inch in height is put 1 inch above the baffle, and the charge particle size is 0.6 mm. $\Delta T = 50$ K is applied on the sidewall and the baffle thickness is 0.28 mm.

Figure 19.19a shows the flow pattern for the growth system with retrograde solubility. The baffle openings are 15% of the cross-sectional area, including 10% in the central hole and 5% in the ring opening between the baffle and the sidewall of the autoclave. Since the GaN charge is put above the baffle, the flow is much stronger below the baffle, as shown in Fig. 19.19a. Highly oscillatory flow is observed across the central hole in the baffle. The flow in the central hole first moves downwards, sending nutrient to the growth zone at the bottom of the autoclave, then the flow moves upwards, sending exhausted fluid back to the dissolving zone in the porous layer. This process repeats with time. The opening of 15% causes very large flow oscillation across the baffle in this case.

The temperature distribution in the case of the retrograde solubility is shown in Fig. 19.19b. The temperature difference between the dissolving zone and growth zone is smaller than the temperature difference applied on the sidewall of the autoclave. A larger baffle opening means more fluid mixing across the baffle and less temperature difference between the two zones. In this case, the temperature across the baffle oscillates with time, and the magnitude of the oscillation of temperature is very large for baffle opening of 15%.

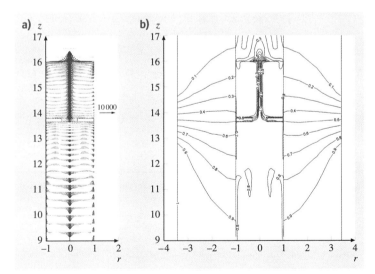

Fig. 19.19 (a) Fluid flow and (b) temperature distribution in an ammonothermal system with a retrograde solubility. Baffle opening is chosen as 15% in cross-sectional area (central opening of 10% and ring opening of 5%) (after [19.65])

19.5 Hydrothermal Growth of ZnO Crystals

19.5.1 Growth Kinetics and Morphology

Figure 19.20 shows the growth planes for hydrothermal zinc oxide crystals. Hydrothermal crystals are highly faceted due to the slow growth rates and lack of confinement during growth. Because each facet has an associated free energy, crystal growth rates can differ for different facets. Hydrothermal ZnO in an alkaline medium grows with the following facets: (0001) and (000$\bar{1}$) monohedra (C^+ and C^- planes, respectively) and the six (10$\bar{1}$0) prismatic faces (M planes). The six (10$\bar{1}$1) pyramid faces (P planes) also can form under certain conditions, which will be describe in Sect. 19.5.2. *Laudise* and *Ballman* first observed the anisotropic growth rate on both spontaneous crystallites and crystals grown on seeds in 1 M NaOH [19.33]. ZnO crystals grown on a C-plane seed and a M-plane seed are shown in Fig. 19.20.

Growth on the C^+ face is always faster than on the C^- face in hydroxide solutions above 1 M. Typical growth rates for 6 M KOH and 1 M LiOH are

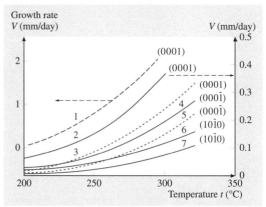

Fig. 19.21 Growth rates of the faces of the monohedra and the {10$\bar{1}$0} prism of ZnO single crystals in alkaline solutions as a function of temperature: (1) 5 M KOH and (2–7) 5.15 M KOH + 1.2 M LiOH as functions of temperature. *Solid lines* correspond to $\Delta t = 75\,°C$; *dashed lines* to $\Delta t = 50\,°C$ (after [19.67])

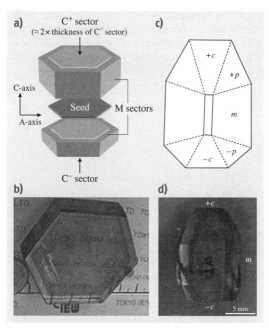

Fig. 19.20a–d Hydrothermal ZnO: (a) schematic and (b) crystal formed by growth on C-plane seed (after [19.13]), (c) schematic and (d) crystal formed by growth on M-plane seed (after [19.66])

0.45 mm/day in the C^+ direction, and 0.22 mm/day in the C^- direction for growth on C-plane seeds [19.24]. Growth rates on M-plane seeds average 0.2 mm/day in the direction normal to the M-plane [19.33]. *Demianets* et al. measured the growth kinetics on the different crystallographic faces by varying the type and concentration mineralizer, growth temperature, and temperature difference between the dissolution and crystallization zones [19.67]. Figure 19.21 shows more detailed kinetics of ZnO growth as a function of temperature. Note the effect that the addition of lithium, which improves the perfection of the ZnO crystal, has in decreasing the growth rate.

The authors went on to determine the elementary surface layers for the possible growth facets of ZnO. The elementary surface layers were then used to determine the relative theoretical growth velocities under ideal conditions (*the absence of any additional components in the crystallization medium*) for the different crystallographic faces of ZnO. The relationships of the velocities are

$$V(10\bar{1}0) < V(000\bar{1}) \sim V(0001)$$
$$< V(10\bar{1}1) < V(10\bar{1}2) < V(1\bar{1}20).$$

The sequence would be reversed to characterize the prevalence of the faces in the formed crystal. The ideal velocities above would form simple shapes such as monohedra and prisms but water, a polar solvent, adds a great deal of complexity to the kinetics of crystal growth.

The anisotropy of the growth rates for the various crystal facets is related to the charge distribution on the facets and the charge of the ions in solution. Several researchers have studied the solubility and thermodynamic parameters of aqueous Zn species for natural hydrothermal systems. *Khodakovsky* and *Yelkin* concluded that $Zn(OH)_4^{2-}$ is the dominant species in alkaline solution at the high temperatures and high pH values at which bulk crystals are grown [19.70]. *Bénézeth* et al. investigated solubility of zinc oxide in 0.03–1.0 M sodium trifluoromethanesulfonate solutions to determine thermodynamic properties of the transport species in dilute acidic and alkaline solutions solutions [19.60, 71]. The Gibbs free energy of formation, entropy, and enthalpy at 25 °C and 1 atm were determined for Zn^{2+}, $Zn(OH^+)$, $Zn(OH)_2^0$, and $Zn(OH)_3^-$ by employing a hydrogen electrode concentration cell and periodic sampling of cell potentials. Solubility data at temperatures up to 200 °C for $Zn(OH^+)$, $Zn(OH)_2^0$, and $Zn(OH)_3^-$, and at temperatures up to 290 °C for Zn^{2+}, were also obtained. The authors concluded that $Zn(OH)_4^{2-}$ was the predominant species in OH^- solutions above 0.1 M NaOH. *Wang* and *Li* et al. performed systematic studies of the morphology and growth rates of zinc oxide powder in alkaline solutions [19.68, 69, 72]. Starting with $Zn(OH)_2$ colloids as nutrient, they investigated morphological changes as a function of pH (Fig. 19.22).

The studies assumed that $Zn(OH)_4^{2-}$ growth units were the predominant species in solution when hydroxide colloids were dissolved. These growth units have a tetrahedral form and charge distribution similar to those of ZnO, which itself is a series of tetrahedra (Fig. 19.23). The corner (point) of a zinc oxide tetrahedron can bind with three hydroxide growth units, the edge with two growth units, and the face of the tetrahedron with only one growth unit. The viscosity of the hydrothermal growth solution is low, so the crystal interface structure plays a large role in kinetics. Thus analysis of Fig. 19.23 would predict the following relative growth rates under ideal conditions

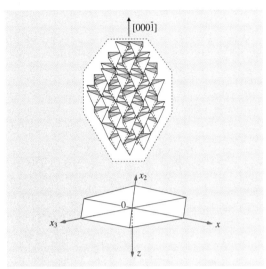

Fig. 19.23 ZnO crystal structure image represented in the form of the coordination tetrahedron along the x-direction (C^- surface made up of flat faces at top) (after [19.69])

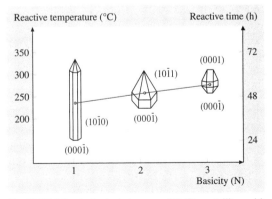

Fig. 19.22 Morphological changes of ZnO crystallites with increasing pH of the growth solution (after [19.68])

$$V\langle 0001\rangle > V\langle 01\bar{1}\bar{1}\rangle > V\langle 0\bar{1}10\rangle$$
$$> V\langle 01\bar{1}1\rangle > V\langle 000\bar{1}\rangle .$$

The hydroxide growth units cluster together by dehydration. In strong alkali solution these clusters are shielded by ions such as Na•O$^-$, shielding the growth units and slowing down growth. Wang et al. contend that these mechanisms account for the growth rates and shape of bulk zinc oxide crystals in strong alkali solutions. *Demianets* and *Kostomarov* proposed a similar mechanism, but argued that $Zn(OH)_4^{2-}$ dissociates into $ZnO_2^{2-} + 2H^+$, and that the ZnO_2^{2-} concentration increases with increasing pH [19.26]. Reaction of one ZnO_2^{2-} with the zinc surface of the crystal allows two ZnO units to form, whereas reaction on the oxygen surface allows only one ZnO unit because of charge compensation.

Addition of lithium to the solution, as hydroxide or carbonate, improves the quality of the bulk zinc oxide crystals but also reduces the growth rate in the (0001) facet while increasing the rate on the ($10\bar{1}0$) facets [19.67, 73]. This may be due to the shielding mechanism mentioned above. *Kuz'mina* et al. grew ZnO crystals grown in KOH solutions that had higher structural quality but more highly faceted than those grown in NaOH solutions [19.27]. *Suscavage* et al. used a 3 M NaOH:1 N KOH:0.1–0.5 N Li$_2$CO$_3$ solution which produced ZnO crystals with low defect densities and less P-plane faceting than KOH-grown crystals [19.21]. The mixed NaOH–KOH solvent had the added benefit of being less corrosive than KOH solutions. The crystal shown on the left-hand side in Fig. 19.20 [19.13] has negligible P-plane faceting and fits the kinetic models of Wang and Demainets discussed before. The low impurity levels in these crystals, high levels of lithium, and growth on fully faceted C-plane seeds may have suppressed formation of the P-plane in these crystals.

Sakagami found that hydrothermal zinc oxide crystals have tens of ppm excess zinc [19.25]. He therefore added H$_2$O$_2$ as an oxidizer; excess zinc was reduced to 1–2 ppm. The addition of an oxidizer such as hydrogen peroxide slows the growth rate on all faces, especially the C$^-$ facet [19.27]. Manganese and nickel had no effect on the kinetics but did color the crystals red and green, respectively. No effect of these dopants on the electrical resistance could be discerned [19.27]. Addition of NH$_4^+$ increased the growth rate on the ($10\bar{1}0$) facets, but crystal quality was degraded [19.27]. *Demianets* et al. published a more detailed paper on the effect of Li$^+$ and several of the divalent and trivalent metals (Co^{2+}, Fe^{2+}, Mn^{2+}, Fe^{3+}, Mn^{3+}, Sc^{3+}, In^{3+}) on growth kinetics and morphology of hydrothermal ZnO [19.67]. Figure 19.20 shows that, as metallic impurities are increased in the ZnO growth medium, P-plane facets are formed and C-axis growth rates decreases. The decrease in growth rates can be explained by the shielding effects impurities can have on the matrix compound (i.e. ZnO) as discussed in Sect. 19.2.

19.5.2 Structural Perfection – Extended Imperfections (Dislocations, Voids, etc.)

Because hydrothermal zinc oxide and quartz are both amphoteric single-component oxides, many insights into zinc oxide hydrothermal can be obtained from studies on hydrothermal quartz growth, which has been intensely investigated during the last 50 years. *Laudise* and *Barnes* [19.74] and *Armington* [19.75] have published excellent reviews on the growth of high-perfection quartz and on dislocation mechanisms. Extended imperfections that can be formed in both zinc oxide and quartz include the following:

- Seed veils and etch channels – small holes or channels filled with voids, water vapor or liquid, caused by etch tracks that form on seeds during initial growth

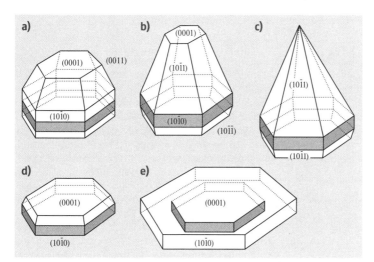

Fig. 19.24a–e Impurity effects on morphology of hydrothermal ZnO crystals (**a**) Li$^+$, (**b,c**) Fe$^+$, (**d**) Mn^{2+}, (**e**) In^{3+} (after [19.67])

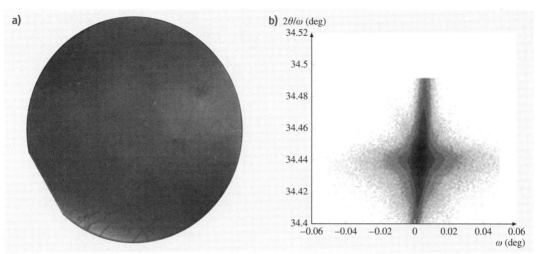

Fig. 19.25 (a) Reflection x-ray topograph of 2 inch ZnO wafer in C-axis projection. (b) (002) reciprocal-space map of corresponding wafer (after [19.13])

- Voids – small holes filled with air, water vapor or liquid; can occur whenever growth conditions change abruptly at the growth surface (impurity clusters, cracks, or crystalline particles from nutrient brought to growth interface by fluid flow)
- Crevice flaws – equivalent to dendritic growth in metals. Uneven or rough growth caused by a change of surface kinetics on growth faces. In extreme cases can cause gaps, cracks, and large numbers of dislocations
- Dislocations – equivalent to those in melt-grown bulk crystals. Strain-induced, because of impurity incorporation or intersection of growth planes, dislocations often propagate from seed into crystal
- Vertical etch channels – equivalent to micropipes that form in vapor-grown crystals. Dislocations decorated with impurities causing cylindrical voids that can reach from the seed to the surface of the crystal

All these imperfections have been observed in hydrothermal zinc oxide bulk crystals. *Laudise* and *Barnes* [19.74] stated that very small nutrient particle size resulted in low growth rates and flawed growth, and an optimal particle of several millimeters was determined, which was confirmed by *Chen* et al. [19.55] by numerical simulation. Addition of lithium, use of low dislocation-density seeds, and use of high purity nutrient also reduces the concentrations of most imperfections. Lithium may reduce imperfections by decreasing the surface free energy when lithium ions incorporate at the growth interface. Lithium may also limit the incorporation of H_2O and OH^- into the crystal lattice at the growth interface [19.76–78].

Figure 19.25 shows a reflection x-ray topograph and the (002) reflection reciprocal-space map for a low-defect commercial 2 inch-diameter (0001) ZnO wafer. Both measurements demonstrate very low defect concentrations in the wafer analyzed [19.13]. Commercial hydrothermal ZnO wafers have measured etch pit densities in the range of $100\,\mathrm{cm}^{-2}$ and the full-width half-maximum (FWHM) of the rocking curve below 20 arcsec, another indication of the high crystallinity of hydrothermal ZnO wafers.

Synchrotron white-beam x-ray topography (SWBXT) in Laue configuration was performed on a series of crystals grown at AFRL-Hanscom [19.79]. To trace the growth history of the crystal, $(10\bar{1}0)$ crystal plates containing both the seed crystal and the bulk region were imaged. One set of topographs showed the usual propagation of edge dislocations from the seeds (Fig. 19.26). The growth sector (GS) boundary can also clearly be seen, marking a change of growth morphology.

On several other crystals the topographs revealed a capping phenomenon similar to that observed in KDP [19.80]. Figure 19.26b shows that the dislocation density is very high near the seed–crystal interface, revealing strain associated with growth initiation rather than dislocations propagating from the seed into the bulk. A growth band, possibly because of a fluctuation

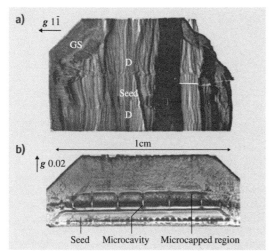

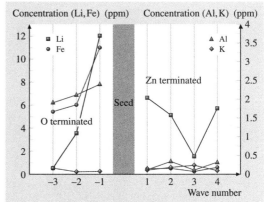

Fig. 19.26a,b Synchrotron white-beam x-ray topographs of ZnO seed–crystal interfaces (after [19.79]): (**a**) normal seed interface (D – dislocations, GS – growth sector boundary); (**b**) ZnO microcapped region, exhibiting microcavities and low dislocation generation above microcapped region in the C^+ direction

Fig. 19.27 Discharge mass spectroscopy (DMS) data for ZnO wafers sliced along the C^- and C^+ growth directions of a commercial 2 inch-diameter boule (after [19.13])

in the growth conditions, stops many of these dislocations. In some cases microcavities originating at the seed–crystal interface were observed; the cavities heal during subsequent growth, nucleating dislocations.

The cause of microcapping has not been established. One possible mechanism is seed etch-back during initial growth, due to fluctuations in the temperature gradient coupled with seed misalignment from C-plane orientation.

Surface studies on hydrothermal crystals showed the C^+ surface to be smooth and specular, with spiral hexagonal growth pyramids [19.27]. The C^- surface is more three-dimensional, with layer-like growth. The M-planes also have layered growth; the P-planes have a series of terraces. No in-depth study has addressed the mechanisms responsible for these various morphologies on the growth surfaces.

19.5.3 Impurities, Doping, and Electrical Properties

Generally the impurity concentrations in hydrothermal ZnO depend not only upon the purity of starting materials, but also upon the growth conditions (solution chemistry, growth temperature, etc.). Fe, Ag, Si, Na, Li, K, and Al are the primary impurities found in hydrothermal ZnO crystals. *Sekiguchi* et al. reported impurity concentration variations among crystals that were grown under various conditions of temperature and pressure but that otherwise were nominally identical [19.66]. Nonetheless, low-ppm impurity and sub-10^{16} cm^{-3} donor/acceptor concentrations have been achieved [19.13, 21], demonstrating that hydrothermal ZnO crystals can have purities that rival or exceed the purities of bulk ZnO grown by other methods – purities, in fact, that rival or exceed those of III–V semiconductors such as InP and GaN. Figure 19.27 shows glow discharge mass spectroscopy (GDMS) data for ZnO wafers sliced along the C^- and C^+ growth directions of a commercial 2 inch-diameter boule [19.13]. High levels of lithium were incorporated in the crystal for growth in both the C^- and C^+ directions. The divalent and trivalent metals levels are two orders of magnitude lower in the wafers grown in the C^+ (Zn-terminated) growth direction.

The resistivity over a 2 inch hydrothermal wafer cut from the C^+ sector (Zn-terminated growth) was measured to be $380\,\Omega\,\text{cm} \pm 15\%$ [19.13].

Lithium is often added to hydrothermal solutions as hydroxide or carbonate because it improves crystallinity and morphology, as stated in the previous section. Li can therefore occur in concentrations of >10 ppm in hydrothermal ZnO [19.13, 21, 25, 66]; it has been employed to achieve resistivities as high as $10^{10}\,\Omega\text{cm}$ by compensating native donors [19.18]. Lithium is anathema to most electronic and optical device fabricators, who fear that Li – typically a fast diffuser – will incorporate into devices and thereby *poison* them. This may not be an insurmountable obstacle for device applications, as Li apparently can be removed from ZnO by annealing in a zinc atmosphere [19.77]. Also, using appropriate mineralizer solutions, it is possible to obtain high-quality as-grown hydrothermal ZnO crystals that have sub-ppm Li concentrations [19.21].

In, Ga, and Al are shallow donors in ZnO [19.82, 83]. As mentioned above, Li occupying the Zn site is believed to be an acceptor (interstitial Li is believed to be a donor [19.77]); addition of Li or Cu increases the resistivity of ZnO after annealing in air or Zn [19.22,24,76,77], probably by compensating donors. The donor/acceptor properties of Fe^{2+} and Fe^{3+} in ZnO are not known. The role of hydrogen in ZnO is controversial: recent theoretical calculations predict it should be a shallow donor [19.84], in overall agreement with experimental measurements performed in the 1950s that associated increases in electrical conductivity with hydrogen incorporation (reviewed in [19.85]); however, in recent work on MOCVD-grown ZnO films, the conductivity increase was attributed to passivation of acceptors [19.86]. Hydrothermal growth of ZnO in an effective overpressure of H_2 was achieved by adding Zn powder to the growth solution; unfortunately only the carrier concentrations after annealing in vacuum or air ($2-5 \times 10^{15}$ cm^{-3}), not carrier concentrations in as-grown ZnO, were reported [19.77].

Native defects such as oxygen vacancies or zinc interstitials have long been regarded as donor centers in ZnO (see, e.g., [19.84, 87] and references therein). In many cases, they may constitute the most numerous donor sites. *Sakagami*'s observation that the ZnO electrical resistance increased when an oxidizing agent was added to the hydrothermal growth solution (equivalent to growth in an oxygen overpressure) [19.25] is indirect evidence that many donor defects result from imperfect zinc–oxygen stoichiometry.

Semi-insulating behavior in hydrothermal ZnO has, as already noted, been achieved through lithium doping [19.83] and growth in an effective oxygen overpressure [19.25]. Semi-insulating behavior, with a net room-temperature free electron concentration of $\approx 2 \times 10^{12}$ cm^{-3}, and electron mobility of ≈ 175 cm^2/V s, was observed [19.21]. The cause of this behavior was found to be a donor center located 340 meV below the conduction band; the microscopic nature of this donor is not understood.

The presence of large growth facets on hydrothermal ZnO crystals facilitates study of electronic properties as a function of crystallographic orientation and surface polarity. *Urbieta* et al. employed scanning

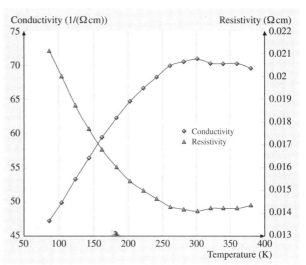

Fig. 19.28 Hall-effect data for indium-doped ZnO wafer (after [19.81])

tunneling microscopy and found clear differences in surface electronic structure that distinguished the C$^+$ (zinc-surface), C$^-$ (oxygen-surface), and m (nonpolar surfaces) [19.88]. *Sakagami* et al. measured I–V and C–V characteristics on the crystallographic C$^+$, C$^-$, and m faces [19.89]. Along all crystallographic axes studied, they observed nearly ohmic behavior when surfaces were zinc-rich and rectifying behavior when surfaces were oxygen rich. The authors' judgment that m-sectors are more suitable than c^+ and c^- sectors for making electrical contacts, if confirmed, has potential significance for device fabrication that could stimulate interest in ZnO crystal growth on nonbasal plane.

Hydrothermal conducting indium-doped ZnO was grown using sintered zinc oxide powder that was mixed with a small percentage indium oxide [19.81]. As previously stated, growth rates were dramatically reduced in the C-axis (Fig. 19.23c). Conductivity measurements for a 5×5 mm^2 In:ZnO sample is shown in Fig. 19.28. The conductivity achieved is adequate for most semiconductor device applications that would benefit from conducting substrates.

19.5.4 Optical Properties

The most sensitive indications of crystal quality often come from optical measurements at near-liquid-helium temperatures, where excitonic and other bands indicate the underlying quality (or lack thereof) of the material.

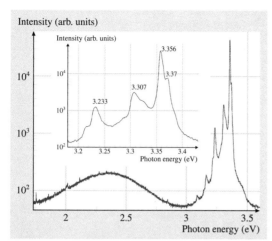

Fig. 19.29 Low-temperature broadband photoluminescence spectrum (after [19.13])

Broadband spectra at both high and low temperatures can provide useful information about impurities and native defects.

The broadband low-temperature photoluminescence (PL) spectrum of a 2 inch ZnO wafer is shown in Fig. 19.29. In addition to the UV band (excitonic and donor–acceptor bands at energies above 3 eV), there is a broad band centered at ≈ 2.3 eV. Similar broad extrinsic *orange* and *green* bands have been reported in several PL [19.90] and cathodoluminescence (CL) [19.91, 92] measurements of hydrothermal grown ZnO. Several authors have shown that the intensity of the UV band compared with the intensity of the broadband can provide qualitative information on the crystal quality of hydrothermal ZnO [19.92, 93].

Conflicting mechanisms have been advanced for the origin of the green band. *Reynolds* et al. compared the ZnO green band to the *yellow band* in GaN (a wurtzitic crystal having a bandgap similar to that of ZnO), whose origin remains a matter of debate, and concluded that both bands arise from a transition between a shallow donor and a deep level [19.94]. However, *Garces* et al. concluded, from electron paramagnetic resonance (EPR) and PL spectra of unannealed and annealed vapor-phase-grown ZnO, that ZnO green bands are caused by emission from Cu^+ and Cu^{2+} ions, respectively [19.82].

Orange and green CL has been observed in the m, c, and p sectors of hydrothermal ZnO crystals. Strong green emission (in addition to near-band-edge emission, which is ignore for the purposes of this discussion) was observed from C^+ (Zn-terminated) surfaces, and weak orange emission from C^- (oxygen-terminated) surfaces; green emission was observed from P^- surfaces (Zn-terminated) and orange emission from P^+ (oxygen terminated) surfaces; and there was weak orange emission from the (nonpolar) M face [19.66]. Combining these and related observations with the growth kinetic models, *Sekiguchi* et al. [19.66] and *Urbieta* et al. [19.95, 96] associated the occurrence of ZnO green and orange CL bands with impurity incorporation efficiencies (during crystal growth) that the growth model attributes to the polarization state of ZnO sectors (c^+, c^-, m, etc.); similarly, UV and broadband luminescence efficiencies were associated with presumed incorporation rates of nonradiative recombination centers. In this vein, *Sekiguchi* et al. noted that orange emission was strongest in their sample that had the highest Li concentration and lower in a flux-grown crystal that contained virtually no Li; they also noted that use of H_2O_2 in the hydrothermal growth solution, which presumably lowered the concentration of oxygen vacancies, significantly reduced the orange emission [19.66]. An overview of visible luminescence in ZnO has more comprehensive information of the possible mechanisms

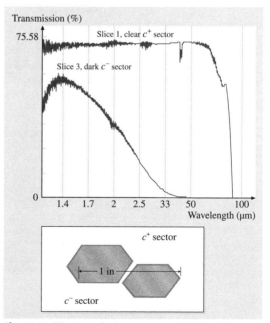

Fig. 19.30 IR transmission spectra of ZnO slices cut from c^+ and c^- growth sectors of a ZnO crystal (after [19.21])

of broadband emission in ZnO bulk crystals and thin films [19.97].

We complete our summary of the optical properties of hydrothermal ZnO by considering transmission spectra (Fig. 19.30) [19.98]. Insulating ZnO is transparent from the near-ultraviolet almost to 10 μm (the spectrum in Fig. 19.30 labeled *clear c^+ sector*). Electrically conducting samples (e.g., the spectrum labeled *dark c^- sector*) exhibit a long-wavelength free-carrier absorption tail. The optical transparency shown in Fig. 19.30, together with the high laser breakdown strength of ZnO [19.98], have made ZnO the leading candidate for transparent conducting electrodes for high-power near-IR laser beam steering devices [19.99].

19.6 Ammonothermal GaN

19.6.1 Alkaline Seeded Growth

Callahan et al. [19.64] synthesized and grew GaN crystals in a retrograde configuration (Fig. 19.8) with 2–3 M KNH_2 concentration. Seeded growth experiments employed autoclaves with 2.2 cm internal diameter and 140 ml volume. Both sets of autoclaves are capable of sustained operation at 600 °C and pressures up to 4 kbar.

Polycrystalline GaN, synthesized by an in-house vapor process [19.100], was suspended in a scaffold near the bottom of the autoclave to obtain solubility data. Single-crystal *free-standing* gallium nitride up to 200 μm thick with a surface area of 1 cm² grown by hydride vapor-phase epitaxy (HVPE) was used as seeds and suspended by a wire scaffold near the bottom (warmer portion) of the autoclave. Seeded growth experiments were carried out for 5–30 days. Transported material nucleates on the walls of the autoclave, the wire scaffold, and on the GaN HVPE seeds. The mass of heterogeneously nucleated materials recovered from the lower walls (crystallization zone) is often eight to ten times the weight gain of the seeds.

Single-crystal growth on the seeds occurred at a rate of up to 40 μm per day. Growth approaching 1 mm in thickness was achieved on multiple seeds. Etching studies were conducted to determine the polarity of the crystalline surfaces. Previous work [19.101] indicates that the C^+ face (gallium face) when exposed to phosphoric acid etchants shows little erosion of the bulk material. The nitrogen polar face (Fig. 19.31a) has flatter surface than the gallium polar face (Fig. 19.31b). A cleaved cross-section of as-grown gallium nitride is shown in Fig. 19.31c. Growth in the C^+ and C^- directions (gallium face and nitrogen face, respectively) exhibits a columnar-type growth, with much more pronounced grain boundaries seen in the growth on the gallium face. Under different experimental conditions growth rates of the C^+ and C^- planes differ, but are approximately equal. Substantial growth in the a-direction has also been observed.

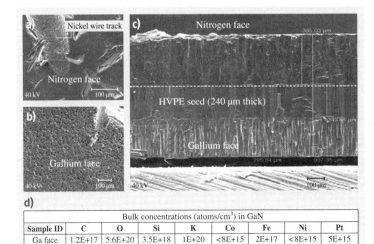

Fig. 19.31a–d Thick ammonothermal GaN growth (after [19.64]) (**a**) SEM image of nitrogen polar face. (**b**) SEM image of gallium polar face. (**c**) SEM image of a cleaved cross-section of gallium nitride grown on HVPE seed. (**d**) Impurity concentrations of Ga and nitrogen faces measured by secondary ion mass spectrometry (SIMS)

	Bulk concentrations (atoms/cm³) in GaN							
Sample ID	C	O	Si	K	Co	Fe	Ni	Pt
Ga face	1.2E+17	5.6E+20	3.5E+18	1E+20	<8E+15	2E+17	<8E+15	5E+15
N face	2.5E+16	2.5E+20	7.9E+17	3E+16	2E+16	4E+16	<8E+15	<5E+16

Secondary ion mass spectrometry (SIMS) were performed on as-grown ammonothermal gallium nitride crystal samples. The results from a characteristic sample are shown in Fig. 19.31d. Impurities do not incorporate into the bulk crystal homogeneously. Concentrations of impurities were different between the gallium and nitrogen surfaces of the bulk crystal. In general metallic impurities incorporated less on the nitrogen face than the gallium face. Figure 19.31d shows that the impurity levels for oxygen and hydrogen are above 10^{19} atoms per cm^3 for both faces. Potassium impurities for the mineralizer were 10^{19} cm^3 atoms per cm^3 on the gallium face and mid 10^{18} cm^3 atoms per cm^3 on the nitrogen face.

The reciprocal-space maps and reflection topographs are shown in Fig. 19.32 for an ammonothermal crystal grown under similar conditions as those in Fig. 19.31 [19.102].

It is evident that the ammonothermal growth has higher defects levels than the HVPE seed. When a thinner film of $\approx 50\,\mu$ was grown [19.103] there was less columnar growth. The levels of impurities in the crystals and the nonuniform crystallinity of the HVPE seeds make quantitative analysis difficult.

The ammonothermal growth of 1 inch-size (0001) GaN crystal in a cylindrical high-pressure autoclave having an internal diameter of 40 mm was reported by *Hashimoto* et al. [19.17, 104]. About 15 μm-thick GaN films were uniformly grown on each side of the GaN seed, which has an oval shape of $3 \times 4\,cm^2$. The applied temperature and pressure were 625–675 °C and about 2.14 kbar. Basic mineralizers were used, resulting in retrograde solubility of GaN in supercritical ammonobasic solutions. The nutrient was placed in the colder region (upper region) and free-standing C-plane HVPE GaN seed crystals were placed in the hotter region (lower

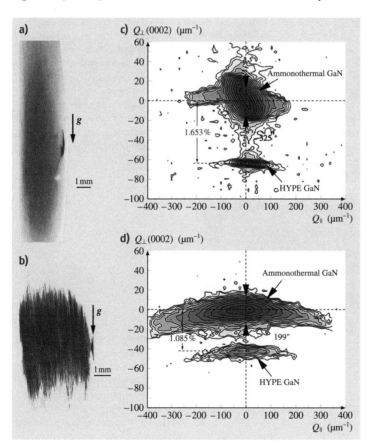

Fig. 19.32a–d Reciprocal-space maps and reflection topographs for ammonothermal GaN and HVPE seed grown with KNH_2 mineralizer (after [19.102])

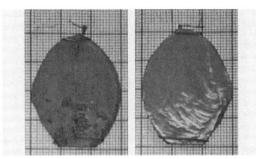

Fig. 19.33 Ammonothermal growth on 40×30 mm^2 HVPE seed (after [19.17, 104])

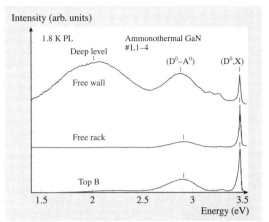

Fig. 19.34 Photoluminescence spectra of ammonothermal GaN (note the difference in broadband emission) (after [19.105])

region). A baffle divides the reactor into an upper region and a lower region to set a temperature difference between the dissolving region and the crystallization region. Uniform growth of GaN films on an over-1 inch oval-shaped seed crystal was achieved through fluid transport of Ga nutrient. Ga nutrient was transformed to GaN on the crucible wall, resulting in abrupt drop of the growth rate in about a day. The crystal is shown in Fig. 19.33.

Polycrystalline GaN nutrient and higher mineralizer concentration was used for longer growth runs. Due to thick wall of the autoclave, the temperature difference between the upper and lower regions of the fluid is estimated to be less than 50 °C. The resulting ammonia pressure was about 1.8–1.9 kbar. The growth rates along C$^+$ (Ga face), C$^-$ (N face) and M direction were 0.8, 3.6, and 6 μm/day. The authors showed a defective growth interface with numerous voids and defects; as growth progressed the structure improved [19.106].

Dwilinski et al. recently reported on GaN crystals grown in alkaline ammonia with a dislocation density on the order of 5×10^3 cm^{-2} [19.107], which is several orders of magnitude lower than what has previously been reported. An x-ray rocking curve FWHM of 17 arcsec was measured on a 1 inch substrate cut from one of the GaN crystals and subsequently polished. There was no evidence of mosaicity or low-angle grain boundaries. X-ray rocking curves under 20 arcsec and radius of curvature on the order of 10^2–10^3 meters were measured on various crystals from separate growth runs. The authors did not disclose how the superior crystallinity of the GaN crystals was obtained.

Figure 19.34 is unpublished data [19.105] that shows photoluminescence (PL) of an alkaline ammonia sample, rack, and wall nucleation. The PL is improved over previous experiments. The broadband emission is reduced over previous runs [19.103] and there is no evidence of a PL signature at 3.4 eV that was identified as stacking-fault-related emission [19.108]. The wall nucleation has a greater number of broadband emissions. This might be evidence of impurities leaching from the autoclave.

19.6.2 Acidic Seeded Growth

Due to the corrosive nature of acidic solvents a platinum or silver liner must be used to contain the experiment. Figure 19.35 shows a scanning electron microscopy (SEM) image of acidic ammonothermal growth at a relatively low pressure of ≤ 170 MPa and temperature in the range 500–550 °C. The film was grown in standard configuration (forward-grade solubility). The authors increase the temperature and obtained columnar growth.

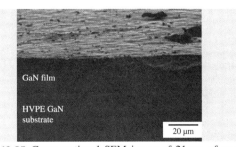

Fig. 19.35 Cross-sectional SEM image of 21 μm of ammonothermal growth on a HVPE seed using NH$_4$Cl as a mineralizer (after [19.43])

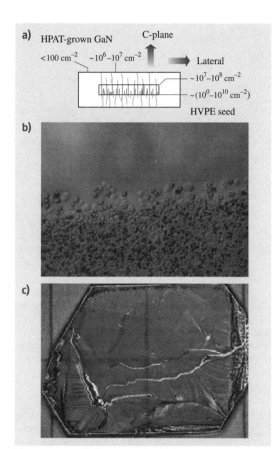

Fig. 19.36a–c GaN grown in high-pressure anvil cell at 700 °C; (**a**) schematic of growth; (**b**) SEM of low-defect/high-defect seed interface in the m-plane direction; (**c**) picture of large HVPE seed with ammonothermal growth showing smooth morphology (after [19.109])

Figure 19.36 shows a GaN crystal grown in a sealed capsule that is placed in a high-pressure cell with a solid pressure medium [19.109]. The cell is similar to the type used for piston-press and belt-press synthesis of diamond. The cell was heated between 600–1000 °C with a pressure in the range 5–20 kbar. The authors show a standard forward-grade solubility configuration for growth. The higher pressure and temperatures that this system can withstand allow for much higher growth rates than those that can be obtained in autoclaves. Figure 19.36a shows a schematic of the dislocation density of a crystal grown by the high-pressure ammonothermal technique (HPAT). Growth in the direction perpendicular to the HVPE seed was almost free of dislocations

and other defects. This shows the advantage of using the ammonothermal technique to obtain very low-defect material. The impurities in the crystals grown were reduced by two orders of magnitude (10^{20}–10^{18}) from a previous publication. The authors fabricated a functional laser diode on one of the HPAT substrates.

19.6.3 Doping, Alloying, and Challenges

There has been synthesis of InGaN nanocrystals [19.110], polycrystalline AlN [19.111], and GaN microcrystals doped with transition-metal ions (Mn, Fe, and Cr) for dilute magnetic semiconductor applications [19.112]. Transition metals could also be used to increase resistance in GaN wafers. Finally, AlGaN microcrystalline balls were synthesized ammonothermally [19.54]. The Al metal nutrient was depleted before the end of the run. The result was two different compositions of AlGaN in the inner and outer portions of the ball, as seen in Fig. 19.37. It is surmised that an AlGaN alloy liquid droplets formed during the heating process and the AlGaN crystallized around the droplets.

Several challenges need to be overcome with ammonothermal technology in order to bring GaN-based crystal growth from research into full production. High-quality seeds, nutrient, equipment, and a complete understanding of the chemistry are required. Ammonothermal growth will leverage other growth technologies such as flux growth and HVPE growth for sources of seeds and nutrient. Electrical and optical characterization will proceed as large-area ammonothermal GaN crystals become available. *Denis* et al. published an excellent review of GaN bulk growth [19.113]. This technology is still in its infancy and there are many unique avenues to be explored.

Fig. 19.37 Ammonothermal AlGaN balls (after [19.54]): 1 – $Al_{0.5}Ga_{0.5}N$; 2 – $Al_{0.25}Ga_{0.75}N$; nutrient – molar ratio of Al : Ga = 3 : 7 (Al nutrient was depleted during course of experiment)

19.7 Conclusion

Hydrothermal growth has produced very high-quality zinc oxide crystals. Nearly dislocation-free growth can be produced in zinc-terminated growth sectors and on nonpolar prism faces. Semi-insulating and conducting crystals have been grown, as well as crystals exhibiting superior low-temperature PL characteristics.

The various crystallographic faces of hydrothermal ZnO exhibit pronounced differences not only in the growth rates, but also in structural, electrical, and optical properties. The growth mechanisms, the incorporation of impurities, and the generation of native defects are (at least in part) connected to the polarity of the growth surfaces; however, the precise mechanisms – and especially the properties of impurities and native defects – have not been fully elucidated. Exploiting the faceted morphology of hydrothermal crystals may facilitate studies of these phenomena, many of which surely have counterparts in ZnO thin-film growth and ZnO bulk growth by other methods. Hydrothermal ZnO technology has progressed rapidly in the last 5 years with the capability to produce hundreds of wafers in a single run. Therefore economical growth of large, low-defect-density ZnO crystals is waiting for innovative strategies in ZnO device design and fabrication to spur demand for large-scale commercialization of ZnO-based devices and therefore ZnO substrates.

Significant progress has been made in acidic and alkaline ammonothermal growth of GaN. Forward solubility in ammonia has been determined when using acidic mineralizers such as NH_4Cl, in contrast with alkaline mineralizers such as KNH_2, where a retrograde solubility was discovered. Several groups have demonstrated that low-defect-density GaN can be produced using the ammonothermal technique, and growth on 1–2 inch-diameter GaN HVPE templates has been shown.

Process modeling and simulation based on physics are necessary in the scale-up of solvothermal systems. Forward solubility has been simulated for an autoclave with an internal diameter of 0.932 cm and internal height of 18.4 cm. The optimum precursor sizes found in the GaN growth experiments have been confirmed by numerical simulations. The fluid field can be significantly increased by an increase of the particle size from 0.6 mm to 3 mm. The optimization of the baffle design has been performed numerically and needs to be carefully selected to achieve a predefined temperature difference across the baffle, to achieve the necessary supersaturations required for the crystallization of single crystals at elevated growth rates.

Ammonothermal technology is a fertile ground for the production of many nitrogenated compounds that are difficult to synthesize by other techniques. The production of gallium nitride by the ammonothermal method requires a large amount of further investigation just as the production of quartz did in the early 20th century. In-depth phase, solubility, growth kinetics, and fluid dynamics studies need to be completed for a multiplicity of mineralizers, and high-strength vessels must be designed for the increased pressure requirements of supercritical ammonia. The cost advantages afforded by growing on multiple seeds simultaneously make this technique an attractive means for meeting the demand for high-quality gallium nitride substrates.

References

19.1 F. Bernardini, V. Fiorentini, D. Vanderbilt: Spontaneous polarization and piezoelectric contants of III-V nitrides, Phys. Rev. B **56**(R10), 24–27 (1997)

19.2 S.J. Pearton, C.R. Abernathy, F. Ren: *Gallium Nitride Processing for Electronics, Sensors and Spintronics* (Springer, Berlin Heidelberg 2006)

19.3 B. Gil: *Low-Dimensional Nitride Semiconductors* (Oxford Univ. Press, Oxford 2002)

19.4 C. Jagadish, S.J. Pearton: *Zinc Oxide, Thin Films and Nanostructions: Processing, Properties, and Applications* (Elsevier Science, Amsterdam 2006)

19.5 Ü. Özgür, Y.I. Alivov, C. Liu, A. Teke, M.A. Reshchikov, S. Doğan, V. Avruntin, S.-J. Cho, H. Morkoç: A comprehensive review of ZnO material and devices, Appl. Phys. Rev. **98**(041301), 1–103 (2005)

19.6 I. Akasaki: Key inventions in the history of nitride-base blue LED and LD, J. Cryst. Growth **300**, 2–10 (2007)

19.7 C. Klingshirn, R. Hauschild, H. Priller, M. Decker, J. Zeller, H. Kalt: ZnO rediscovered – once again!?, Superlattice Microstruct. **38**, 209–222 (2005)

19.8 H.J. Scheel: Historical aspects of crystal growth technology, J. Cryst. Growth Technol. **211**, 1–12 (2000)

19.9 J. Nause, B. Nemeth: Pressurized melt growth of ZnO boules, Semicond. Sci. Technol. **20**, S45–S48 (2005)

19.10 J. Karpiński, J. Jun, S. Porowski: High pressure thermodynamics of GaN, J. Cryst. Growth **66**, 1–10 (1984)

19.11 H.D. Sun, Y. Segawa, M. Kawasaki, A. Ohtomo, K. Tamura, H. Koinuma: Phonon replicas in ZnO/ZnMgO multiquantum wells, J. Appl. Phys. Lett. **91**(10), 6450–6457 (2002)

19.12 D.C. Look, D.C. Reynolds, J.R. Sizelove, R.L. Jones, C.W. Litton, G. Cantwell, W.C. Harsch: Electrical properties of bulk ZnO, Solid-State Commun. **105**, 399–401 (1998)

19.13 D. Ehrentraut, H. Sato, Y. Kagamitani, H. Sato, A. Yoshikawa, T. Fukuda: Solvothermal growth of ZnO, Prog. Cryst. Growth Charact. Mater. **52**, 280–335 (2006)

19.14 D.A. Kramer: Nitrogen (fixed) Ammonia. In: *US Geological Survey*, ed. by US Department of the Interior (United States Government Printing Office, Washington 2005) pp. 116–117

19.15 K. Byrappa, M. Yoshimura: *Handbook of Hydrothermal Technology* (William Andrew, New York 2001)

19.16 M.T. Harris, J.J. Larkin, J.J. Martin: Low-defect colorless $Bi_{12}SiO_{20}$ grown by hydrothermal techniques, Appl. Phys. Lett. **60**, 2162–2163 (1992)

19.17 T. Hashimoto, K. Fujito, M. Saito, J.S. Speck, S. Nakamura: Ammonothermal growth of GaN on an over-1-inch seed crystal, Jpn. J. Appl. Phys. **44**, L1570–L1572 (2005)

19.18 R.A. Laudise: *Hydrothermal Growth in The Growth of Single Crystals* (Prentice-Hall, New Jersey 1970) pp. 275–293

19.19 H. Jacobs, D. Schmidt: High-pressure ammonolysis in solid-state chemistry. In: *Current Topics in Materials Science*, Vol. 8, ed. by E. Kaldis (North Holland, Amsterdam 1982) pp. 381–427

19.20 B. Wang, M.J. Callahan: Ammonothermal synthesis of III-nitride crystals, Cryst. Growth Des. **6**(6), 1227–1246 (2006)

19.21 M. Suscavage, M. Harris, D. Bliss, P. Yip, S.Q. Wang, D. Schwall, L. Bouthillette, J. Bailey, M. Callahan, D.C. Look, D.C. Reynolds, R.L. Jones, C.W. Litton: High quality ZnO crystal, Mater. Res. Soc. Symp. Proc. **537**, 294–299 (1999)

19.22 R.R. Monchamp, R.C. Puttbach, J.W. Nielson: Hydrothermal growth of ZnO crystals (Airtron Division of Litton Industries, Morris Plains, technical report AFML-TR-67-144 1967)

19.23 R.A. Laudise, E.D. Kolb: The solubity of zincite in basic hydrothermal solvents, Am. Mineral. **48**(3), 642–648 (1963)

19.24 D.F. Croxall, R.C.C. Ward, C.A. Wallace, R.C. Kell: Hydrothermal growth and investigation of Li-doped zinc oxide crystals of high purity and perfection, J. Cryst. Growth **22**, 117 (1974)

19.25 N. Sakagami: Hydrothermal growth and characterization of ZnO single crystals of high purity, J. Cryst. Growth **99**, 905–909 (1990)

19.26 L. Demianets, D. Kostomaro: Mechanism of zinc oxide single crystal growth under hydrothermal conditions, Ann. Chim. Sci. Mater. **26**(1), 193–198 (2001)

19.27 I.P. Kuz'mina, A.N. Lobachev, N.S. Triodina: *Synthesis of Zincite by the Hydrothermal Method in Crystallization Process Under Hydrothermal Conditions* (Nauka, Moscow 1973) pp. 27–41

19.28 T. Fukuda, D. Ehrentraut: Prospects for the ammonothermal growth of large GaN crystals, J. Cryst. Growth **305**, 304–310 (2007)

19.29 E.V. Kortunova, P.P. Chvanski, N.G. Nikolaeva: The first attempts of industrial manufacture of ZnO single crystals, J. Phys. IV France **126**, 39–42 (2005)

19.30 L.E. McCandlish, R. Urhin: Mild conditions for hydrothermal growth of ZnO with potential for p-type semiconductor behavior, Poster Presentation at 5th Int. Conf. Solvotherm. React. Conf (East Brunswick, 2002), image supplied directly by L.E. McCandlish

19.31 G.F. Hüttig, H. Möldner: The specific heat of crystallized zinc hydroxide and calculation of the affinities between zinc oxide and water, Z. Anorg. Chem. **211**, 368–378 (1933)

19.32 C.H. Lu, C.H. Yeh: Influence of hydrothermal conditions on the morphology and particle size of zinc oxide powder, Ceram. Int. **26**, 351–357 (2000)

19.33 R.A. Laudise, A.A. Ballman: Hydrothermal synthesis of zinc oxide and zinc sulfide, J. Phys. Chem. **64**(5), 688–691 (1960)

19.34 M.M. Lencka, R.E. Riman: Synthesis of lead titanate: thermodynamic modeling and experimental verification, J. Am. Ceram. Soc. **76**, 2649–2659 (1993)

19.35 M.M. Lencka, A. Anderko, R.E. Riman: Hydrothermal precipitation of lead zirconate titanate solid solutions: thermodynamic modeling and experimental synthesis, J. Am. Ceram. Soc. **78**, 2609–2618 (1995)

19.36 M.M. Lencka, R.E. Riman: Themodynamic modeling of hydrothermal synthesis of ceramic powders, Chem. Mater. **5**, 61–70 (1993)

19.37 R. Dwilinski, R. Doradzinski, J. Garczynski, L. Sierzputowski, J.M. Baranowski, M. Kaminska: AMMONO method of GaN and AlN production, Diam. Relat. Mater. **7**, 1348–1350 (1998)

19.38 A.P. Purdy: Ammonothermal sytheis of cubic gallium nitride, Chem. Mater. **11**, 1648–1651 (1999)

19.39 D.R. Ketchum, J.W. Kolis: Crystal growth of gallium nitride in supercritical ammonia, J. Cryst. Growth **222**, 431–434 (2001)

19.40 A. Yoshikawa, E. Ohshima, T. Fukuda, H. Tsuji, K. Oshima: Crystal growth of GaN by ammonothermal method, J. Cryst. Growth **260**, 67–72 (2004)

19.41 Y.C. Lan, X.L. Chen, M.A. Crimp, Y.G. Cao, Y.P. Xu, T. Xu, K.Q. Lu: Single crystal growth of gallium nitride in supercritical ammonia, Phys. Status Solidi (c) **2**(7), 2066–2069 (2005)

19.42 B. Wang, M.J. Callahan, K. Rakes, D.F. Bliss, L.O. Bouthillette, S.-Q. Wang, J.W. Kolis: Am-

19.43 monothermal growth of GaN crystals in alkaline solutions, J. Cryst. Growth **287**, 376–380 (2006)

19.43 Y. Kagamitani, D. Ehrentraut, A. Yoshikawa, N. Hoshino, T. Fukuda, S. Kawabata, K. Inaba: Ammonothermal epitaxy of thick GaN film using NH_4Cl mineralizer, Jpn. J. Appl. Phys. **45**(5A), 4018–4020 (2006)

19.44 D. Peters: Ammonothermal synthesis of aluminium nitride, J. Cryst. Growth **104**, 411–418 (1990)

19.45 B. Wang, M.J. Callahan: Transport growth of GaN crystals by the ammonothermal technique using various nutrients, J. Cryst. Growth **291**, 455–460 (2006)

19.46 A.P. Purdy, R.J. Jouet, F.G. Clifford: Ammonothermal recrystallization of gallium nitride with acidic mineralizers, Cryst. Growth Des. **2**(2), 141–145 (2002)

19.47 D. Ehrentraut, N. Hoshino, Y. Kagamitani, A. Yoshikawa, T. Fukuda, H. Itoh, S. Kawabata: Temperature effect of ammonium halogenides as mineralizers on the phase stability of gallium nitride synthesized under acidic ammonothermal conditions, J. Mater. Chem. **17**, 886–893 (2007)

19.48 B. Raghothamachar, W.M. Vetter, M. Dudley, R. Dalmau, R. Schlesser, Z. Sitar, E. Michael, J.W. Kolis: Synchrontron white beam topography charctrization of physical vapor transport grown AIN and ammonothermal GaN, J. Cryst. Growth **246**, 271–280 (2002)

19.49 A.P. Purdy, S. Case, N. Murastore: Synthesis of GaN by high-pressure ammonolysis of gallium triiodide, J. Cryst. Growth **252**, 136–143 (2003)

19.50 T. Hashimoto, K. Fujito, R. Sharma, E.R. Letts, P.T. Fini, J.S. Speck, S. Nakamura: Phase selection of microcrystalline GaN synthesized in supercritical ammonia, J. Cryst. Growth **291**, 100–106 (2006)

19.51 A. Purdy: Growth of cubic GaN crystals from hexagonal GaN feedstock, J. Cryst. Growth **281**, 355–363 (2005)

19.52 A.N. Mariano, R.E. Hanneman: Crystallographic polarity of ZnO crystals, J. Appl. Phys. **34**, 384–389 (1963)

19.53 B. Wang, M.J. Callahan, L.O. Bouthillette: Hydrothermal growth and photoluminescence of $Zn_{1-x}Mg_xO$ alloy crystals, Cryst. Growth Des. **6**, 1256–1260 (2006)

19.54 M. J. Callahan, B. Wang, unpublished results

19.55 Q.-S. Chen, V. Prasad, W.R. Hu: Modeling of ammonothermal growth of nitrides, J. Cryst. Growth **258**, 181–187 (2003)

19.56 M. Carr: Penetrative convection in a superposed porous-medium-fluid layer via internal heating, J. Fluid Mech. **509**, 305–329 (2004)

19.57 V. Prasad: Convective flow interaction and heat transfer between fluid and porous layers. In: *Convective Heat and Mass Transfer in Porous Media*, ed. by S. Kakaç, B. Kilkiş, F.A. Kulacki, F. Arinç (Kluwer, Netherlands 1991) pp. 563–615

19.58 Q.-S. Chen, V. Prasad, A. Chatterjee, J. Larkin: A porous media-based transport model for hydrothermal growth, J. Cryst. Growth **198/199**, 710–715 (1999)

19.59 Q.-S. Chen, V. Prasad, A. Chatterjee: Modeling of fluid flow and heat transfer in a hydrothermal crystal growth system: use of fluid-superposed porous layer theory, J. Heat Transf. **121**, 1049–1058 (1999)

19.60 H. Zhang, V. Prasad, M.K. Moallemi: Numerical algorithm using multizone adaptive grid generation for multiphase transport processes with moving and free boundaries, Num. Heat Transf. **29**(B), 399–421 (1996)

19.61 H. Zhang, V. Prasad: An advanced numerical scheme for materials process modeling, Comput. Model. Simul. Eng. **2**, 322–343 (1997)

19.62 Q.-S. Chen, S. Pendurti, V. Prasad: Effects of baffle design on fluid flow and heat transfer in ammonothermal growth of nitrides, J. Cryst. Growth **266**, 271–277 (2004)

19.63 A.J. Chapman: *Heat Transfer* (Macmillan, New York 1984)

19.64 M. Callahan, B.-G. Wang, K. Rakes, D. Bliss, L. Bouthillette, M. Suscavage, S.-Q. Wang: GaN single crystals grown on HVPE seeds in alkaline supercritical ammonia, J. Mater. Sci. **41**, 1399–1407 (2006)

19.65 Q.-S. Chen, S. Pendurti, V. Prasad: Modeling of ammonothermal growth of gallium nitride single crystals, J. Mater. Sci. **41**, 1409–1414 (2006)

19.66 T. Sekiguchi, S. Miyashita, K. Obara, T. Shishido, N. Sakagami: Hydrothermal growth of ZnO single crystals and their optical characterization, J. Cryst. Growth **214/215**, 72–76 (2000)

19.67 L.N. Demianets, D.V. Kostomarov, I.P. Kuz'mina, S.V. Pushko: Mechanism of growth of ZnO single crystals from hydrothermal alkali solutions, Cryst. Rep. **47**, S86–S98 (2002), Supp 1

19.68 B.G. Wang: Understanding and controlling the morphology of ZnO crystallites under hydrothermal conditions, Cryst. Res. Technol. **32**, 659–667 (1997)

19.69 W.J. Li, E.W. Shi, W.Z. Zhong, Z.W. Yin: Growth mechanism and growth habit of oxide crystals, J. Cryst. Growth **203**, 186–196 (1999)

19.70 I.L. Khodakovskiy, A.Y. Yelkin: Measurement of the solubility of zincite in aqueous NaOH at 100, 150, and 200 °C, Geokhimiya **10**, 1490–1498 (1975)

19.71 P. Bénézeth, D. Palmer, D. Wesolowski: The solubility of zinc oxide in 0.03 m NaTr as a function of temperature with in-situ pH measurement, Geochim. Cosmochi. Acta **63**, 1571–1586 (1999)

19.72 B.G. Wang, E.W. Shi, W.Z. Zhong: Twinning morphologies and mechanisms of ZnO crystallites under hydrothermal conditions, Cryst. Res. Technol. **33**, 937–941 (1998)

19.73 M.M. Lukina, M.V. Lelekova, V.E. Khadzhi: Effect of lithium on the growth rate of zincite and quartz under hydrothermal conditions, Sov. Phys. Crystallogr. **15**, 530–531 (1970)

19.74 R.A. Laudise, R.L. Barnes: Perfection of quartz and its connection to crystal growth, IEEE Trans. Ultrasonics Ferroelectr. Freq. Control. **35**, 277–287 (1998)

19.75 A.F. Armington: Recent advances in the growth of high quality quartz, Prog. Cryst. Growth Charact. **21**, 97–111 (1990)

19.76 E.D. Kolb, R.A. Laudise: Hydrothermally grown ZnO crystals of low and intermediate resistivity, J. Am. Ceram. Soc. **49**, 302–305 (1966)

19.77 E.D. Kolb, S. Coriell, R.A. Laudise, A.R. Hutson: The hydrothermal growth of low carrier concentration ZnO at high water and hydrogen pressures, Mater. Res. Bull **2**, 1099–1106 (1967)

19.78 I.P. Kuz'mina: Crystallization kinetics of zincite under hydrothermal conditions, Sov. Phys. Crystallogr. **13**(5), 803–805 (1969), translated from Kristallogr., Vol. 13, No.5

19.79 G. Dhanaraj, M. Dudley, D. Bliss, M. Callahan, M. Harris: Growth and process induced dislocation in zinc oxide crystals, J. Cryst. Growth **297**, 74–79 (2006)

19.80 H. Youping, Z. Jinbo, W. Dexang, S. Genbo, Y. Mingshan: New technology of KDP crystal growth, J. Cryst. Growth **169**, 196–198 (1996)

19.81 B. Wang, M.J. Callahan, C. Xu, L.O. Bouthillette, N.C. Giles, D.F. Bliss: Hydrothermal growth and characterization of indium-doped-conducting ZnO crystals, J. Cryst. Growth **304**, 73–79 (2007)

19.82 N.Y. Garces, L. Wang, L. Bai, N.C. Giles, L.E. Halliburton, G. Cantwell: Role of copper in the green luminescence from ZnO crystals, Appl. Phys. Lett. **81**, 622–624 (2002)

19.83 E.D. Kolb, R.A. Laudise: Hydrothermally grown ZnO crystals of low and intermediate resistivity, J. Am. Ceram. Soc. **49**, 302–305 (1966)

19.84 C.G. Van de Walle: Hydrogen as a cause of doping in zinc oxide, Phys. Rev. Lett. **85**(5), 1012–1015 (2000)

19.85 R. Littbarski: Carrier concentration and mobility. In: *Current Topics in Materials Science*, Vol. 7, ed. by E. Kaldis (North-Holland, Amsterdam 1981) pp. 212–225

19.86 B. Theys, V. Sallet, F. Jomard, A. Lusson, J. Rommeluère, Z. Teukam: Effects of intentionally introduced hydrogen on the electric properties of ZnO layers grown by metalorganic chemical vapor deposition, J. Appl. Phys. **91**, 3922–3924 (2002)

19.87 D.C. Look, J.W. Hemsky, J.R. Sizelove: Residual native shallow donor in ZnO, Phys. Rev. Lett. **82**, 2552–2555 (1999)

19.88 A. Urbieta, P. Fernández, J. Piqueras, T. Sekiguchi: Scanning tunneling spectroscopy characterization of ZnO single crystals, Semicond. Sci. Technol. **16**, 589–593 (2001)

19.89 N. Sakagami, M. Yamashita, T. Sekiguchi, S. Miyashita, K. Obara, T. Shishido: Variation of electrical properties on growth sectors of ZnO single crystals, J. Cryst. Growth **229**, 98–103 (2001)

19.90 M. Yoneta, K. Yoshino, M. Ohishi, H. Saito: Photoluminscense studies of high-quality ZnO single crystals by hydrothermal method, Phys. B **376–377**, 745–748 (2006)

19.91 J. Mass, M. Avella, J. Jiménez, M. Callahan, E. Grant, K. Rakes, D. Bliss, B. Wang: Cathodoluminescence characterization of hydrothermal ZnO crystals, Superlattice Microstruct. **38**, 223–230 (2005)

19.92 L.N. Dem'yanets, V.I. Lyutin: Status of hydrothermal growth of bulk ZnO: latest issues and advantages, J. Cryst. Growth **310**, 993–999 (2008)

19.93 J. Mass, M. Avella, J. Jiménez, A. Rodriquez, T. Rodriquez, M. Callahan, D. Bliss, B. Wang: Cathodoluminescence study of ZnO wafer cut from hydrothermal crystals, J. Cryst. Growth **310**, 1000–1005 (2008)

19.94 D.C. Reynolds, D.C. Look, B. Jogai, H. Morkoç: Simililarities in the bandedge and deep-centre photoluminescence mechanisms of ZnO and GaN, Solid State Commun. **101**, 643–646 (1997)

19.95 A. Urbieta, P. Fernández, J. Piqueras, C. Hardalov, T. Sekiguchi: Cathodoluminescence microscopy of hydrothermal and flux grown ZnO single crystals, J. Phys. D Appl. Phys. **34**, 2945–2949 (2001)

19.96 A. Urbieta, P. Fernández, C. Hardalov, J. Piqueras, T. Sekiguchi: Cathodoluminescense and scanning tunneling spectroscopy, Mater. Sci. Eng. **B91–92**, 345–348 (2002)

19.97 J. Mass, M. Avella, J. Jiménez, M. Callahan, E. Grant, K. Rakes, D. Bliss, B. Wang: Visable luminescence in ZnO. In: *New Materials and Proceesses for Incoming Semiconductor Technologies*, ed. by S. Dueñas, H. Castán (Transworld Research Network, Kerala 2006)

19.98 D. Bliss: Zinc oxide. In: *Encyclopedia of Advanced Materials*, ed. by D. Bloor, M.C. Flemings, R.J. Brook, S. Mahajan, R.W. Cahn (Pergamon, Oxford 1994) pp. 9888–9891

19.99 C. Woods, A.J. Drehman: Presentation, Natl. Space Missile Mater. Symp. (Monterey, 2001)

19.100 B. Wang, M. Callahan, J. Bailey: Synthesis of dense polycrystalline GaN of high purity by the chemical vapor reaction process, J. Cryst. Growth **286**, 50–54 (2005)

19.101 K. Lee, K. Auh: Dislocation density of GaN grown by hydride vapor phase epitaxy, MRS Int. J. Nitride Semicond. Res. **6**, 9 (2001)

19.102 B. Raghothamacher, J. Bai, M. Dudley, R. Dalmau, D. Zhuang, Z. Herro, R. Schlesser, Z. Sitar, B. Wang, M. Callahan, K. Rakes, P. Konkapaka, M. Spencer: Characterization of bulk-grown GaN and AlN single-crystals materials, J. Cryst. Growth **287**, 349–353 (2006)

19.103 M.J. Callahan, B. Wang, L. Bouthillette, S.-Q. Wang, J.W. Kolis, D. Bliss: Growth of GaN crystals under ammonothermal conditions, MRS Fall Meet. Symp. Proc. **798**, Y2.10 (2004)

19.104 T. Hashimoto, M. Saito, K. Fujito, F. Wu, J.S. Speck, S. Nakamura: Seeded growth of GaN by the basic ammonothermal method, J. Cryst. Growth **305**, 311–316 (2007)

19.105 Images provided by Prof. Brian Skromme's group, Arizona St. Univ.

19.106 T. Hashimoto, K. Fujito, F. Wu, B.A. Haskell, P.T. Fini, J.S. Speck, S. Nakamura: Structural characterization of thick GaN films grown on free-standing GaN seeds by the ammonothermal method using basic ammonia, Jpn. J. Appl. Phys. **44**(25), L797–L799 (2005)

19.107 R. Dwilinski, R. Doradzinski, J. Garzynski, L.P. Sierzputowski, A. Puchalski, Y. Kanaba, K. Yagi, H. Minakuchi, H. Hayashi: Excellent crystallinity of truly bulk ammonothermal GaN, J. Cryst. Growth **310**, 3911–3916 (2008)

19.108 J. Bai, M. Dudley, B. Raghothamachar, P. Gouma, B.J. Skrome, L. Chen, P.J. Hartlieb, E. Michaels, J. Kolis: Correlated structural and optical characterization of ammonothermally grown bulk GaN, Appl. Phys. Lett. **84**(17), 3289–3291 (2004)

19.109 M.P. D'Evelyn, H.C. Hong, D.-S. Park, H. Lu, E. Kaminsky, R.R. Melkote, P. Perlin, M. Lesczynski, S. Porowski, R.J. Molnar: Bulk GaN crystal growth by th high-pressure ammonothermal method, J. Cryst. Growth **300**, 11–16 (2007)

19.110 S.V. Bhat, K. Biswas, C.N.R. Rao: Synthesis and optical properties of In-doped GaN nanocrystals, Solid State Commun. **141**, 325–328 (2007)

19.111 B.T. Adekore, K. Rakes, B. Wang, M.J. Callahan, S. Pendurti, Z. Sitar: Ammonothermal synthesis of aluminum nitride crystals on group III-nitride templates, J. Electron. Mater. **35**, 1104–1111 (2006)

19.112 M. Zajac, J. Gosk, E. Grzanka, S. Stelmakh, M. Palczewska, A. Wysmołek, K. Korona, M. Kamińska, A. Twardowski: Ammomonothermal sythesis of GaN doped with transition metal ions (Mn, Fe, Cr), J. Alloys Compd. **456**, 324–338 (2008)

19.113 A. Denis, G. Goglio, G. Demazeau: Gallium nitride bulk crystal growth processes: a review, Mater. Sci. Eng. R **50**, 167–194 (2006)

20. Stoichiometry and Domain Structure of KTP-Type Nonlinear Optical Crystals

Michael Roth

In recent years the growth technologies of only a few inorganic oxide crystals, such as BBO (BaB_2O_4), LBO (LiB_3O_5), and the KTP ($KTiOPO_4$) group of isomorphic compounds, have matured to a degree allowing their extensive integration into commercial laser systems in the form of nonlinear optical (NLO) and electrooptic (EO) devices. The KTP-type crystals are ferroelectrics at room temperature. They are also well known for their large birefringence, high NLO and EO coefficients, wide acceptance angles, thermally stable phase-matching properties, and relatively high damage threshold. These properties make them especially useful for high-power wavelength-conversion applications, such as the second-harmonic generation (SHG) and optical parametric oscillations (OPO), as well as for electrooptic phase modulation and Q-switching. Lately, a great deal of effort has been put into the development of periodically poled KTP (PPKTP) devices based on quasi-phase-matched (QPM) wavelength conversion. However, both birefringent and QPM properties of KTP crystals depend on their structural characteristics, such as morphology, chemical composition, point defect distribution (stoichiometric and impurities), and particularly the ferroelectric domain structure, which are closely related to the specific crystal growth parameters. Current research includes studies of nonstoichiometry and distribution of point defects, e.g., vacancies and impurities, as well as the basic mechanisms underlying ferroelectric domain formation during

20.1	**Background** 691
	20.1.1 KTP Crystal Structure 692
	20.1.2 Crystal Growth 694
20.2	**Stoichiometry and Ferroelectric Phase Transitions** 697
	20.2.1 $KTiOPO_4$ Crystals........................ 697
	20.2.2 $RbTiOPO_4$ Crystals....................... 700
	20.2.3 Other KTP Isomorphs.................... 702
20.3	**Growth-Induced Ferroelectric Domains** 703
	20.3.1 Domains in Top-Seeded Solution-Grown KTP 704
	20.3.2 Domain Boundaries.................... 705
	20.3.3 Summary of Ferroelectric Domain Structures ... 707
	20.3.4 Single-Domain Growth 707
20.4	**Artificial Domain Structures** 708
	20.4.1 Electric Field Poling 708
	20.4.2 As-Grown Periodic Domain Structure 711
20.5	**Nonlinear Optical Crystals** 713
	20.5.1 Optical Nonuniformity 713
	20.5.2 Gray Tracks 715
References 716

KTP crystal growth and cool-down. By controlling the stoichiometry and achieving single-domain growth of bulk crystals it is also possible to create as-grown periodic domain structures useful for nonlinear QPM applications.

20.1 Background

Potassium titanyl phosphate (KTP) belongs to the family of isomorphic compounds with generic composition of $MTiOXO_4$, where X = {P or As} and M = {K, Rb, Tl, NH_4 or Cs (for X = As only)} [20.1]. All crystals of this family, including their solid solutions, are orthorhombic and belong to the noncentrosymmetric point group *mm*2

Table 20.1 Selected physical properties of KTP-family crystals

Crystal	KTP	RTP	KTA	RTA	CTA	References
Optical transparency (μm)	0.35–4.3	0.35–4.3	0.35–5.3	0.35–5.3	0.35–5.3	[20.2–4]
Birefringence, $n_z - n_x$ at 1.064 μm	0.0921	0.0884	0.0863	0.0782	0.0700	[20.4–8]
NLO susceptibilities (pm/V) at 1.064 μm						
d_{33}	16.9	17.1	16.2	15.8	18.1	[20.9, 10]
d_{32}	4.4	4.1	4.2	3.8	3.4	
d_{31}	2.5	3.3	2.8	2.3	2.1	
d_{24}	3.6	–	3.2	–	–	
d_{15}	1.9	–	2.3	–	–	
d_{eff} for type-II SHG	3.35	2.51	–	–	–	[20.4, 11]
SHG cutoff (μm) along						
x-direction	1.082	1.147	1.134	1.243	1.548	[20.11]
y-direction	0.994	1.038	1.074	1.138	1.280	
EO coefficients (pm/V) at 0.633 μm						
r_{33}	36.3	39.6	37.5	40.5	38.0	[20.10, 12]
r_{23}	15.7	17.1	15.4	17.5	18.5	
r_{13}	9.5	12.5	11.5	13.5	14.2	

at room temperature. The unique ferroelectric, EO, and NLO properties of the KTP-family crystals are stipulated by this specific structure. Table 20.1 gives a partial list of the important physical properties exhibited by the five more extensively studied KTP group members: KTiOPO$_4$ (KTP), RbTiOPO$_4$ (RTP), KTiOAsO$_4$ (KTA), RbTiOAsO$_4$ (RTA), and CsTiOAsO$_4$ (CTA).

All these crystals exhibit a broad optical transparency range allowing for NLO [20.13] and EO [20.14] interactions from the visible to the near- and mid-infrared. The larger birefringence and suitable dispersion of the orthophosphates, KTP and RTP, brings their phase-matching directions into the x–y plane for efficient second-harmonic generation (SHG) of the principal Nd:YAG laser radiation at 1.064 μm. Although not suitable for frequency doubling into the green, CTA includes the 1.32 μm line of the Nd:YAG laser in its phase-matchable wavelength interval within the x–y plane [20.9] allowing doubling into the red. In general, the transparency range of the orthoarsenates extends by 1 μm further into the mid-infrared (Table 20.1) and contains no orthophosphate overtone absorption at wavelengths shorter than 3.5 μm, which makes them particularly attractive for noncritically phase-matched (NCPM) eye-safe OPO (signal at ≈ 1.5 μm) with negligible absorption of the idler radiation at ≈ 3.3 μm [20.15]. It is noteworthy that a more detailed insight into the idler absorption problem in KTP, namely its reduced value for z-polarization, demonstrates that orthophosphates may also generate high-average-power eye-safe OPO [20.16]. All KTP-family crystals also feature large EO coefficients necessary for electrooptic modulation and switching [20.14], but mainly KTP, and especially RTP, are widely used for these applications [20.17, 18]. This may be associated with the fact that the orthoarsenates are apt to ferroelectric multidomain formation [20.19, 20], and their crystal growth is a costly process [20.21]. Therefore, the more technologically important KTP and RTP materials will be primarily discussed in terms of their defect structure and the structure–properties relationships versus the main processing parameters.

20.1.1 KTP Crystal Structure

The five KTP-family members identified above have identical crystal structures at room temperature. The orthorhombic structure ($a \neq b \neq c$, $\alpha = \beta = \gamma = 90°$) of KTP belongs to the space group $Pna2_1$, as initially determined by *Tordjman* et al. in 1974 [20.22]. There are 64 atoms in a unit cell in the KTP-type lattice. This 64-atom group separates into four subgroups of 16 atoms each, and within each such subgroup there are two inequivalent K (Rb) sites, two inequivalent titanium sites, two inequivalent P (As) sites, and ten inequivalent oxygen sites. Two of the latter oxygen sites represent bridging ions located between titanium ions, while the other eight are contained in PO$_4$ (AsO$_4$) groups where they link one Ti and one P (As) ion.

Table 20.2 Crystal lattice parameters of KTP-family crystals

Lattice constants ()	KTP [20.23]	RTP [20.24]	KTA [20.25]	RTA [20.26]	CTA [20.27]
a	12.819	12.952	13.125	13.264	13.486
b	6.399	6.4925	6.5716	6.682	6.682
c	10.584	10.555	10.786	10.7697	10.688

Lattice Parameters and Site Occupation

The room-temperature lattice constants of KTP-family crystals are given in Table 20.2. They change relatively little in going from KTP even to CTA. The values for KTP and RTP are very close, thus verifying that the PO_4 tetrahedra have much more influence on the lattice constants than the monovalent cations K and Rb. In KTA, the a and b lattice constants are approximately 1.2% larger than in KTP and the c lattice constant is about 2.1% larger than in KTP, thus reflecting the slightly larger size of the AsO_4 tetrahedron compared with PO_4. One 16-atom subgroup can be transformed into one of the other three subgroups by simple transformations within the unit cell. The [001] projection of the KTP structure is shown in Fig. 20.1.

The physical structure of the KTP-type crystal contains helical chains of distorted TiO_6 octahedra running parallel to the $\langle 011 \rangle$ crystallographic directions and forming a two-dimensional network in the b–c plane. The TiO_6 octahedra are linked at two corners by alternately changing long and short Ti(1)−O bonds which are commonly assumed as primarily responsible for the optical nonlinearity [20.28]. Other sources of optical nonlinearities have been invoked as well. Bond-polarizability calculations of KTP [20.29] show that the nonlinearity derives from the various K−O bonds and $P(2)O_4$ groups rather than from the TiO_6 groups. Nuclear magnetic resonance (NMR) studies of KTP [20.30] suggest that the large electric field gradients at the Ti sites and associated charge-transfer mechanisms are important factors establishing the necessary conditions for high nonlinear response. However, very recent stimulated Raman measurements show that, in KTP, the strongest vibrations occur in the direction collinear with the distortion of TiO_6 octahedra [20.31]. The other four oxygen ions around a titanium ion are parts of the PO_4 (or AsO_4) tetrahedra. These tetrahedra bond the −Ti−O−Ti−O− networks into a three-dimensional covalent ($TiOPO_4$) framework. The crystal structure is completed by potassium (or rubidium) ions occupying cavities, or cages, within this framework. These monovalent ions are either 8- or 9-coordinated with respect to oxygen, and they are denoted as K(1) and K(2), respec-

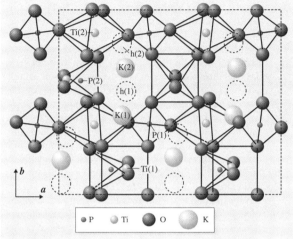

Fig. 20.1 A view of the KTP crystal structure along the c-axis direction

tively, in Fig. 20.1 representing the KTP structure. Each K atom is asymmetrically coordinated by O atoms in the two inequivalent sites, and at a short distance along the b-direction from each K atom there is a void similar in size and coordination to the K site. The voids are termed *hole sites*, h(1) and h(2) [20.32], and they are pseudosymmetrically related to the K(2) and K(1) sites respectively, as is apparent from Fig. 20.1. At high temperatures, above the transition to the ferroelectric ($Pna2_1$) to paraelectric (centrosymmetric) $Pnan$ phase, the K(1) and h(1) as well as K(2) and h(2) sites merge along the b-direction into positions halfway between the room-temperature K sites and their associated holes sites. More refined studies of the crystallographically unique K(1) and K(2) sites reveal additional subsplittings of their positions [20.33, 34]. Also, the K(1) cage has a volume 25% smaller than the K(2) cage [20.35]. This may explain why the larger Rb atoms substituting for K atoms in mixed $K_xRb_{1-x}TiOPO_4$ crystals occupy preferentially the larger K(2) sites, independently of the Rb incorporation mechanism, by crystal growth or ion exchange [20.36].

Ionic Conductivity

Finally, we refer again to Fig. 20.1 and conclude that there are additional consequences to the quite different environments for the K(1) and K(2) atoms. The K(2)O$_9$ cage forms a channel, parallel to the c-axis, which runs through the entire crystal structure and along which the K(2) atoms are expected to move relatively freely. On the other hand, the K(1) atom is constrained from a similar motion by the confinement of the P(1)O$_4$−Ti(2)O$_6$ chain which restricts its movement along the c-axis. Although the K(1)O$_8$ cage forms a channel along the a-axis, the K(1) atom does not show any significant movement along this axis even at high pressures [20.35]. Therefore, when KTP is described as a quasi-one-dimensional superionic conductor [20.37], with the conductivity σ_{33} being about three orders of magnitude higher than σ_{11} and σ_{22} [20.38], mainly K(2) ions can diffuse through cavities combined into channels along the c-axis direction via a vacancy mechanism [20.39]. Similar ionic conductivity anisotropy is valid also for KTA [20.2]. In RTP and RTA, σ_{33} is relatively smaller, since the activation energy for hopping of larger ions is higher [20.40, 41]. The concentration of potassium/rubidium vacancies, or the degree of deviation from stoichiometry, depends on the crystal growth method and on specific growth parameters that will be discussed throughout the chapter.

20.1.2 Crystal Growth

KTP-family crystals decompose before melting (e.g., KTP decomposes at 1172 and 1158 °C in air and argon atmospheres, respectively [20.42]) and may be thus grown only from solutions. Relatively small but high-quality crystals can be grown from aqueous solutions by the hydrothermal process. Larger crystals, yet requiring meticulous quality control, can be grown from high-temperature tungstate and molybdate fluxes or self-fluxes. All these methods are being used for commercial production of KTP-family crystal in their seeded versions, although unseeded small crystals are occasionally processed for research purposes. The typical morphology of a KTP crystal [20.43, 44] grown on a seed fully immersed into the solution (by either of the methods mentioned above) is shown in Fig. 20.2.

The crystal exhibits 14 facets belonging to four families of crystallographic planes, namely: {100}, {110}, {011} and {201}. Accordingly, 14 growth sectors develop simultaneously on the submerged seed. The morphology can be altered depending also on the seed shape and the solution composition [20.45, 46].

Hydrothermal Growth

A detailed description of the first attempts to grow a few millimeter-sized optical-quality KTP-family crystals hydrothermally was given by *Bierlein* and *Gier* in the patent literature [20.47, 48]. Typically, the hydrothermal process involved growing the crystal in an autoclave having a growth region and a nutrient region, and an aqueous mineralizer solution was employed. For example, *Laudise* et al. [20.49] used a Pt-lined low-carbon steel autoclave (75% filled) at a constant pressure of about 1360 atm with baffle-separated flux-grown KTP crystals as a nutrient (at 435 °C) and a (2 M K$_2$HPO$_4$ + 0.5 M KPO$_3$) mineralizer and TiO$_2$ to grow KTP crystals at around 400 °C. Better quality crystals were obtained by growing on (201)- and (010)-oriented seeds at $\approx 0.07-0.14$ mm/day rates. A typical commercial production of KTP crystals uses a potassium phosphate mineralizer at temperatures in the range of about 520−560 °C and pressures in the range of about 1700−2000 atm [20.50]. The relatively high temperatures and pressures employed in this process makes scale-up difficult and expensive, yet the use of more moderate conditions can lead to a problem of anatase (TiO$_2$) coprecipitation [20.51], making the process less useful. In principle, it has been shown that growth temperatures (< 500 °C) and pressure (< 1000 atm) can be reduced while using potassium-rich mineralizer solutions, but the growth rate becomes limited (< 0.13 mm/day) due to the relatively low solubility of KTiOPO$_4$ in the mineralizer [20.52].

Other types of mineralizers adopted for the hydrothermal growth of KTP involve the use of KF solutions, as suggested by *Jia* et al. [20.53, 54]. By utilizing KF as a mineralizer, relatively lower temperature

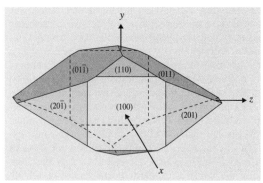

Fig. 20.2 Typical habit of an immersion-seeded KTP crystal

and pressure (≈ 1000 atm) can be employed for a hydrothermal KTP crystal growth process. In addition, the use of pure KF mineralizer provides a solubility of about 2% under the stated growth conditions, but like the process described by *Laudise* et al. [20.49], the higher solubility occurs near the phase-stability boundary (with respect to temperature, pressure, and mineralizer concentration), so the possibility of co-precipitation of an undesirable non-KTP phase exists. To avoid this disadvantage, *Cheng* [20.55] suggested complex aqueous solutions as mineralizers comprising both KF and potassium (rubidium) phosphates (arsenates) for hydrothermal production of KTP-family crystals, in amounts effective to assure solubility of the KTP-type crystals of at least about 1% by weight. Although grown under higher temperatures and pressures, optical-quality KTP and KTA crystals of at least 1 mm^3 volume could be recovered after a few hours rather than days and weeks. In the most recent report on hydrothermal KTP growth a process utilizing a (2 M K_2HPO_4 + 0.1 M KH_2PO_4 + 1 wt% H_2O_2) solution as a mineralizer and crushed flux-grown KTP crystals as a nutrient was described [20.56]. Crystals with dimensions up to $14.5 \times 28 \times 17$ mm^3 have been grown on (011) and spherical seeds at a rate of 0.15 mm/day in the 400–540 °C temperature range and 1200–1500 atm pressure range. They exhibited an exceptionally high (for KTP) optical damage threshold of 9.5 GW/cm^2.

Growth from Low-Viscosity Melts

In the early days of optical-grade KTP development, the only commercial process reported was the hydrothermal crystal growth technique requiring cumbersome high-pressure vessels with noble-metal liners and yielding relatively small crystals. Initial attempts to grow KTP crystals from phosphate fluxes at atmospheric pressure conditions [20.42] seemed to encounter the problem of high melt viscosity. *Ballman* et al. [20.57] were the first to report on the growth of up to 1 cm^3 KTP single crystals from a low-melting low-viscosity water-soluble tungstate flux. Tungstic anhydride was added to the potassium phosphate-titanium oxide melt, and KTP as a single stable phase formed according to the following reaction

$$4K_2HPO_4 + 2TiO_2 + 3WO_3 \xrightarrow{1000\,°C}$$
$$2KTiOPO_4 + 3K_2WO_4 \cdot P_2O_5 + 2H_2O \,. \quad (20.1)$$

Iliev et al. [20.58] found that with a $K_2O : P_2O_5$ ratio of 0.9 : 0.1 and 40 mol % WO_3 in the melt, the flux viscosity was as low as 10–15 cP in the 1050–850 °C temperature range. High-optical-quality KTP crystals up to $30 \times 20 \times 20$ mm^3 in size could be grown from the K_2O-P_2O_5-TiO_2-WO_3 system in the 930–700 °C temperature range at a rate of 0.9 mm/day [20.59]. Not only KTP, but also RTP, KTA, and RTA crystals were subsequently grown from both tungstate and molybdate fluxes [20.44]. However, optical inhomogeneities associated with tungsten-containing (0.5–1.0 wt %) striations parallel to the {011} faces were identified [20.60, 61]. Indeed, electron paramagnetic resonance (EPR) measurements show that tungsten enters KTP as an impurity occupying both the Ti(1) and Ti(2) sites as W^{5+} and, partly, the Ti(2) site as W^{4+} [20.62]. The tungsten impurity is found to cause detrimental multiplication of growth sectors in the case of $\langle 011 \rangle$-seeded (and less $\langle 010 \rangle$-seeded) KTP crystal growth from $3K_2WO_4 \cdot P_2O_5$ fluxes [20.63].

When small amounts of MoO_3 are added to the phosphate flux ($K_6P_4O_{13}$) in the case of spontaneous nucleation growth, only trace amounts of Mo are detected in the KTP crystal, but they prevent spurious nucleation and increase the crystal hardness necessary for improved polishing [20.64]. In a different attempt to reduce the melt viscosity, potassium halides were added to the phosphate flux to grow KTP [20.65] or RbI to grow RTP [20.66] crystals.

Growth from Self-Fluxes

The main crystal growth problems pointed out above were that the hydrothermal process yielded high-quality but small single crystals, and the use of tungstate fluxes resulted in incorporation of W impurities into the crystals, increasing the optical loss [20.60] and lowering their laser damage resistance [20.67]. Therefore, the development of most of the practical NLO devices unfolded in parallel with constant advancement of the seeded growth of large and ever-improving quality KTP-family crystals from self-fluxes, namely high-temperature solutions in alkali phosphates. In the early work [20.42, 68, 69], KTP crystals up to 1 cm^3 in size were grown from flux compositions within the KPO_3-$K_4P_2O_7$ system. The end components obtained by thermal decomposition of K_2HPO_4 and KH_2PO_4 together with TiO_2 were used for flux preparation, e.g., according to the following high-temperature reactions

$$KH_2PO_4 + TiO_2 \rightarrow KTiOPO_4 + H_2O \uparrow \,,$$
$$2KH_2PO_4 + 2K_2HPO_4 \rightarrow K_6P_4O_{13} + 3H_2O \uparrow \,.$$
$$(20.2)$$

More chemical routes for preparing self-fluxes of KTP and its isomorphs are given by *Iliev* et al. [20.70] and *Cheng* et al. [20.10]. $K_6P_4O_{13}$ (denoted as K6 in the literature), with the K-to-P ratio $R = 1.5$, is the most popular solvent for KTP growth due to the moderate viscosity [20.71] and volatility [20.70] of the KTP/K6 flux and relatively high solubility of KTP. Other solvents, such as $K_4P_2O_7$ (K4), $K_5P_3O_{10}$ (K5), $K_8P_6O_{19}$ (K8), $K_{10}P_8O_{25}$ (K10), $K_{15}P_{13}O_{40}$ (K15), and $K_{18}P_{16}O_{49}$, have been also studied, mainly in terms of KTP solubility [20.46, 72] in the 900–1100 °C temperature range. The solubility decreases down this sequence due to the relative decrease of the concentration of $(P_2O_7)^{-4}$ anions in the flux, since all the above solvents can be viewed as composed of $(xKPO_3 + yK_4P_2O_7)$ [20.72]. In the case of KTP isomorphs, their solubility in similar self-fluxes (RTP/R6 [20.73, 74], KTA/K6, RTA/R5, and CTA/C5 [20.10], KTA/K5 [20.75], and RTP/R4 [20.76]) has been also reported.

Reproducible and controllable seeded growth of large-size KTP crystals from the KTP/K6 flux has been repeatedly reviewed [20.77–79]. In view of the small undercooling required for KTP nucleation, close to nongradient conditions in the growth furnace are appropriate. However, it is necessary to provide for a high degree of spatial temperature uniformity in order to avoid spurious nucleation. *Bordui* et al. [20.80] used a heat-pipe-based furnace for this purpose with submerged seeding. Others used three-zone [20.81] or five-zone [20.82] furnaces to produce large (200–300 g) and almost inclusion-free KTP crystals by the top-seeded solution growth (TSSG) method. *Angert* et al. [20.83] suggested a way to improve the yield of KTP crystal growth by combining the TSSG technique with seed pulling. Figure 20.3 shows the growth schemes typical for the submerged seed, top-seeded, and top-seeded with pulling configurations.

The convection-dominated mass transfer in submerged seed growth of KTP (left image in Fig. 20.3) was studied theoretically by *Vartak* and *Derby* [20.84], and some modeling problems were pointed out. An interesting approach to the submerged growth was put forward earlier by *Bordui* and *Motakef* [20.85] (modeling and experiment), who suggested an asymmetric 90° seed orientation and rotation at 50 rpm with periodic reversal in order to reduce the amount of inclusions. Their model was initially supported by numerical analyses performed by *Vartak* et al. [20.86], but a more recent boundary-layer analysis for flow and mass transfer [20.87] showed that the straightforward computational approach, such as the use of a spinning disk approximation, could provide satisfactory results only under experimental conditions not matching the realistic setups employed in the field. It is noteworthy that surface kinetics must be also considered in the growth of KTP crystals [20.46, 88].

Although the submerged seed method is still being used for commercial production by a few vendors, clearly the development of up to 14 growth sectors (Fig. 20.2) limits the use of such KTP crystals to fab-

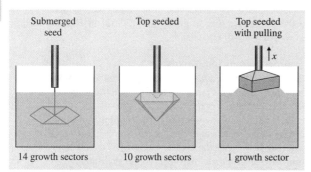

Fig. 20.3 Three configurations for seeded high-temperature solution growth of KTP-family crystals

Fig. 20.4a–c KTP crystals grown by the top-seeded solution growth method with pulling in the (**a**) [100]-, (**b**) [010]-, and (**c**) [001]-directions

ricating only small optical elements, since elements cut from a single sector are required for high-quality performance [20.89]. Multiplication of growth sectors results in nonuniform growth rates [20.46, 63] and incorporation of trace impurities at the sector boundaries and may affect the ferroelectric domain structure to be discussed below. The same is true for the conventional top-seeding method, shown in the middle sketch of Fig. 20.3. In contrast, TSSG with pulling reduces the number of growth sectors formed. *Bolt* et al. [20.81] suggest the use of a [001]-oriented seed, since maximal linear growth rate can be achieved in this direction. (Note that the low growth rate in the [100] direction can be overcome by using seeds elongated in this direction [20.45] to grow larger crystals.) However, in the case of [001] seeding, four growth sectors form, and the crystal is bound by the {201} and {011} facets. The use of [100]-oriented seeds is, therefore, favorable [20.83], since growth of a single growth sector crystal can be achieved. The corresponding configuration is shown on the right side of Fig. 20.3. With the largest (100) facet (Fig. 20.2) parallel to the growth front, as is typical for growth from the KTP/K6 flux, a planar solid–liquid interface bounds the crystal, assuring the best possible transverse uniformity of its physical properties. Photographs of three KTP crystals grown by the TSSG method with pulling on [100]-, [010]-, and [001]-directed seeds are shown in Fig. 20.4 to illustrate their distinct morphologies.

20.2 Stoichiometry and Ferroelectric Phase Transitions

Regardless of the growth method of KTP-family crystals from self-fluxes, say of MTiOXO$_4$ (M = {K, Rb} and X = {P, As}) from MTiOXO$_4$/M$_6$X$_4$O$_{13}$, the M-to-X ionic ratio increases as the crystal grows out, leaving behind a solution that becomes gradually enriched in M cations. Obviously, the M-metal content in the growing crystal increases accordingly, presumably improving the associated stoichiometry; for example, in the initially solidifying portion of the KTP crystal grown from the KTP/K6 flux, large concentrations of potassium vacancies (500–800 ppm) have been detected [20.90]. These vacancies can be charge-compensated by holes trapped at bridging oxygen ions between two titanium ions [20.91] and by positively charged oxygen vacancies [20.92]. This is reflected in the high-temperature defect formation reaction [20.93]

$$\text{KTiOPO}_4 \rightarrow \text{K}_{1-x}\text{TiOPO}_{4(1-x/8)} + \tfrac{x}{2}\text{K}_2\text{O}, \quad (20.3)$$

where x is the concentration of potassium vacancies.

Most of the KTP-family crystals are high-temperature ferroelectrics [20.9] which undergo a second-order phase transition changing its point group symmetry from *mmm* in the high-temperature paraelectric phase to the *mm*2 polar symmetry class in the low-temperature ferroelectric phase [20.94]. The ferroelectric phase transition (Curie) temperature, T_C, is very sensitive to the crystal stoichiometry and ionic substitution, which is manifested in the large spread of published Curie temperatures for all KTP-family crystals [20.95]. In this section we will discuss the correlation between the Curie temperatures and the defect structure, mainly stoichiometry, in KTP and RTP and present the limited results available also on other related crystals.

20.2.1 KTiOPO$_4$ Crystals

In KTP, addition of dopants, such as Ga, Al, and Ba, in the 400–2000 ppm concentration range is known to reduce the Curie temperature by several tens of degrees and modifies the Curie constant in the Curie–Weiss law [20.96, 97]. Isomorphic substitution of K and P ions by Rb, Cs or Tl and As reduces the Curie temperature by hundreds of degrees [20.97, 98]. Partial substitution of Na for K has been shown to initially increase the T_C from 944 °C for pure KTP to 956 °C and then decrease it to 914 °C for [Na] = 4 and 47 mol %, respectively [20.99]. A large diversity of Curie temperatures was reported for pure KTP crystals. The highest T_C values, 955–959 °C [20.97, 99, 100], were measured on hydrothermally grown crystals. A variety of lower T_C values was given for flux-grown KTP crystals: 952 °C [20.95] (second-harmonic generation (SHG) measurements), 947–951 °C [20.100], 946 °C [20.97], 944 °C [20.99], 934 °C [20.101], 910 °C [20.102] (dielectric measurements), and 892 °C [20.103] (birefringence measurements). Such scatter of data is beyond any imaginable experimental error and can be explained only in terms of variable stoichiometry of the KTP crystals grown from different solutions under variable growth conditions. *Angert* et al. [20.72] have studied the dependency of T_C on the composition of self-fluxes in great detail, and the main results will be presented below. The influence of high-temperature annealing on T_C has been addressed as well.

Curie Temperature Versus Composition

Curie temperatures of numerous crystals grown from different self-fluxes with varying initial concentrations were measured by a standard dielectric technique [20.72]. The results are reproduced in Fig. 20.5, where the measured T_C values are given as a function of weight concentration of KTP in each of the seven fluxes used, namely corresponding to KTP dissolved in the K4, K5, K6, K8, K10, K15, and K18 solvents with [K]/[P] ratios (R) varying from 2 to 1.125. The lowest T_C value found was 898 °C and the highest was 960 °C, but these limits could be stretched even further upon broadening the KTP concentration and solvent composition ranges. Within the part of the crystallization field studied (of the K_2O-TiO_2-P_2O_5 ternary system) the T_C dependencies can be approximated by straight lines for all fluxes employed. This provides the possibility of a quantitative representation of the T_C dependences on two parameters: R and the KTP concentration in the flux (C), which can be fitted by the equation

$$T_C(C, R) = 980 + (C + 0.24) \times (98R^3 - 555.6R^2 + 1074R - 733), \quad (20.4)$$

for $1 < R \leq 2$. Equation (20.4) allows to forecast or to choose any desirable initial T_C for the crystal, since it holds for any arbitrary self-flux composition. In principle, the higher the T_C value, the more perfect the crystal in terms of stoichiometry or content of impurities, which may be important for many physical applications of this type of crystals.

It is apparent from Fig. 20.5 that, the lower the KTP concentration in any particular self-flux, the higher the Curie temperature of the solid. Also, for any given concentration, the Curie temperatures are higher for self-fluxes exhibiting higher values of R. These two observations can be combined into one conclusion, namely that higher overall concentrations of K in the solution result in higher Curie temperatures of KTP crystals. Moreover, this implies a gradual increase in T_C during growth of a large KTP crystal, since for any given self-flux the solution becomes richer in K while the crystal grows out. T_C increase of up to 20 °C along the growth direction in some large KTP crystals grown from the $K_6P_4O_{13}$ solvent are reported [20.72]. Such spatial variation of T_C in the crystal is obviously associated with compositional (mainly stoichiometry) gradients resulting in nonuniformity of the crystal physical properties. Indeed, *Miyamoto* et al. [20.104] have observed appreciable gradients in the refractive indices, $\Delta n_x/\Delta x = 1.2 \times 10^{-5}$ mm^{-1} and $\Delta n_z/\Delta z = -2.0 \times 10^{-5}$ mm^{-1}, in a large TSSG KTP crystal.

Another a priori observation from Fig. 20.5 is that growth above the dotted line is supposed to yield single-ferroelectric-domain crystals, while growth below this line is apt to result in multidomain crystals, which is occasionally claimed [20.105]. However, in reality, multidomain KTP crystals are frequently obtained for growth below T_C, and single-domain crystals may form when growth is initiated above T_C [20.106]. This may be understood after a more detailed insight into the nature of compositional gradients arising from the variable potassium stoichiometry. Even a relatively large concentration of potassium vacancies of up to 800 ppm [20.90] is small in comparison with the overall amount of potassium ions. Therefore, the gradients of the latter are small in absolute terms. The diffusion coefficient of the K$^+$ ions at typical growth temperatures is small as well, e.g., 8×10^{-10} cm^2/s at 965 °C [20.107]. This explains why the growth-induced potassium concentration gradients essentially *freeze in* along the crystal. The associated gradients of ionized oxygen vacancies are, in contrast, relatively high, and they are responsible for the production of a built-in electric field according to the model proposed earlier by *Kugel* et al. [20.92]. This built-in field may be signifi-

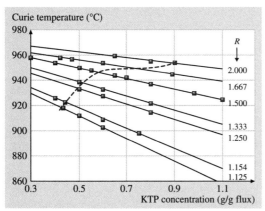

Fig. 20.5 Curie temperatures as a function of KTP concentration in the flux for seven different solvents ($K_4P_2O_7$, $K_5P_3O_{10}$, $K_6P_4O_{13}$, $K_8P_6O_{19}$, $K_{10}P_8O_{25}$, $K_{15}P_{13}O_{40}$, and $K_{18}P_{16}O_{49}$) corresponding to different potassium-to-phosphorus ratios (R). The *dotted line* crosses the slopes at compositions of equal Curie and crystallization temperatures (deduced from the measured solubilities of each flux)

cantly larger than the coercive field [20.92] depending on the magnitude of the existing concentration gradients of charged species. The resulting *memory* effects may have an impact on the ferroelectric domain formation.

Effect of Thermal Treatment

Direct proof for the existence of a close relationship between potassium nonstoichiometry and the Curie temperatures of KTP crystals has been found while monitoring the T_C variation in samples exposed to high-temperature annealing in air. A typical example of the results is shown in Fig. 20.6. A set of 1.35–1.5 mm thick z-cut samples exhibiting an initial T_C value of 950 °C has been maintained at 1000 °C in air for variable long periods of time. Exposure to high temperatures causes partial decomposition of the crystal surface, which manifests itself in the appearance of a translucent outer layer, a few tens of micrometer thick. This layer has been removed upon cooling to room temperature, leaving behind a transparent, colorless crystal. Figure 20.6 shows that the Curie temperature can be reduced by prolonged high-temperature annealing in air down to 883 °C, i.e., even beyond the lowest T_C values obtained in as-grown KTP crystals. This effect can be explained in terms of gradual potassium escape, or increasing potassium deficiency in the crystal, upon heat treatment [20.92, 107, 108]. A moderate increase in the Curie temperature following 400 h of heat treatment is not an artifact and has been repeatedly observed with a number of samples. This may result from as-yet unknown kinetics of stoichiometry variation in high-temperature-annealed KTP crystals associated with the surface layer decomposition [20.108].

The existence of potassium nonstoichiometry in KTP crystals was realized a long time ago [20.109], but there was disagreement about the limits of its range. According to some reports, this range is narrow, less than 0.1 mol % [20.90], while others claim that it may exceed 10 mol % [20.108]. Our preliminary electron microprobe measurements indicate that the potassium deficiency may be of the order of a few mol. % and that phosphorus nonstoichiometry may exist as well [20.110]. An attempt to evaluate the KTP composition based on T_C measurements has been made only using ceramic KTP compounds synthesized at 800 and 900 °C [20.89], since the single crystal decomposes before melting. It is common to think that stoichiometry of as-grown KTP depends on the crystallization temperature alone [20.90]. However, this is true only for a single kind of self-flux (Fig. 20.5), when the saturable KTP concentration (solubility) is a direct function of temperature. A generally more correct statement is that KTP stoichiometry, and thus the Curie temperature, depend on the self-flux composition; for example, two crystals grown from 1.1 g/g $K_6P_4O_{13}$ and 0.4 g/g $K_{15}P_{13}O_{40}$ fluxes have an identical T_C value of 925 °C, yet their crystallization temperatures (1080 and 926 °C, respectively) are completely different.

Effect of Impurities

We have referred above to the gradients of potassium and oxygen vacancies V_K and V_O, respectively. However, the latter can be superimposed by the gradients of residual impurities on the various KTP cation and anion sites. Some impurities may create a charge-compensating effect and alter the magnitude of the built-in electric field. Depending on the valence and the nature of a substitutional site captured, impurity ions may enhance or diminish the concentration of oxygen vacancies that are normally charge-compensating the potassium vacancies. Table 20.3 gives a chart of the influence of various impurities located at different sites, assuming that the effective distribution coefficients (k_{eff}) of all impurities are less than one. The parallel arrows indicate enhancement of the V_O concentration gradients, while antiparallel arrows indicate their compensation with a consequent reduction in the value of the built-in electric field in the z-direction.

It should be noted that miscellaneous dopants are frequently introduced into the KTP crystals for a variety of reasons, such as alteration of their resistivity [20.90]

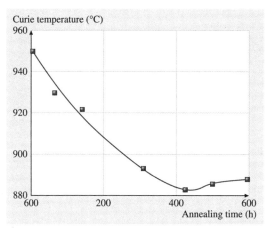

Fig. 20.6 Dependence of Curie temperature on time for long-term high-temperature annealing (at 1000 °C) of 1.35–1.5 mm thick KTP crystals grown from a 0.5 (g/g) KTP/$K_6P_4O_{13}$ flux (after [20.72])

Table 20.3 Concentration gradients of oxygen vacancies in KTP crystals grown from self-fluxes and solutions with charge-compensating dopants ($k_{\text{eff}} < 1$)

Case	Type of doping	Direction of V_O concentration gradient induced by		
		V_K in self-flux	Doping	Result
1	Self-flux	⇓	None	⇓
2	A^{3+}/Ti^{4+}	⇓	⇓	⇓ ⇓
3	A^{5+}/Ti^{4+}	⇓	⇑	⇓ or ⇑
4	A^{6+}/T^{4+}	⇓	⇑	⇓ or ⇑
5	B^{2+}/K^+	⇓	⇑	⇓ or ⇑
6	Si^{4+}/P^{5+}	⇓	⇓	⇓ ⇓
7	F^-/O^{2-}	⇓	⇑	⇓ or ⇑

⇓ Direction from the seed to a growing facet, ⇑ direction from a growing facet to the seed,
$A^{3+} = Sc^{3+}, Ga^{3+}, Al^{3+}, In^{3+}, Cr^{3+}$; $A^{5+} = Nb^{5+}, Ta^{5+}$; $A^{6+} = W^{6+}, Mo^{6+}$; $B^{2+} = Pb^{2+}, Ca^{2+}, Mg^{2+}$

or nonlinear optical properties [20.111, 112]. Lead oxide is deliberately added to the K6 self-flux [20.106] in order to enhance the crystal growth rate and to reduce the probability of spurious nucleation at the crucible wall. As a result, several hundred ppm of Pb^{2+} ions have been introduced into the crystal [20.113]. They cause a beneficial effect of significantly reducing the concentration of color centers responsible for the detrimental *gray-tracking* phenomenon [20.67] in frequency-doubling the 1.06 μm Nd:YAG laser. Most stable *gray-track* defects are attributed to the presence of oxygen-vacancy-associated Ti^{3+} centers in the KTP crystal. Addition of small amounts of Pb^{2+} ions substituting for K^+ ions provides charge compensation for potassium vacancies, and oxygen vacancies are no longer needed. (Gray-tracking will be discussed in more detail at the end of this chapter.) Reduction in the concentration of oxygen vacancies and their gradients has the effect of diminishing the built-in electric field and is, therefore, influencing the ferroelectric domain formation in KTP crystals during growth and cooling through the Curie temperature.

20.2.2 RbTiOPO$_4$ Crystals

Presently, RTP is viewed as a particularly useful member of the KTP family of crystals for electrooptic applications, such as high-frequency Q-switching and light modulation, due to its large electrooptic coefficients, high damage threshold, and the absence of piezoelectric ringing [20.14, 114]. The low-ionic-conductivity RTP crystals are required for low-leakage-current operation of the devices. Although the Rb^+ ion is larger than the K^+ ion, RTP is also a classical ionic conductor [20.39], and vacancy-assisted one-dimensional ionic conductivity is expected. It is thus important to monitor the concentration of Rb vacancies, or the degree of Rb stoichiometry, which may and usually does depend on the crystal growth conditions. The fundamental ideas about such dependence for growing RTP crystals from self-fluxes can be based on the knowledge accumulated with KTP crystals as described above. In the present section, we will review the recent results on the variable stoichiometry of RTP crystals as assessed using the Curie temperature measurements by *Roth* et al. [20.76]. Some peculiar results distinct from those observed with KTP were revealed.

Differentiation of Growth Sectors

In similarity with KTP, an RTP crystal growing out from a particular flux, e.g., $Rb_6P_4O_{13}$ (R6), and becoming enriched in rubidium in course of the process is expected to be gradually more stoichiometric in terms of Rb^+ ions. To verify this, a 300 g RTP crystal (of Fig. 20.4a type) was grown from an R6 solvent by pulling on a [100]-oriented seed, and the Curie temperatures of the top and bottom parts of the boule were measured using the dielectric (capacitance) anomaly technique. The top part exhibited $T_C = 782\,°C$, and the bottom part showed a higher T_C value by 5 °C. This result was in accord with the expected improvement of rubidium stoichiometry. The absolute T_C values depend, of course, on the specific solute content and solvent composition in the flux, to be discussed below.

Before engaging in the analysis of Curie temperature dependencies on the flux chemical composition, we will address the peculiar effect revealed in the course of taking the capacitance versus temperature characteristics in some RTP samples. These characteristics frequently feature two peaks, pointing to the fact that

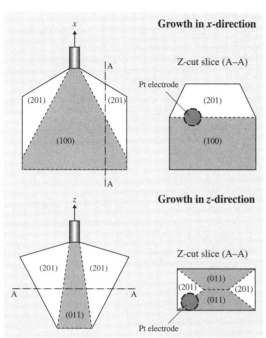

Fig. 20.7 Schematic representation of growth sectors developed in RTP crystals grown on x- and z-oriented seeds; z-cut slices may contain two growth sectors, which can be partially included in the area below the Pt electrode

two different Curie temperatures exist in the same sample. A close examination of such samples revealed that they were cut from crystal areas comprising two different growth sectors, and that the Pt electrodes covered areas containing both sectors. We recall that RTP crystals (like other KTP-family crystals) contain 14 growth sectors, displayed through four types of well-developed facets: $2 \times \{100\}$, $4 \times \{110\}$, $4 \times \{011\}$, and $4 \times \{201\}$, in its typical morphological habit (Fig. 20.2). The number of growth sectors and the volume and geometry of each particular sector depend on the growth direction and type (submerged, TSSG, with or without pulling). Figure 20.7 shows schematically the structure of main growth sectors in two RTP crystals grown by the TSSG method from R6 fluxes, one with pulling on an x-oriented [100] seed and the other without pulling and grown on a z-oriented [001] seed. Samples for T_C measurements were machined from z-cut slices along the A–A lines shown in the figure for both cases. Apparently, Pt electrodes could cover either a single growth sector or two sectors, and a double-peaked $C = C(T)$ characteristic was obtained in the latter case.

Curie Temperature Versus Composition

The intriguing result described above implies that the variation of the Curie temperature, or of the stoichiometry of RTP crystals, as a function of the flux chemical composition must be studied for each growth sector separately. In Fig. 20.8, the results for three differen-

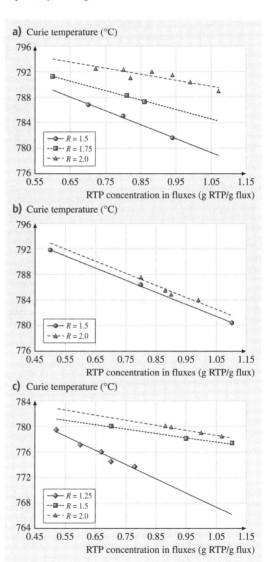

Fig. 20.8a–c Curie temperatures as a function of RTP concentration in the solution for various self-fluxes, measured separately for three growth sectors: (**a**) $\{100\}$, (**b**) $\{011\}$, and (**c**) $\{201\}$

tial growth sectors, of the {100}, {011}, and {201} types, are presented. For each sector, T_C as a function of RTP concentration (in weight units) is given for self-fluxes with different R ([Rb]/[P]) values. The nearly linear dependencies obtained are in complete similarity with the KTP case: (1) the lower the RTP concentration, in any self-flux, the higher the T_C value of the crystal, and (2) for any given RTP concentration, T_C is higher for self-fluxes with higher values of R. The combined result is that a higher Curie temperature corresponds to a higher overall concentration of the Rb ions in the solution, and therefore to higher rubidium content in the crystal. Naturally, the solution becomes gradually enriched in rubidium in the course of RTP crystal growth. The important practical consequence of this behavior is that a rubidium concentration gradient builds up in the as-grown crystal. This gradient is not averaged out during the cool-down stage, since the diffusion coefficient of the larger Rb^+ ions is presumably even smaller than that of K^+ ions.

The observed span of T_C values in RTP, from 770 to 800 °C, is much narrower than the corresponding range of Curie temperatures in KTP (880–960 °C). In addition, the slopes of the linear dependencies are shallower. We presume, therefore, that the overall extent of change of the Rb stoichiometry in RTP crystals is essentially smaller than the corresponding variation of K stoichiometry in KTP crystals. Another distinctive feature of RTP crystals is that they exhibit abrupt *jumps* in the Curie temperature over boundaries between any pair of simultaneously solidifying growth sectors of different types, as can be deduced from Fig. 20.8. It is noteworthy that a sign of a double peak in the $C(T)$ curve has been reported for the isomorphic CTA crystal [20.20], but has never been observed in nominally pure (undoped) KTP crystals. Only deliberately doped KTP may show double peaks due to the different distribution coefficients of dopants at various growth faces along the crystal–melt interfaces [20.76].

Extensive studies of nominally pure and deliberately doped RTP crystals exhibiting double peaks in their $C(T)$ curves, with a separation of over 10 °C between the T_C values on a single sample, cannot be explained by the presence of trace impurities. A different explanation must be invoked. It starts with the recognition that the RTP growth temperatures are typically 100–200 °C higher than the T_C values, and the crystals solidify in the pseudosymmetric *mmm* phase [20.116], in which the R(1) and R(2) sites are symmetrically identical [20.24]. However, the likely diverse formation mechanisms of native defects within the various growth facets at high temperatures may cause a variation in the statistical distribution of the Rb ions between the Rb(1) and Rb(2) sites during cooling through the ferroelectric phase transition. The defects may be associated not only with the rubidium and charge-compensating oxygen vacancies, but also with other stoichiometric components, namely titanium and phosphorus ions. However, initial attempts to identify any deviation from the stoichiometric composition of these components using the electron-microprobe technique did not contribute positive results.

20.2.3 Other KTP Isomorphs

Only a limited attempt has been made to study the crystal stoichiometry versus flux composition dependencies in KTP isomorphs other than RTP. This is regrettable, since the arsenates, especially KTA, are very useful for OPO applications in the near-infrared [20.15], and optically uniform long *x*-cut elements are required. The same applies to RTA crystals used for periodically poled waveguide frequency-doubling devices [20.117]. In this section, some preliminary results on the arsenate isomorphs are given.

Curie Temperature Variation

In a recent work [20.118], a series of relatively small ($< 1 \text{ cm}^3$) KTP, RTP, KTA, and RTA crystals were grown by the TSSG method from high-temperature K6 and R6 self-fluxes containing different starting concentrations of the (KTP, KTA) and (RTP, RTA) solutes, respectively. Figure 20.9 shows the Curie tempera-

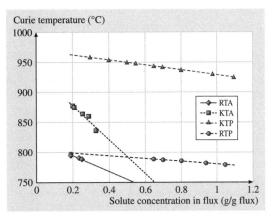

Fig. 20.9 Curie temperature dependence on solute (KTP, RTP, KTA, and RTA) concentration in the respective K6 or R6 flux for four KTP isomorphs (after [20.115])

ture dependencies of all the corresponding crystals on the solute concentration in the flux for comparison. The trend lines expanded for the two arsenate crystals do not represent accurate slopes, since only a few growth experiments have been carried out, at small solute concentrations. There is a difficulty in obtaining such crystals with reliably changing compositions from highly concentrated solutions, namely at higher growth temperatures, due to arsenic evaporation. However, the common rule for all isomorphs is quite apparent, namely, the more dilute the solution, the higher the Curie temperature, and the more stoichiometric the crystal. This reflects the fact that, at smaller solute concentrations, the crystal solidifies from solutions richer in (primarily) potassium or rubidium ions with respect to titanium.

In the preceding discussion, we have referred only to the potassium or rubidium stoichiometry. The reasons for such approach are that (1) titanium (oxide) alone is left behind during prolonged high-temperature annealing of KTP, while the potassium and phosphorus compounds evaporate [20.93], and (2) there is a general belief that PO_4 tetrahedra are the basic building blocks of the KTP structure [20.115], making phosphorus nonstoichiometry less likely. On the contrary, the existence of potassium [20.90] and rubidium [20.119] vacancies in KTP and RTP, respectively, has been demonstrated.

Concluding Remarks

To summarize this section, large crystals of KTP and its isomorphs, such as RTP, KTA, and RTA, are found to exhibit gradual compositional changes in their stoichiometric components (primarily potassium or rubidium ions) during TSSG from self-fluxes. This has been verified by measuring the ferroelectric transition (Curie) temperatures of the various crystals grown from fluxes of different initial solute concentrations. Higher Curie temperatures of crystals solidified from dilute solutions indicate their improved stoichiometry, which translates into better performance in optical devices; for example, more stoichiometric KTP becomes highly resistant to the detrimental gray-tracking in SHG [20.89]. Better stoichiometry implies also lower coercive fields necessary for efficient processing of KTP-based periodically poled nonlinear optical devices (see Sect. 20.4.1 for more detail) and orders of magnitude higher electrical resistivity, which is particularly important for the use of RTP crystals in electrooptic devices [20.14].

20.3 Growth-Induced Ferroelectric Domains

Single-domain KTP-family crystals can be readily grown from high-temperature solutions well below the Curie temperature. However, high-temperature growth is of more practical interest due to the lower viscosity or significantly larger growth rates involved [20.83]. When the high-temperature growth results in multidomain crystals, the latter can be converted into single-domain structures by complex thermal annealing [20.105, 107]. The presence of multiple domains degrades the optical uniformity and, thus, the device performance of the crystals. Numerous questions associated with the domain structure and their formation during crystal growth, cooling through the Curie temperature, and thermal annealing are still not understood to a full extent. Historically, *Zumsteg* et al. [20.116] were the first to predict the existence of a ferroelectric domain structure in KTP-family crystals. *Bierlein* and *Ahmed* [20.120] used piezoelectric, electrooptic, and pyroelectric techniques to reveal the domain structure in hydrothermally grown KTP crystals. They found domains oriented parallel to the (100) crystallographic plane and zigzag domain walls characteristic of a head-to-head domain configuration. A more complex domain structure was observed by *Loiacono* and *Stolzenberger* [20.121] in flux-grown KTP crystals. This structure, termed *dark line defects*, could not be ascribed any definite crystallographic orientation. Moreover, the domain pattern could not be altered by thermal cycling to above the Curie temperature and, thus, the domains were identified as possible nonferroelectric Dauphiné twins. *Shi* et al. [20.122] reported on the existence of a ferroelectric domain twin boundary in KTP parallel to the crystallographic (001) plane. They concluded, based on X-ray double-crystal diffraction measurements, that the (100) facet of KTP was formed by two symmetric vicinal planes with the interface between the vicinal planes being the 180° twin boundary. However, other authors [20.10, 83] pointed out that the (100) vicinal plane boundary and the corresponding 180° twinned domain wall might not coincide with the (001) plane but rather exhibit a considerably more complicated structure.

20.3.1 Domains in Top-Seeded Solution-Grown KTP

KTP crystals grown from highly concentrated solutions usually crystallize in the paraelectric phase above the Curie temperature (Fig. 20.5). The ferroelectric domain structure is formed upon postgrowth cooling, at a rate of $\approx 25-40\,°\text{C}$ in a $2-5\,\text{K/cm}$ temperature gradient. In order to visualize the domain structure, diverse techniques are employed, such as etching and piezoelectric measurements [20.107] or pyroelectric toning [20.123]. Below, visualization of the various KTP ferroelectric domain structures will be presented as obtained by simple etching in a $2:1$ (molar ratio) $KOH:KNO_3$ solution at $220\,°\text{C}$. The main parameters governing the kinetics of the diverse domain shape formation are elucidated in this section as well.

Bidomain Structure

TSSG of KTP crystals from the K6 self-flux by pulling on x-oriented seeds and initiated above the T_C (Fig. 20.4a) frequently results in a bidomain structure, namely the boule contains two large ferroelectric domains of opposite sign in a head-to-head configuration. Figure 20.10a shows the details of this structure through a crystal cross-section cut perpendicularly to the growth direction. Two central domains with vectors of spontaneous polarization P_S pointing towards the domain boundary, or inwards, usually exist. The domain boundary coincides with the z-plane or has an orientation very close to it (since this is not always a coherent boundary [20.107]) and crosses vertically the entire crystal. Additionally, the boule is enveloped by a thin domain layer, $0.5-2\,\text{mm}$ thick, and a fragment of such envelope is shown in Fig. 20.10b. Its polarization is always opposite in sign to the neighboring main domain and positive towards the outer surface. The surface appears to be the Ti (or K) positive face of the crystal. The en-

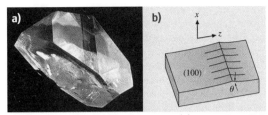

Fig. 20.11a,b Photograph of a hillock (a) on the flat (100) facet of a TSSG KTP and sketch of the associated edge-like perturbation (b)

velope domain is clearly formed during the slow crystal cooling above the melt. As a result, the outer crystal surface is charged positively, however can be reversed by an appropriate thermal treatment [20.83].

The origin of the bulk bidomain structure is of particular importance and interest. We recall that in the course of pulling KTP crystals on x-oriented seeds the growth interface is formed entirely by the large flat (100) facet. However, *Bolt* and *Enckevort* [20.124] have reported on the existence of one or several hillocks on the (100) facet, or the formation of growth surfaces vicinal to the (100) crystallographic plane at an angle of about $30'$. In fact, they have observed such hillocks on top of all facets parallel to the z-direction. Our experimental observations are consistent with their reports as well. A sample photograph of a hillock on the (100) facet of a top-seeded KTP crystal is shown in Fig. 20.11a. These linear hillocks are essentially edge-like perturbations on flat facets, described by *Chernov* [20.125] over three decades ago, and they are due to temperature fluctuations at the growth interface. A sketch of the edge-like perturbation is given in Fig. 20.11b.

Kinetics of Domain Formation

As discussed above, potassium concentration gradients are developed in the growing KTP crystal perpendicularly to the growth interface. With the existence of vicinal surfaces inclined at small angles to the z-direction, small components of the concentration gradients in this direction inevitably emerge. If the associated built-in electric field is larger than the coercive field, the direction of the concentration gradient's z-component defines the domain direction when the crystal undergoes the ferroelectric transition upon cooling. A bidomain structure is thus formed, as shown in Fig. 20.12a. Subsequent temperature fluctuations may cause additional edge-like perturbations resulting in formation of new domains by a similar mechanism. Local

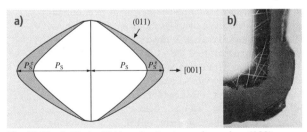

Fig. 20.10a,b Typical bidomain structure scheme in a (100) cross-section of an as-grown (above Curie temperature) KTP crystal (a) and a photograph ($\times 30$) of an envelope domain fragment (b)

variations in the magnitude of the built-in field may cause an opposite effect, namely healing of the main domain (also shown in Fig. 20.12a) leaving behind domain islands of opposite sign. Examples of such islands are shown in Fig. 20.12b, and their existence has been reported elsewhere [20.126] as well. Incorporation of residual impurities may play a role in this process, as follows from Table 20.3. Each half of the bidomain crystal can be monodomainized using high-temperature thermal treatment [20.83].

If seeded growth above the Curie temperature is enforced in such a way that the crystal is always submerged in the solution, no envelope domain is formed, but the bidomain structure can be also readily obtained. Figure 20.13a shows a y-cut cross-section of such a crystal, which generally acquires the typical morphology sketched in Fig. 20.2. The (100)- and (201)-type growth sectors are defined by dotted lines in the drawing. Clearly, the (201)-type facet forms a large angle (about 31°) with the z-direction, which results in a large z-component of the concentration gradient and a large built-in electric field. Relatively small temperature fluctuations cannot reverse the domain sign in this case, and single-domain (201)-type sectors always grow. The same applies to the (011)-type growth sectors not shown in the picture. The bidomain formation within the (100)-type growth sector is also reflected in Fig. 20.13a; similar structure occurs in the (110)-type growth sectors, since the (110) crystallographic plane is also parallel to the z-axis. Figure 20.13b demonstrates that, when the growth is initiated below the Curie temperature, the compositional gradient may be so strong across the {201} and {011} facets that a double-domain structure rather than a single-domain occasionally occurs, with a configuration as shown.

20.3.2 Domain Boundaries

We have already demonstrated elsewhere [20.107] that the domain boundary within the (100) growth sector is diffuse rather than a sharp 180° twin boundary [20.10]. Similar diffuse boundaries exist within the (110) growth sector. Of special interest are the domain boundaries at the borders of growth sectors, which are formed at the edges of two meeting facets during growth. In the following, the various types and subtypes of ferroelectric domain boundaries will be described in more detail.

Classification of Boundaries
It should be pointed out that not only the intrasector boundaries, but also the edge boundaries between

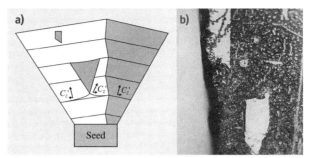

Fig. 20.12a,b Formation of ferroelectric domains in the (100) growth sector of KTP due to edge-like perturbations: (a) schematic drawing, (b) image of etched surface parallel to the [100]-direction

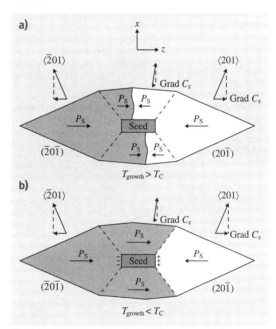

Fig. 20.13a,b Ferroelectric domain structure in immersion-seeded KTP crystals (central cuts in the y-plane) grown above (a) and below (b) the Curie temperature

(100)- and (110)-type sectors are diffuse, while those between (011)- and (01$\bar{1}$)-type sectors (Fig. 20.2) are always sharp. A boundary between (110)- and (011)-type growth sectors may not exist at all or appear in its special form shown in Fig. 20.14a. Such apex-like perturbations of the edge between the corresponding growth facets may arise [20.125] due to temperature fluctuations, and a z-component of the potassium con-

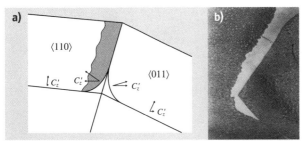

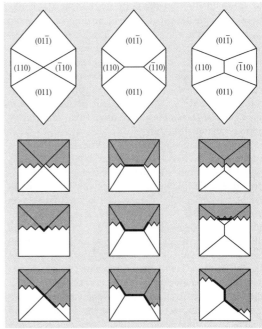

Fig. 20.14a,b Complex ferroelectric domain boundary between (110) and (011) growth sectors: (**a**) schematic drawing, (**b**) metallographic image (×40)

Fig. 20.15 Ferroelectric domain structure and types of domain boundaries in TSSG KTP crystals terminating with edges or sharp spikes in the z-direction

centration gradient is created on the (110) facet. The associated built-in electric field may be large enough to reverse the domain sign near the boundary, but only within the (110) growth sector. This results in a special type of edge domain boundary, which is sharp on the (011) sector side and diffuse on the (110) sector side. Such complex boundaries are frequently observed experimentally, as shown in Fig. 20.14b.

The variety of domain boundaries and their extent depend directly on the number of growth sectors formed and on the kinetics of growth of each sector, which in turn depend on the growth method employed and on the chemical composition of the flux. According to *Loiacono* et al. [20.46], in the case of K6 flux, complex ions of $(PO_4)^{3-}$, $(P_2O_5)^{4-}$, and $(P_3O_{10})^{5-}$ are present in the solution in a ratio $3:8:1$, leading to the ratio of $1:2:3$ for growth rates in the x, y, and z crystallographic directions, respectively. They report on a relatively close to the former growth rate ratio of $1:1:3$ for K8 and K15 fluxes, but a quite distinct ratio of $1:18:17$ (extremely slow growth of the x facet) for the K4 flux. The differences in growth rates are reflected in the variation of morphological shapes of KTP crystals [20.43]. Certain facets may be overdeveloped or underdeveloped or not exit at all. A good example is the (001) facet that is expected to exist

Table 20.4 Classification of ferroelectric domain boundaries in KTP-type crystals*

Type of domain boundary		Details and comments
Edge boundary		Traces of intersection of growing planes; they are always sharp boundaries
	– Full-size edge	{011}/{011}; {201} types never intersect
	– Partial edge	{110}/{110}, {110}/{201}, {110}/{100}
Intrasector boundary		
	– Central	Inside {100} and {110} growth sectors (diffuse boundaries)
	– Streak	Adjacent to {011}/{110} edge (sharp on the {011} side, diffuse inside the {110} sector)
	– Dispersed	Contours of small domains inside the {100} and {110} growth sectors (irregular shape)
	– Envelope	External crystal surface (for TSSG – on top of the crystal, for SSSG – the entire surface)

* {201}/{100} and {201}/{011} boundaries *do not* exist

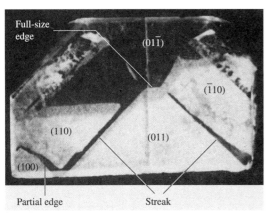

Fig. 20.16 Metallographic image of a full (011)-cut slice of a TSSG KTP crystal showing a real structure of ferroelectric domains and domain boundaries

theoretically [20.43], but which is rarely observed experimentally. Usually, only edges parallel to the x- (as in Fig. 20.2) or y-directions or just sharp spikes develop instead of this facet. These three cases are presented schematically in Fig. 20.15, where the central cut (perpendicularly to the y-axis) reveals the specific domain structure and the various domain boundaries of TSSG KTP crystals.

Similar types of diffuse and sharp domain boundaries can be found in crystals cut also along other crystallographic planes, but the y-cut slice reveals this variety sufficiently, allowing to suggest their presumably complete classification, which is summarized in Table 20.4. In this classification, the main distinction is made between domain boundaries forming at the borders of different growth sectors and within the growth sectors. Boundaries associated with the envelope domains discussed above and the irregular shape domains within the {100} growth sectors are added as a special case of intrasector boundaries. Figure 20.16 shows a real domain structure as revealed on the etched surface of a (011)-cut slice of a TSSG KTP crystal as well as part of the domain boundaries shown in Fig. 20.15 (for a y-cut surface) and listed in Table 20.4.

20.3.3 Summary of Ferroelectric Domain Structures

The foregoing analysis of the domain structure and boundaries applies to all KTP-type crystals grown by the TSSG method, both above and below the corresponding Curie temperatures. The main features observed are as follows:

- The ferroelectric domain structure of high-temperature solution grown KTP crystals is defined by the nature of growth sector development.
- Domain formation within the (100)- and (110)-type growth sectors (containing central and streak boundaries) is governed by the elementary crystal growth mechanisms.
- (100)- and (110)-type growth sectors exhibit predominantly bidomain and rarely multidomain structures; (011)- and (201)-type growth sectors are single domain.
- The bidomain, or 180° twin domain, is not a twin in the crystallographic sense.
- The diverse domain boundaries can be classified into edge type (partial and full size) forming at intersections of growth sectors and intrasector type (central, streak, dispersed, envelope).
- Domain orientation is defined by the direction of the built-in electric field originating from the concentration gradients of stoichiometric components (mainly potassium in KTP) and cation impurities.

20.3.4 Single-Domain Growth

The kinetics of ferroelectric domain formation during growth of KTP crystals described above allows to suggest a number of avenues for growing single-domain crystals. A few examples are given below:

- Meniscus growth with pulling on (201)- or (011)-oriented seeds yields single-domain growth immediately, which becomes clear from the observation of Fig. 20.13. Large compositional gradients formed upon crystallization on {201} and {011} facets imply high built-in electrical fields, preventing domain inversion and keeping the body of the crystal enclosed within these facets' single domain. The drawback of this method is that *boot-like* growth occurs due to the lack of symmetry in these directions, and the crystals may hit the crucible wall before a large chunk of crystal is grown.
- TSSG with pulling on a z-oriented seed generally starts with a multidomain structure, but eventually converts into a single domain when the main body of the crystal is formed solely by the (201)- and (011)-type growth sectors as well.
- Bidomain growth is frequently also acceptable for practical applications, when the crystals are large

enough. Such growth has already been discussed above for the case of TSSG with pulling on x-oriented seeds. Naturally, the same applies to growth on $\langle 110 \rangle$-oriented seeds. Hereby, there is always a danger of multidomain formation, which can be prevented by minimizing temperature fluctuations at the growth interface.

- TSSG with pulling in the y-direction behaves as follows: the (110)-type facets grow out fast, and the growth is governed by two (011)-type sectors, resulting in a bidomain growth with a sharp $(011)/(01\bar{1})$ domain boundary.

We reiterate that growth of KTP or KTA below the Curie temperature, even by several tens of degrees, may result in similar domain formation features due to the large potassium concentration gradients, or large built-in electric fields involved. The rubidium isomorphs follow these rules very closely as well.

20.4 Artificial Domain Structures

Single-domain KTP-family crystals are indispensable for most nonlinear optical and electrooptic applications; for example, their excellent birefringent phase-matching properties, such as appreciable effective nonlinear coefficients, broad temperature tolerance, and large angular bandwidth, make them highly suitable for frequency doubling of near-infrared lasers. Unfortunately, in the useful type II phase-matching scheme, the optical birefringence is too small to allow for frequency doubling into the green by KTA or into the blue by KTP [20.127]. In the latter case, it is possible to obtain the important blue coherent radiation with KTP using sum-frequency generation [20.128] or broadening its birefringence range by doping [20.129] or with the thermally unstable $KNbO_3$ crystal [20.130]. However, all these methods are not attractive due to their technological complexity. An appropriate alternative is the use of quasi-phase-matching (QPM) based on periodic domain structures (PDS). The QPM technique, first suggested by *Armstrong* et al. [20.131], corrects the relative phase of the fundamental and secondary waves at regular intervals by means of structural periodicity built into the nonlinear medium. Phase-matching is thus achieved by periodic spatial modulation of the nonlinear coefficient along the direction of propagation. The most efficient implementation of QPM occurs when the sign of the nonlinear coefficient is periodically reversed, which can be achieved by periodic domain inversion in a ferroelectric material. This has been initially performed by means of fabricating several-micrometer-deep KTP waveguides by the ion-exchange method [20.132–134], in which Rb^+ and Ba^{2+} ions replace K^+ ions. Attempts to obtain deeper periodic domain structures involve scanned electron beams [20.135] in KTP and, more recently, atomic force microscopy (AFM) cantilever tips [20.136] in RTP. Well-controlled domain inversion in more than 1 mm thick KTP-type crystals with down to a few μm small period structures has been obtained reproducibly using electrical field poling, and we will review this technique and structures in more detail below. As-grown periodic domain structures allowing for larger apertures and, thus, for higher-average-power laser applications will be discussed as well.

20.4.1 Electric Field Poling

Fabrication of QPM PDS by external electrical field poling has been initially implemented in $LiNbO_3$ [20.137, 138], where the coercive field is $\approx 21\,kV/mm$ (congruent crystals) and the wafer thickness is limited to 0.5 mm due to the high field (24 kV/mm) needed to produce domain inversion. The coercive field of KTP and its isomorphs is about an order of magnitude lower and, unlike $LiNbO_3$, they do not suffer from photorefractive damage. The main problem of KTP-type crystals is the relatively high ionic conductivity, which may complicate the periodic poling. Among the techniques employed to overcome this problem are ion exchange in the surface layer and low-temperature poling. Alternatively, more stoichiometric (in potassium or rubidium) crystals exhibiting lower ionic conductivities can be used. These approaches will be discussed in more detail below.

Periodically Poled KTP

KTP is the most readily available material within the family, and the main bulk of published work in the field of QPM structures describes periodically poled KTP (PPKTP). The conventional process of periodic domain structure fabrication [20.137–139] is based on the application of an electric field to a photolithographically patterned electrode on the z-cut crystal surface, com-

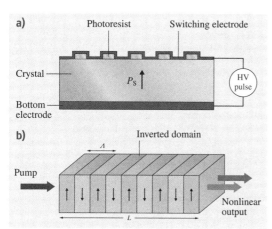

Fig. 20.17 (a) Conventional experimental setup for fabrication of domain gratings; (b) basic structure of electrically poled crystal for frequency conversion. The SHG efficiency, $\eta \propto d_{\text{eff}}^2 L$ [20.140], with L the crystal length, and the effective nonlinear coefficient for SHG of an order of m is given by [20.141] $d_{\text{eff}} = (2/\pi m) \sin(\pi m D) d_{33}$, where D is the duty cycle (ratio between inverted and uninverted domain widths). With $m = 1$ for the first-order interaction and $D = 0.5$, the optimal $d_{\text{eff}} = 2d_{33}/\pi$. Most importantly, the z-direction of light propagation with the highest nonlinear coefficient d_{33} can be chosen

prising a periodic array of metal and insulating stripes, as shown in Fig. 20.17a, or forming a grating. For SHG, the grating period (Λ) shown in Fig. 20.17b is related to the refractive indices of the fundamental (pump) and second-harmonic waves, $n_{2\omega}$ and n_ω, respectively, by

$$\Lambda = \frac{\lambda_\omega}{2(n_{2\omega} - n_\omega)}, \qquad (20.5)$$

where λ_ω is the pump wavelength. For $\lambda_\omega = 0.84\,\mu\text{m}$, frequency doubling into the blue requires PPKTP with $\Lambda = 4\,\mu\text{m}$ [20.139], while for $\lambda_\omega = 1.06\,\mu\text{m}$ (Nd:YAG laser) doubling into the green $\Lambda \approx 9\,\mu\text{m}$ [20.142]. Much larger periods are sufficient for OPO, where they are deduced from the first-order ($m = 1$) phase-matching condition [20.138]

$$\frac{1}{\Lambda} = \frac{n_\text{p}}{\lambda_\text{p}} - \frac{n_\text{s}}{\lambda_\text{s}} - \frac{n_\text{i}}{\lambda_\text{i}}, \qquad (20.6)$$

where n_p, n_s, and n_i are the pump, signal, and idler refractive indices, respectively. The signal and idler outputs are shown by arrows in Fig. 20.17b. For example, for an OPO interaction with $\lambda_\text{p} = 1.06\,\mu\text{m}$, the period required is $\Lambda = 37.8\,\mu\text{m}$ in PPKTP [20.143].

As shown in Fig. 20.17, the pulsed switching voltage is applied to the polar z-faces of the crystal through metal electrodes. The required periodic domain inversion is obtained if spontaneous polarization switching occurs only under the micrometer-wide metal stripes (top electrode) that are in intimate contact with the crystal and does not occur below the dielectric photoresist layers of the same width. If polarization switching below the insulating stripes is not prevented, domain broadening takes place below them [20.144], which degrades the duty cycle of the device (Fig. 20.17b). The resulting inverted domain narrowing is due to high tangential fields arising immediately below the dielectric barrier layer and causing depolarization by internal charge carriers [20.144]. The kinetics of depolarization thus depends strongly on the crystal electrical conductivity. In insulating ferroelectric crystals, e.g., LiNbO$_3$, the polarization relaxes very slowly ($\tau \approx 10^5$ s), while in KTP and its isomorphs exhibiting high ionic conductivity the relaxation process is very fast ($\tau \approx 0.2\,\mu\text{s}$) [20.144]. One way of decreasing the relaxation time, or screening of depolarization, is the reduction of electrical conductivity by ion exchange.

A detailed experimental account of ion-exchange-assisted poling of KTP crystals was given by *Karlsson* and *Laurell* [20.145]. They made use of the fact that the larger Rb$^+$ ions had a lower mobility than K$^+$ ions in the KTP lattice and indiffused Rb$^+$ ions, by immersing flux-grown KTP crystals in a RbNO$_3$ melt, to obtain low conductivity ($\sim$ three orders of magnitude lower than the bulk) thin (2–5 μm) layers of Rb$_x$K$_{1-x}$TiOPO$_4$ at the two z-faces, where x varies gradually from 100% at the surface to 0% in the bulk. When voltage is applied to such crystal, the field in the exchanged layer is significantly higher than in the bulk, due to the difference in conductivity. At a certain voltage the coercive field is exceeded in the exchanged layer and an inverted domain nucleates. It continues growing towards the opposite side of the crystal under the influence of the remaining field in the bulk. The periodicity of the field near the surface is well defined, and the domains formed there can grow in a low field (lower than the coercive field) without significant broadening. Up to 1 mm thick PPKTP wafers could be fabricated using this method, and efficient first and higher-order green SH signals were obtained with 9.01 and 31 μm domain gratings [20.145] and blue SH with 6.09 μm gratings [20.146].

Another method of poling under reduced KTP ionic conductivity and requiring no chemical treatment of the crystal is the low-temperature polarization switching suggested by *Rosenman* et al. [20.147]. They used 0.5 mm thick KTP wafers with a room-temperature conductivity of $10^{-7}\,\Omega^{-1}\mathrm{cm}^{-1}$. The latter reduced to about $10^{-12}\,\Omega^{-1}\mathrm{cm}^{-1}$ at $T=170\,\mathrm{K}$ due to the freeze-in of K^+-ion migration. This increased the relaxation time τ by five orders of magnitude as well, and polarization switching under the dielectric layer stipes (Fig. 20.17) was prevented. One negative feature of the low-temperature poling method is that it causes a strong increase in the coercive field, but even at $T=170\,\mathrm{K}$ it is still twice as low as in LiNbO$_3$ [20.147]. High-quality PPKTP elements with 9 μm domain grating period were fabricated using this method for efficient frequency doubling into the green [20.142] and OPO [20.148], while no optical damage due to gray tracking (green-induced infrared absorption) was observed up to continuous pump power of 4.5 kW/cm^2.

In spite of the tangible technological progress in PPKTP processing, many aspects of the field-induced ferroelectric domain formation still remain unclear, and additional research efforts have to be made for their understanding. Clearly, the domain propagation speed along the polar z-axis is orders of magnitude larger than that in the x–y plane [20.149] in consonance with the anisotropy of ionic conductivity. In addition, anisotropy in the domain wall propagation velocity between the x- and y-directions exists as well, being 6–30 times larger in favor of the latter [20.149, 150]. Room-temperature polarization reversal in conventional KTP superionic conductor results in spontaneous nucleation of domains in the crystal bulk, since the mobile potassium ions can screen the depolarization field of the nucleating domain [20.151]. Additional observations related to room-temperature poling include the widely known features, such as domain nucleation at the electrode edges, domain broadening due to charge deposition underneath the dielectric layer and overpoling by multiple pulses, lateral domain growth and merging before reaching the opposite electrode, and also nonuniformity of the domain growth kinetics across the wafer. The latter is attributed to the variable stoichiometry of the KTP crystal [20.152]. Some attempts to return to the doping schemes for increasing the KTP resistivity for poling have been made lately as well [20.153].

High-Potassium KTP

Improvement of the morphology of PPKTP domain structures by ion exchange or low-temperature poling discussed above incorporates also certain disadvantages. Ion exchange is a two-step method involving chemical treatment of the crystal and resulting in variable success. Low-temperature poling requires application of external electric fields above the coercive field, which is as high as 12 kV/mm [20.147]. Therefore, a better alternative is to use more stoichiometric (in potassium) KTP crystals, exhibiting lower ionic conductivities and allowing poling at higher temperatures and lower voltages. Such an attempt has been made by *Rosenman* et al. [20.154], who have studied polarization reversal in three different KTP crystals denoted as HK, IK, and LK for high, intermediate, and low potassium content, since it is quite difficult to determine the exact potassium stoichiometry. The conductivities of the samples varied from 10^{-12} to $10^{-8}\,\Omega^{-1}\mathrm{cm}^{-1}$ at room temperature. Both HK and IK samples exhibited a relatively long polarization switching relaxation time, $\tau \approx 0.18\,\mathrm{s}$, which is sufficient for preventing domain broadening. Their measured room-temperature coercive field was less than 3 kV/mm, namely four times lower than that of LK at 170 K. The crystallographically more perfect 0.5 mm thick IK sample was used for actual poling at 248 K under a 4.6 kV/mm electric field. The duty cycle obtained was 50%, which indicated a complete lack of domain broadening. The element with a grating period of 24.7 μm was subsequently used for successful doubling of 1.55 μm radiation from a 18 mW diode laser into the deep red.

In a follow-up paper, *Jiang* et al. [20.155] grew HK-grade KTP crystals from a highly potassium-containing K4 flux for electric poling application. Their results confirmed that crystals with better potassium stoichiometry exhibited significantly lower ionic conductivities and lower coercive fields, allowing successful room-temperature poling. Application of a less than 3 kV/mm electric field to a 1 mm thick HK-grade KTP crystal resulted in polarization reversal throughout the sample thickness, and PPKTP elements with grating periods of 60 μm could be obtained. A surprising detail pointed out by the authors was that periodic poling for the grating vectors along [100]- and [010]-directions gave similar results, which was presumably sample dependent, since it was never observed with the usual smaller 9 μm gratings.

KTP Isomorphs

Very few attempts have been made to produce PPKTA crystals, since in addition to the difficulties of arsenate growth the dielectric and ferroelectric properties of KTA are hardly better than those of KTP crystals. In fact, the arsenate crystallographic framework is even more open [20.25, 26], facilitating the enhanced chiral mobility of the K$^+$ ions in KTA. Consequently, the superionic-to-insulating transition occurs at even lower temperature than in KTP, and the crystal temperature must be reduced to 150 K for low-temperature poling [20.156]. The reported coercive field is as high as 90 kV/cm at $T = 150$ K, and 0.5 mm thick KTA wafers have been poled at this temperature under 5 kV switching bias to obtain PPKTA elements with grating periods of 37.4 and 39 μm. It was suggested that domain inversion hereby occurred by twinning across the (100) domain walls with As(1) acting as linking atoms [20.157]. KTA has better transparency in the infrared than KTP [20.15] and is preferred for OPO applications, hence the large grating periods prepared [20.156].

Periodically poled rubidium isomorphs, namely 1 mm thick PPRTP [20.158, 159] and 3 mm thick PPRTA [20.143, 160, 161], have been prepared as well. Unlike KTP and KTA, commercially available RTP and RTA crystals have low electrical conductivities in the 10^{-12}–10^{-8} Ω^{-1}cm^{-1} range [20.14, 143]. This is due to the larger size of Rb$^+$ ions in comparison with K$^+$ ions and thus their lower mobility along the polar axis. The much narrower span of the Curie temperatures in RTP crystals (Fig. 20.8) than in KTP (Fig. 20.5) also indicates a smaller deviation from stoichiometry, which is likely in the case of RTA as well. As a result, PDS in RTP and RTA can be obtained by electrical poling at room temperature. The reported room-temperature coercive fields of RTP vary from 2.65 to 6 kV/mm, obviously depending on the specific resistivity of the crystals used [20.40, 149]. Typically, ≈ 4 kV voltages are used for creating PPRTP homogeneous structures over more than 80% of the 1 mm depth [20.158]. Both HK KTP and RTP crystals can be successfully poled at room temperature for a variety of applications. Unfortunately, the commercial production of PPKTP devices existing to date is based mostly on the low-temperature poling technology. Voltage pulses as low as 5.3 kV are sufficient for obtaining 3 mm thick PPRTA elements, also at room temperature [20.143]. PPRTA, like PPKTA, is used mainly in OPO applications [20.143, 162–164] due to the lower absorption in the mid-infrared. The detailed study of the thermal dependences of refractive indices performed for RTP and RTA shows that PPRTP is more appropriate for applications that require immunity to thermal lensing, whereas PPRTA is suitable for realizing temperature-tuned nonlinear devices [20.165]. Of great current interest are PPRTP-based submicrometer-size ($\Lambda = 1.18$ μm) domain structures produced in a 200 μm thick RTP crystal by applying a voltage of 650 V from a high-voltage atomic force microscope [20.136]. Such PPRTP nanodomain gratings are used, for example, in backward-propagating quasi-phase-matched converters [20.136] and noncollinear SHG [20.166]. It is noteworthy that, concurrently, 0.5 mm thick flux-grown KTP crystals have been poled by the ion-exchange-aided method with application four 1 ms long 1 kV pulses to produce submicrometer PPKTP gratings with a period as small as 800 nm [20.167]. The latter have been used for demonstrating electrically adjustable Bragg reflectivity.

20.4.2 As-Grown Periodic Domain Structure

Domain miniaturization in electrically poled PDS, discussed at the end of the previous paragraph, is of great interest for the development of novel photonic devices. However, the ability to produce large-aperture PDS for high-power QPM nonlinear optical devices is of equal technological importance. The straightforward way of generating periodic domains in the bulk of large ferroelectric crystals is to introduce appropriate temperature oscillations during their growth to induce striations associated with impurities. This has been attempted with other ferroelectric crystals, such as rare-earth-doped LiNbO$_3$ [20.168, 169] and Li-doped KTa$_x$Nb$_{1-x}$O$_3$ [20.170]. In the case of KTP, the influence of the various impurities on the direction of impurity-related concentration gradients and, therefore, domain formation has been discussed in Sect. 20.2.1. An additional advantage of KTP over LiNbO$_3$ is that the latter grows (by Czochralski pulling from the melt) with a curved interface, whereas KTP exhibits a flat facet morphology. Until now, only a single attempt to grow a large KTP crystal with a periodic domain structure has been made [20.106], and it will be described comprehensively below.

Kinetics of Impurity Incorporation

It has been pointed out in Sect. 20.3.4 that TSSG growth of KTP crystals by pulling on z-oriented seeds eventually results in stable single-domain growth due to large potassium concentration gradients, or their vacancies, in the ⟨201⟩ and ⟨011⟩ growth directions. However, the presence of charge-compensating impurities, such as divalent ions (Mg^{2+} or Pb^{2+}) on K$^+$-sites, may minimize or eliminate these gradients and the associated built-in electric fields. In this case, periodic variation in the domain sign during crystal cooling through the Curie temperature can be achieved by imprinting a periodic variation in the impurity effective distribution coeffi-

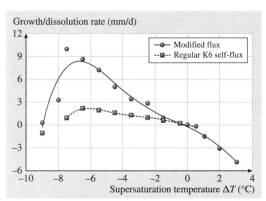

Fig. 20.18 Growth and dissolution rate of KTP in z-direction

cient [20.171]

$$k_{\text{eff}} = \frac{k^*}{k^* + (1-k^*)\exp(-f^*\delta_D/D_L)}, \quad (20.7)$$

where k^* is the interface distribution coefficient, f^* is the interfacial linear growth rate, δ_D is the diffusion boundary layer, and D_L is the diffusion coefficient. Since $\delta_D \sim \sqrt{\nu/\omega}$, where ν and ω are the kinematic viscosity and crystal rotation rate, respectively, one way of altering k_{eff} is to introduce a variable rotation rate. This is complex, since the hydrodynamic flow may be altered and is then difficult to control. The second approach is to vary the growth rate by controlling the interface temperature.

The average linear growth rate can be expressed by [20.172]

$$\langle f \rangle = \frac{V}{A\rho}\left(\frac{dC_e}{dT}\right)\frac{\Delta T}{\Delta t}, \quad (20.8)$$

where V is the volume of the solution, A is the growing crystal area, ρ its density, C_e is the equilibrium concentration of the solute, dC_e/dT is the change of solubility per degree (known from the solubility curve), ΔT is the undercooling, and $\Delta T/\Delta t$ is the cooling rate. The magnitude of undercooling is defined by the width of the metastable zone, which has been studied experimentally for the regular K6 and modified K6 (addition of PbO) fluxes [20.106]. The results are reproduced in Fig. 20.18, showing the growth or dissolution rate as a function of undercooling.

Temperature Oscillations and Crystal Growth

It is apparent from Fig. 20.18 that the width of the metastable zone almost does not depend on the type of flux, and its average value is 6.6 K. This is remarkably close to the theoretically estimated value of 6.67 K [20.173] for the K6 self-flux. The average undercooling, $\langle\Delta T\rangle$, must be of the order of half of the width of the metastable zone, or about 3.3 K, for normal KTP growth. If an additional periodic, e.g., sinusoidal, temperature oscillation is superimposed on the average undercooling, the actual undercooling can be expressed by

$$\Delta T = \langle\Delta T\rangle + B\sin(2\pi t/\tau), \quad (20.9)$$

where τ is the time period of temperature oscillation and the constant B must be chosen so that the undercooling does not exceed the limits of the metastable zone. As a result, growth rate oscillations will occur followed by variable (positive and negative) concentration gradients of the solute and production of a periodic domain structure. The domain metric period, Λ, depends on the

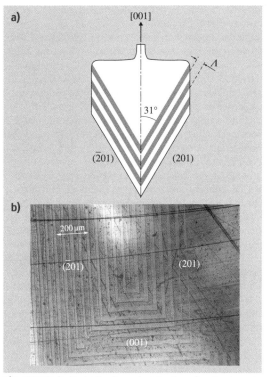

Fig. 20.19a,b Periodic ferroelectric domain structure in as-grown KTP crystal pulled in the z-direction: (a) schematic drawing of domain formation; (b) etched z-cut slice showing the rectangular periodic domain pattern within {201} and {011} growth sectors

average growth rate and the time period of temperature oscillations through the straightforward relation

$$\langle f \rangle \tau = \Lambda \,. \tag{20.10}$$

Growth experiments involving an undercooling oscillatory regime by multistep programming of the growth system's temperature controller yield large KTP crystals with a well-defined periodic domain structure. The geometry of such structure for TSSG with pulling in z-direction is presented schematically in Fig. 20.19a. When the {201} and {011} facets are fully developed, temperature oscillation is turned on, and the periodic domain structure is formed with period Λ. This is not the true period for nonlinear interactions; the latter is $\Lambda/\cos(31°)$, since wafers are cut perpendicularly to the [001] axis in order to utilize the largest d_{33} nonlinear coefficient.

Figure 20.19b shows an etched surface of a fraction of a real z-cut slice, where clearly defined domain structures are visible within the {201} and {011} growth sectors. The domain contours are straight since both the {201} and {011} are perfectly flat. The grating period in this picture is $50\,\mu\text{m}$, but structures with Λ of $25-40\,\mu\text{m}$ have been easily obtained. They can be effectively used for a variety of OPO applications. Frequency doubling requires smaller periods, such as $9\,\mu\text{m}$, for SHG of the $1.06\,\mu\text{m}$ radiation of the Nd:YAG laser, and there is no principal limitation in creating the narrower domain as-grown structures. The quality of the PDS is determined by the value of the duty cycle, $D = a/\Lambda$, where a is the actual inverted domain width. The optimal effective NLO coefficient and, thus, frequency conversion efficiency is obtained at $D = 0.5$ [20.142]. The D values in Fig. 20.19b are in the $0.2-0.5$ range along $10\,\text{mm}$ of the wafer in the [100] direction (within a single growth sector), which are still useful for practical implementation of the as-grown PDS. It is noteworthy that in $LiNbO_3$ reasonable quality PDS can be obtained only along $1\,\text{mm}$ due to the curved Czochralski growth interface where the inverted domains are obtained.

20.5 Nonlinear Optical Crystals

The combined knowledge of KTP-type crystal morphology development, compositional variations, and ferroelectric domain formation kinetics during growth allows one to adapt the crystal growth method to the requirements of particular applications. It has been pointed out in the previous section that TSSG of KTP crystals with pulling in the [001]-direction is most suitable for creating as-grown PDS. Crystals grown by top-seeding with pulling from K6-forming fluxes in the [100]-direction are the best choice for extracting x-oriented (for highest electrooptic coefficients) Q-switching elements, since the large flat (100) facet (Fig. 20.2) assures excellent transverse optical homogeneity needed for maximum extinction ratio [20.118]. This approach is specifically implemented in producing RTP Q-switching devices [20.14, 114]. Fortunately, TSSG with pulling in both [001]- and [100]-directions allows to obtain large volumes of single-domain crystals, as summarized in Sect. 20.3.4. In addition, establishment of compositional gradients during growth and poor potassium stoichiometry in general can not only influence the domain formation mechanisms (Sect. 20.3.1) but also affect the optical properties of KTP crystals. In the following, we will analyze the nonuniformity in the distribution of refractive indices associated with compositional gradients [20.118] and generation of detrimental point defects related to potassium vacancies [20.89].

20.5.1 Optical Nonuniformity

Every application of nonlinear optical and electrooptic crystals requires variable degrees of optical uniformity. However, one has to make a distinction between the relative importance of transverse and longitudinal uniformity with respect to the direction of propagation of the laser beam. While x-oriented crystal pairs of electrooptic Q-switches are not too sensitive to longitudinal distribution of refractive indices [20.14], their inhomogeneity along the OPO interaction path may lead to phase mismatch and, consequently, to degradation of the OPO interaction efficiency. In this section, we will mainly focus on this latter example.

Distribution of Refractive Indices
The variation of refractive indices in large solution-grown KTP crystals, from the seed to the crystal periphery, has been studied experimentally by transmission interferometry and reported in a series of works by *Sasaki* et al. [20.60, 82, 104]. They have discovered the

existence of constant gradients of the refractive indices in different crystallographic directions of TSSG-grown KTP crystals, which are ascribed to compositional variation in the solute, in agreement with the Curie temperature variation described in Sect. 20.2.1. Particularly useful are the results of [20.82], where growth and characterization details of a large TSSG-grown crystal are given including the initial flux composition and saturation temperature, final crystal weight, and the end values of the refractive indices measured at 633 nm. With the knowledge of KTP solubility in K6 [20.10], their data allow to calculate the average variation of the refractive indices per degree for growing out crystal volumes upon the solution temperature reduction. The values obtained are [20.118]

$$\frac{\Delta n_{x,y}}{\Delta T} \approx 10^{-7}\,\mathrm{K}^{-1} \quad \text{at 633 nm},$$
$$\frac{\Delta n_z}{\Delta T} \approx 2 \times 10^{-6}\,\mathrm{K}^{-1} \quad \text{at 633 nm}. \quad (20.11)$$

Assuming, as a simplifying approximation, that Δn varies linearly with the increase of the solidified mass and depends weakly on wavelength, (20.11) are a universal result, independent of the specific method of solution growth (Fig. 20.3). In particular, this result can be applied to crystals obtained by TSSG with pulling, and their optical uniformity in specific crystallographic directions can be assessed if certain growth parameters are known.

Figure 20.20 shows a photograph of an almost ideal TSSG (meniscus-pulled in the [100]-direction) KTP crystal of ≈ 850 g weight. A nearly constant growth interface area has been maintained after developing the shoulders of the crystal by careful monitoring the temperature reduction program and the pulling rate. The height of the nearly rectangular part of the crystal l is about 50 mm, and the temperature drop during its growth was 75 K. Using these parameters and the data of (20.11), the following average gradients of the refractive indices along the crystal length can be calculated

$$\frac{\Delta n_z}{\Delta l} = 3 \times 10^{-5}\,\mathrm{cm}^{-1},$$
$$\frac{\Delta n_{x,y}}{\Delta l} = 1.5 \times 10^{-6}\,\mathrm{cm}^{-1}. \quad (20.12)$$

OPO Interaction in KTP

[100]-Cut KTA and KTP crystals are widely used in frequency conversion of Nd-doped solid-state lasers to longer wavelengths by OPO, primarily for eye-safe lidar systems operating in the 1550 nm spectral region [20.15, 16]. One of the major advantages of KTP-family crystals is the ability to operate with noncritical phase matching (NCPM), allowing large acceptance angles of the incident beam. The basic configuration of the type II process in a KTP-type crystal pumped in the x-direction is shown in Fig. 20.21. In this interaction, a y-polarized pump (p) generates a y-polarized signal (s) and z-polarized idler (i) beams, and the corresponding frequencies are bound by the following equations:

$$n_y^p \omega^p = n_y^s \omega^s + n_z^i \omega^i,$$
$$\omega^p = \omega^s + \omega^i, \quad (20.13)$$

which are the perfect NCPM phase-matching conditions, namely the momentum and energy conservation laws, respectively. (The frequencies can be recalculated from the known wavelengths, which for the Nd:YAG-

Fig. 20.20 TSSG KTP crystal meniscus pulled on a [100]-oriented seed

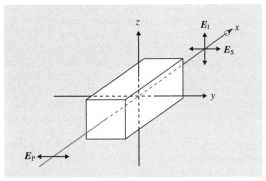

Fig. 20.21 Noncritical phase-matching scheme for type II OPO in the x–z plane of a KTP crystal ($\boldsymbol{E}_p$, $\boldsymbol{E}_s$, and $\boldsymbol{E}_i$ are the appropriate electric field polarization vectors)

induced NCPM OPO in KTP are $\lambda^p = 1064.2$ nm, $\lambda^s = 1576$ nm, and $\lambda^i = 3277$ nm [20.16].)

It is well known that the inherently present variation of the refractive indices due to gradual compositional changes inevitably causes a phase mismatch during the interaction. This mismatch can be evaluated by deriving the associated bandwidth of the generated signal wave $\Delta\omega^s$. If we disregard the negligible change in n_y [20.104] and assume that only $\Delta n_z \neq 0$, the signal bandwidth is given by

$$\Delta\omega^s = \frac{\Delta n_z^i}{n_z^i - n_y^s} \omega^i. \qquad (20.14)$$

Keeping in mind that Δn_z does not vary dramatically with wavelength, the value of $\Delta n_z^i = 3 \times 10^{-5}$ for a 1 cm long crystal, from the first of (20.12), can be used as a rough approximation. The relevant refractive indices can be calculated from the Sellmeier equations for KTP [20.10] as $n_z^i = 1.7717$ and $n_y^s = 1.7362$. Final calculation using these parameters gives $\Delta\omega^s = 2.6$ cm^{-1}, which is well within the KTP/OPO acceptance bandwidth of 9.02 cm^{-1} for a 1 cm interaction length [20.174]. This explains the high NCPM OPO conversion efficiency achieved even by pumping 20 mm long KTP crystals with a multimode laser [20.16].

It is also important to remember that, with a proper choice of the growth method, KTP-type crystals of sufficient optical uniformity can be obtained for both nonlinear optical and electrooptic applications. We recall that the main cause of inhomogeneity in the distribution of refractive indices along the growing crystal is the composition gradient, mainly of potassium ions, building up in the course of growth from self-fluxes. Such gradients are considerably smaller in the case of one-dimensional meniscus growth with pulling in the x-direction than in the case of three-dimensional submerged-seeded growth (Fig. 20.3).

20.5.2 Gray Tracks

In some applications where KTP is used to generate the second harmonic of the 1064 nm Nd:YAG laser radiation [20.175–179] or OPO pumped at 532 nm [20.180], detrimental optical absorption often builds up in the crystal bulk following intense high-repetition pulse or continuous-wave (CW) pumping. This phenomenon is termed gray-tracking, since darkening is observed along the laser beam path, which is attributed to formation of color centers absorbing in most of the visible 400–700 nm range [20.67] and emitting in the 700–1200 nm range under green laser excitation [20.181]. The induced absorption diminishes the nonlinear output power and causes significant local heating, leading to severe beam distortion. It is imperative, therefore, to understand the nature of color centers involved and to find the means of their suppression.

Origin of Gray Track Centers

Gray tracks have been unambiguously related to the presence of Ti^{3+} ions [20.67, 178, 182], which are electron color centers induced photochemically via electron trapping by the constitutional Ti^{4+} ions. The latter can be reduced to Ti^{3+} also by high-temperature annealing in vacuum or by charge transfer (e.g., $Fe^{2+} \rightarrow Ti^{4+}$) from adjacent transition-metal impurities [20.108]. Yet, in thermally untreated and nominally pure KTP crystals, EPR and electron-nuclear double resonance (ENDOR) measurements have revealed four distinct types of Ti^{3+} centers as well [20.183, 184]. Two of them, appearing only in hydrothermal KTP, have a Ti^{3+} ion occupying the inequivalent Ti(1) and Ti(2) sites with a proton (in the form of an OH$^-$ ion) bonded to an adjacent oxygen ion as a stabilizing entity for the electron trapping. The remaining two types of centers are present only in flux-grown crystals. One of them represents a self-trapped electron at the Ti(1) site stabilized by a neighboring divalent cation impurity, according to a tentative suggestion [20.184], but it is unstable above 140 K. The other is a Ti^{3+} ion located at the Ti(2) site adjacent to an oxygen vacancy and is stable at room temperature for days or weeks.

It is now well established that the reason for the existence of electron-trap-stabilizing OH$^-$ ions (in hydrothermal crystals) and oxygen vacancies (in flux-grown crystals) is charge compensation of K$^+$ vacancies present to variable degrees in all KTP crystals [20.185], as also discussed in Sect. 20.2. In an attempt to reduce the concentration of oxygen vacancies, PbO was added to the KTP growth flux to form the modified flux referred to in Sect. 20.4.2. A Pb^{2+} ion, with its 6s^2 closed shell configuration, has an ionic radius approximately 0.2 Å smaller than a K$^+$ ion, and thus easily substitutes for the K$^+$ ion to provide charge compensation for one potassium vacancy. Oxygen vacancies are no longer needed to stabilize the Ti^{3+} color center if a sufficient amount of the Pb^{2+} dopant is present. Instead, [Ti^{3+}-Pb^{2+}] centers may form under intense laser irradiation. Four types of these centers have been revealed by EPR and ENDOR measurements [20.113], but they are all unstable above 250 K and thus short-lived at room temperature. Of course, operation under the extreme conditions of high-power, high-repetition-rate or CW pumping will cause the transient formation of large concentrations of Ti^{3+} centers with the detrimental optical absorption in the visible. Therefore, doping with lead is only a partial remedy. A better way to eliminate gray-tracking is the reduction of potassium vacancies, or improvement of potassium stoichiometry, in as-grown KTP crystals by using K-rich solvents and/or more dilute solutions as discussed in Sect. 20.2.

Kinetics of Gray Track Formation

Figure 20.5 shows explicitly that KTP crystals grown at lower temperatures, especially from self-fluxes with higher [K]/[P] ratios, are more stoichiometric, exhibiting higher Curie temperatures. In order to test the influence of potassium stoichiometry, a comparative study of the kinetics of gray track formation in high- and low-T_C KTP crystals has been conducted [20.89]. This has been done using the green-radiation-induced infrared absorption (GRIIRA) method which has been initially developed for periodically poled $LiNbO_3$ [20.186]. In the quoted experiment [20.89], a $10\,kW/cm^2$ CW green (532 or 514 nm) excitation beam was used to induce the gray track damage, while the resulting infrared (1064 nm) absorption was monitored. At the initial stage of gray track formation, usually lasting from several seconds to several minutes, an increase in the infrared absorption can be observed. This increase (after the onset of green irradiation) is clearly seen in Fig. 20.22.

During the studied initial stage (80 s) the gray track is located within the green beam region and disappears fast when the green light is turned off. However, the first minute of the test is sufficient to determine the quality of the sample. The results of Fig. 20.22 clearly show that the infrared absorption of the HK KTP crystals with a higher potassium content (high-T_C) saturates deep below the level of $10^{-4}\,cm^{-1}$ while the similar absorption of a low-T_C crystal (LK) continues to rise linearly. Obviously, crystals with an improved potassium stoichiometry reveal a higher resistance to gray track formation. Such crystals may compete successfully with hydrothermal KTP (where the concentration of gray track centers is higher, but

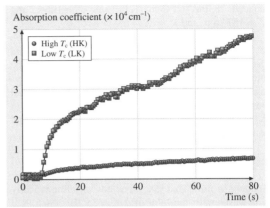

Fig. 20.22 Initial stages of gray-track formation in potassium-deficient (LK, low-T_C) and potassium-rich (HK, high-T_C) grown from self-fluxes

they are less stable [20.183, 184]) in high-power frequency doubling. Crystals grown from pure self-fluxes have an advantage also with respect to low-power applications, since their initial absorption never exceeds $10^{-5}\,cm^{-1}$ (Fig. 20.22). The known initial absorption of most hydrothermal KTP crystals is typically higher than $10^{-4}\,cm^{-1}$, presumably due to the presence of impurities associated with the hydrothermal growth.

RTP crystals are seldom used for frequency doubling into the green [20.4] and, therefore, their susceptibility to gray track formation has not been investigated directly. However, very recent EPR and ENDOR studies have shown that room-temperature-stable Ti^{3+} centers stabilized by oxygen vacancies can be created also in this crystal [20.187]. The narrow range of Curie temperature variation in RTP (Fig. 20.8) and the usually observed high electrical resistivity of these crystals [20.14] imply that their concentration of stable Ti^{3+} electron trap centers is generally low. Nevertheless, current attempts to develop Yb and Nb co-doped RTP crystals as new self-doubling laser media [20.188] necessitates an examination of RTP crystals using the GRIIRA test, similarly to the one described above for KTP.

References

20.1 J.D. Bierlein, H. Vaherzeele: Potassium titanyl phosphate: Properties and new applications, J. Opt. Soc. Am. B **6**, 622–633 (1989)

20.2 G.M. Loiacono, D.N. Loiacono, J.J. Zola, R.A. Stolzenberger, T. McGee, R.G. Norwood: Optical properties and ionic conductivity of $KTiOAsO_4$

20.3 D.E. Spence, C.L. Tang: Characterization and applications of high repetition rate, broadly tunable, femtosecond optical parametric oscillators, IEEE J. Sel. Top. Quantum Electron. **1**, 31–43 (1995)

20.4 Y.S. Oseledchik, A.I. Pisarevsky, A.L. Prosvirin, V.V. Starshenko, N.V. Svitanko: Nonlinear optical properties of the flux grown RbTiOPO$_4$ crystal, Opt. Mater. **3**, 237–242 (1994)

20.5 T.Y. Fan, C.E. Huang, B.Q. Hu, R.C. Eckardt, R.L. Byer, R.S. Feigelson: Second harmonic generation and accurate index of refraction measurements in flux-grown KTiOPO$_4$, Appl. Opt. **26**, 2390–2394 (1987), .

20.6 C.-S. Tu, R.S. Katiyar, V.H. Schmidt, R.-M. Chien, R. Guo, A.S. Bhalla: Hypersonic anomalies and optical properties of RbTiOAsO$_4$ and KTiOPO$_4$ single crystals, Phys. Rev. B **59**, 251–256 (1999)

20.7 L.T. Cheng, L.K. Cheng, J.D. Bierlein, F.C. Zumsteg: Properties of doped and undoped single domain KTiOAsO$_4$, Appl. Phys. Lett. **62**, 346–348 (1993)

20.8 L.T. Cheng, L.K. Cheng, J.D. Bierlein, F.C. Zumsteg: Nonlinear optical and electro-optical properties of single crystal CsTiOAsO$_4$, Appl. Phys. Lett. **63**, 2618–2620 (1993)

20.9 L.K. Cheng, J.D. Bierlein: KTP and isomorphs – Recent progress in device and material development, Ferroelectrics **142**, 209–228 (1993)

20.10 L.K. Cheng, L.T. Cheng, J. Galperin, P.A. Morris Hotsenpiller, J.D. Bierlein: Crystal growth and characterization of KTiOPO$_4$ isomorphs from the self-fluxes, J. Cryst. Growth **137**, 107–115 (1994)

20.11 L.K. Cheng, L.T. Cheng, J.D. Bierlein: Phase-matching property optimization using birefringence tuning in solid solutions of KTiOPO$_4$ isomorphs, Appl. Phys. Lett. **64**, 1321–1323 (1994)

20.12 M. Roth, N. Angert, M. Tseitlin, G. Wang, T.P.J. Han, H.G. Gallagher, N.I. Leonyuk, E.V. Koporulina, S.N. Barilo, L.A. Kurnevich: Recent developments in crystal growth and characterization of nonlinear optical borate and phosphate materials, Proc. 3rd Int. Conf. Single Crystal Growth, Strength Probl. Heat Mass Transf., Vol. 2, ed. by V.P. Ginkin (Uch.-Izd., Obninsk 2000) pp. 416–426

20.13 D.N. Nikogosyan: *Nonlinear Optical Crystals: A Complete Survey* (Springer, New York 2005)

20.14 M. Roth, M. Tseitlin, N. Angert: Oxide crystals for electro-optic Q-switching of lasers, Glass Phys. Chem. **31**, 86–95 (2005)

20.15 M.S. Webb, P.F. Moulton, J.J. Kasinski, R.I. Burnham, G. Loiacono, R. Stolzenberger: High-average-power KTiOAsO$_4$ optical parametric oscillator, Opt. Lett. **23**, 1161–1163 (1998)

20.16 L.S. Lingvay, N. Angert, M. Roth: High-average-power KTP ring OPO, Proc. SPIE **3928**, 52–56 (2000)

20.17 X.D. Wang, P. Basseras, R.J.D. Miller, H. Vanherzeele: Investigation of KTiOPO$_4$ as an electro-optic amplitude modulator, Appl. Phys. Lett. **59**, 519–521 (1991)

20.18 S. Pearce, C.L.M. Ireland: Performance of a CW pump Nd:YVO$_4$ amplifier with kHz pulses, Opt. Laser Technol. **35**, 375–379 (2003)

20.19 M.G. Jani, J.T. Murray, R.R. Petrin, R.C. Powell: Pump wavelength tuning of optical parametric oscillations and frequency mixing in KTiOAsO$_4$, Appl. Phys. Lett. **60**, 2327–2329 (1992)

20.20 G.M. Loiacono, D.N. Loiacono, R.A. Stolzenberger: Crystal growth and characterization of ferroelectric CsTiOAsO$_4$, J. Cryst. Growth **131**, 323–330 (1993)

20.21 J. Nordborg, G. Svensson, R.J. Bolt, J. Albertsson: Top seeded solution growth of [Rb,Cs]TiOAsO$_4$, J. Cryst. Growth **224**, 256–268 (2001)

20.22 I. Tordjman, R. Masse, J.C. Guitel: Crystalline structure of monophosphate KTiPO$_5$, Z. Krist. **139**, 103–115 (1974)

20.23 P.A. Thomas, A.M. Glazer, B.E. Watts: Crystal structure and nonlinear optical properties of KSnOPO$_4$ and their comparison with KTiOPO$_4$, Acta Crystallogr. B **46**, 333–343 (1990)

20.24 P. Delarue, C. Lecomte, M. Jannin, G. Marnier, B. Menaert: Evolution towards centrosymmetry of the nonlinear-optical material RbTiOPO$_4$ in the temperature range 293–973 K: Alkaline displacements and titanyl deformations, Phys. Rev. B **58**, 5287–5295 (1998)

20.25 S.C. Mayo, P.A. Thomas, S.J. Teat, G.M. Loiacono, D.N. Loiacono: Structure and nonlinear-optical properties of KTiOAsO$_4$, Acta Crystallogr. B **50**, 655–662 (1994)

20.26 P.A. Thomas, S.C. Mayo, B.E. Watts: Crystal structures of RbTiOAsO$_4$, KTiO(P$_{0.58}$As$_{0.42}$)O$_4$, RbTiOPO$_4$ and (Rb$_{0.465}$K$_{0.535}$)TiOPO$_4$ and analysis of pseudosymmetry in crystals of the KTiOPO$_4$ family, Acta Crystallogr. B **48**, 401–407 (1992)

20.27 J. Protas, G. Marnier, B. Boulanger, B. Menaert: Crystal structure of CsTiOAsO$_4$, Acta Crystallogr. C **45**, 1123–1125 (1989)

20.28 M. Munowitz, R.H. Jarman, J.F. Harrison: Theoretical study of the nonlinear optical properties of KTiOPO$_4$: Effects of Ti-O-Ti bond angles and oxygen electronegativity, Chem. Mater. **5**, 1257–1267 (1993)

20.29 D. Xue, S. Zhang: The origin of nonlinearity in KTiOPO$_4$, Appl. Phys. Lett. **70**, 943–945 (1997)

20.30 P.A. Thomas, A. Baldwin, R. Dupree, P. Blaha, K. Schwartz, A. Samoson, Z.H. Gan: Structure-property relationships in the nonlinear optical crystal KTiOPO$_4$ investigated using NMR and ab initio DFT calculations, J. Phys. Chem. B **108**, 4324–4331 (2004)

20.31 V. Pasiskevicius, C. Canalias, F. Laurell: Highly efficient stimulated Raman scattering of picosecond pulses in KTiOPO$_4$, Appl. Phys. Lett. **88**, 041110 (2006)

20.32 P.A. Thomas, A.M. Glazer: Potassium titanyl phosphate, KTiOPO$_4$. II. Structural interpretation of

20.33 P. Delarue, C. Lecomte, M. Jannin, G. Marnier, B. Menaert: Behavior of the non-linear optical material KTiOPO$_4$ in the temperature range 293–973 K studied by x-ray diffractometry at high resolution: Alkaline displacements, J. Phys. Condens. Matter **11**, 4123–4134 (1999)

20.34 S.T. Norberg, N. Ishizawa: K-site splitting in KTiOPO$_4$ at room temperature, Acta Crystallogr. C **61**, i99–i102 (2006)

20.35 D.R. Allan, J.S. Loveday, R.J. Nelmes, P.A. Thomas: A high-pressure structural study of potassium titanyl phosphate (KTP) up to 5 GPa, J. Phys. Condens. Matter **4**, 2747–2760 (1992)

20.36 P.A. Thomas, R. Duhlev, S.J. Teat: A comparative structural study of a flux-grown crystal of K$_{0.86}$Rb$_{0.14}$TiOPO$_4$ and an ion-excganged crystal of K$_{0.84}$Rb$_{0.16}$TiOPO$_4$, Acta Cryst. B **50**, 538–543 (1994)

20.37 S. Furusawa, H. Hayasi, Y. Ishibashi, A. Miyamoto, T. Sasaki: Ionic conductivity of quasi-one-dimensional superionic conductor KTiOPO$_4$ (KTP) single crystal, J. Phys. Soc. Jpn. **62**, 183–195 (1993)

20.38 J.D. Bierlein, C.B. Arweiler: Electro-optic and dielectric properties of KTiOPO$_4$, Appl. Phys. Lett. **49**, 917–919 (1986)

20.39 B. Mohamadou, G.E. Kugel, F. Brehat, B. Wyncke, G. Marnier, P. Simon: High-temperature vibrational spectra, relaxation and ionic conductivity effects in KTiOPO$_4$, J. Phys. Condens. Matter **3**, 9489–9501 (1991)

20.40 Q. Jiang, A. Lovejoy, P.A. Thomas, K.B. Hutton, R.C.C. Ward: Ferroelectricity, conductivity, domain structure and poling conditions of rubidium titanyl phosphate, J. Phys. D Appl. Phys. **33**, 2831–2836 (2000)

20.41 P. Urenski, N. Gorbatov, G. Rosenman: Dielectric relaxation in flux grown KTiOPO$_4$ and isomorphic crystals, J. Appl. Phys. **89**, 1850–1855 (2001)

20.42 J.C. Jacco, G.M. Loiacono, M. Jaso, G. Mizell, B. Greenberg: Flux growth and properties of KTiOPO$_4$, J. Cryst. Growth **70**, 484–488 (1984)

20.43 R.J. Bolt, P. Bennema: Potassium titanyl phosphate KTiOPO$_4$ (KTP): Relation between crystal structure and morphology, J. Cryst. Growth **102**, 329–340 (1990)

20.44 L.K. Cheng, J.D. Bierlein, A.A. Ballman: Crystal growth of KTiOPO$_4$ isomorphs from tungstate and molybdate fluxes, J. Cryst. Growth **110**, 697–703 (1991)

20.45 X. Wang, X. Yuan, W. Li, J. Qi, S. Wang, D. Shen: Flux growth of large potassium titanyl phophate crystals and their electro-optical applications, J. Cryst. Growth **237–239**, 672–676 (2002)

20.46 G.M. Loiacono, T.F. McGee, G. Kostecky: Solubility and crystal growth of KTiOPO$_4$ in polyphosphate solvents, J. Cryst. Growth **104**, 389–391 (1990)

20.47 J.D. Bierlein, T.E. Gier: Crystals of (K,Rb,Tl,NH$_4$)TiO(P,As)O$_4$ and their use in electrooptic devices, US Patent 3949323 (1976)

20.48 T.E. Gier: Hydrithermal process for growing a single crystal with an aqueous mineralizer, US Patent 4305778 (1981)

20.49 R.A. Laudise, R.J. Cava, A.J. Caporaso: Phase relations, solubility and growth of potassium titanyl phosphate, KTP, J. Cryst. Growth **74**, 275–280 (1986)

20.50 R.F. Belt, G. Gashurov, R.A. Laudise: Low temperature hydrothermal growth of KTiOPO$_4$ (KTP), Proc. SPIE **968**, 100–106 (1988)

20.51 R.A. Laudise, W.A. Sunder, R.F. Belt, G. Gashurov: Solubility and P-V-T relations and the growth of potassium titanyl phosphate, J. Cryst. Growth **102**, 427–433 (1990)

20.52 A. Ferretti, T.E. Gier: Hydrothermal process for growing optical-quality crystals, US Patent 5066356 (1991)

20.53 S.Q. Jia, P.Z. Jiang, H.D. Niu, D.Z. Li, X.H. Fan: The solubility of KTiOPO$_4$ (KTP) in KF aqueous solution under high temperature and high pressure, J. Cryst. Growth **79**, 970–973 (1986)

20.54 S.Q. Jia, H.D. Niu, J.G. Tan, Y.P. Xu, Y. Tao: Hydrothermal growth of KTP crystals in the medium range of temperature and pressure, J. Cryst. Growth **99**, 900–904 (1990)

20.55 L.K. Cheng: Hydrothermal aqueous mineralizer for growing optical-quality single crystals, US Patent 5500145 (1996)

20.56 C. Zhang, L. Huang, W. Zhou, G. Zhang, H. Hou, Q. Ruan, W. Lei, S. Qin, F. Lu, Y. Zuo, H. Shen, G. Wang: Growth of KTP crystals with high damage threshold by hydrothermal method, J. Cryst. Growth **292**, 364–367 (2006)

20.57 A.A. Ballman, H. Brown, D.H. Olson, C.E. Rice: Growth of potassium titanyl phosphate (KTP) from molten tungsten melts, J. Cryst. Growth **75**, 390–394 (1986)

20.58 K. Iliev, P. Peshev, V. Nikolov, I. Koseva: Physicochemical properties of high-temperature solutions of the K$_2$O-P$_2$O$_5$-TiO$_2$-WO$_3$ system, suitable for the growth of KTiOPO$_4$ (KTP) single crystals, J. Cryst. Growth **100**, 225–232 (1990)

20.59 C.G. Chao, Z.Q. Qiang, T.G. Kui, S.W. Bao, T.H. Gao: Top seeded growth of KTiOPO$_4$ from molten tungsten solution, J. Cryst. Growth **112**, 294–297 (1991)

20.60 A. Yokotani, A. Miyamoto, T. Sasaki, S. Nakai: Observation of optical inhomogeneities in flux grown KTP crystals, J. Cryst. Growth **110**, 963–967 (1991)

20.61 D.P. Shumov, M.P. Tarassov, V.S. Nikolov: Investigation of optical inhomogeneities in KTiOPO$_4$ (KTP) single crystals grown from high-temperature tungsten-containing solutions, J. Cryst. Growth **129**, 635–639 (1993)

20.62 D. Bravo, X. Ruiz, F. Díaz, F.J. López: EPR of tungsten impurities in KTiOPO$_4$ single crystals grown from

20.63 J.H. Kim, J.K. Kang, S.J. Chung: Effects of seed orientation on the top-seeded solution growth of $KTiOPO_4$ single crystals, J. Cryst. Growth **147**, 343–349 (1995)

20.64 F.J. Kumar, S.G. Moorthy, C. Subramanian, G. Bocelli: Growth and characterization of $KTiOPO_4$ single crystals, Mater. Sci. Eng. B **49**, 31–35 (1997)

20.65 G. Marnier: Process for the flux synthesis of crystals of the $KTiOPO_4$ potassium titanyl monophosphate type, US Patent 4746396 (1988)

20.66 Y. Guillien, B. Ménart, J.P. Fève, P. Segonds, J. Douady, B. Boulanger, O. Pacaud: Crystal growth and refined Sellmeier equations over the complete transparency range of $RbTiOPO_4$, Opt. Mater. **22**, 155–162 (2003)

20.67 G.M. Loiacono, D.N. Loiacono, T. McGee, M. Babb: Laser damage formation in $KTiOPO_4$ and $KTiOAsO_4$ crystals: Grey tracks, J. Appl. Phys. **72**, 2705–2712 (1992)

20.68 T.E. Gier: Method for flux growth of $KTiOPO_4$ and its analogues, US Patent 4231838 (1980)

20.69 D. Shen, C. Huang: A new nonlinear optical crystal KTP, Prog. Cryst. Growth Charact. **11**, 269–274 (1985)

20.70 K. Iliev, P. Peshev, V. Nikolov, I. Koseva: Physicochemical properties of high-temperature solutions of the $K_2O-P_2O_5-TiO_2$ system suitable for the growth of $KTiOPO_4$ (KTP) single crystals, J. Cryst. Growth **100**, 219–224 (1990)

20.71 P.F. Bordui, J.C. Jacco: Viscosity and density of solutions used in high-temperature growth of $KTiOPO_4$ (KTP), J. Cryst. Growth **82**, 351–355 (1987)

20.72 N. Angert, M. Tseitlin, E. Yashchin, M. Roth: Ferroelectric phase transition temperatures of $KTiOPO_4$ crystals grown from self-fluxes, Appl. Phys. Lett. **67**, 1941–1943 (1995)

20.73 Y.S. Oseledchik, S.P. Belokrys, V.V. Osadchuk, A.L. Prosvirin, A.F. Selevich, V.V. Starshenko, K.V. Kuzemchenko: Growth of $RbTiOPO_4$ single crystals from phosphate systems, J. Cryst. Growth **125**, 639–643 (1992)

20.74 C.V. Kannan, S. Ganesa Murthy, V. Kannan, C. Subramanian, P. Ramasamy: TSSG of $RbTiOPO_4$ single crystals from phosphate flux and their characterization, J. Cryst. Growth **245**, 289–296 (2002)

20.75 L.I. Isaenko, A.A. Merkulov, V.I. Tyurikov, V.V. Atuchin, L.V. Sokolov, E.M. Trukhanov: Growth and real structure of $KTiOAsO_4$ crystals from self-fluxes, J. Cryst. Growth **171**, 146–153 (1997)

20.76 M. Roth, N. Angert, M. Tseitlin, G. Schwarzman, A. Zharov: Ferroelectric phase transition temperatures of self-flux-grown $RbTiOPO_4$ crystals, Opt. Mater. **26**, 465–470 (2004)

20.77 M.E. Hagerman, K.R. Poeppelmeier: Review of the structure and processing-defect-property relationships of potassium titanyl phosphate: A strategy for novel thin-film photonic devices, Chem. Mater. **7**, 602–621 (1995)

20.78 M.N. Satyanarayan, A. Deepthy, H.L. Bhat: Potassium titanyl phosphate and its isomorphs: growth, properties and applications, Crit. Rev. Solid State **24**, 103–189 (1999)

20.79 J. Wang, J. Wei, Y. Liu, X. Yin, X. Hu, Z. Shao, M. Jiang: A survey of research on KTP and its analogue crystals, Prog. Cryst. Growth Charact. **40**, 3–15 (2000)

20.80 P.F. Bordui, J.C. Jacco, G.M. Loiacono, R.A. Stolzenberger, J.J. Zola: Growth of large single crystals of $KTiOPO_4$ (KTP) from high-temperature solution using heat pipe based furnace system, J. Cryst. Growth **84**, 403–408 (1987)

20.81 R.J. Bolt, M.H. van der Mooren, H. de Haas: Growth of $KTiOPO_4$ (KTP) single crystals by means of phosphate and phosphate/sulfate fluxes out of a three-zone furnace, J. Cryst. Growth **114**, 141–152 (1991)

20.82 T. Sasaki, A. Miyamoto, A. Yokotani, S. Nakai: Growth and optical characterization of large potassium titanyl phosphate crystals, J. Cryst. Growth **128**, 950–955 (1993)

20.83 N. Angert, L. Kaplun, M. Tseitlin, E. Yashchin, M. Roth: Growth and domain structure of potassium titanyl phosphate crystals pulled from high-temperature solutions, J. Cryst. Growth **137**, 116–122 (1994)

20.84 B. Vartak, J.J. Derby: On stable algorithms and accurate solutions for convection-dominated mass transfer in crystal growth modeling, J. Cryst. Growth **230**, 202–209 (2001)

20.85 P.F. Bordui, S. Motakef: Hydrodynamic control of solution inclusion during crystal growth of $KTiOPO_4$ (KTP) from high-temperature solution, J. Cryst. Growth **96**, 405–412 (1989)

20.86 B. Vartak, Y. Kwon, A. Yeckel, J.J. Derby: An analysis of flow and mass transfer during the solution growth of potassium titanyl phosphate, J. Cryst. Growth **210**, 704–718 (2000)

20.87 B. Vartak, A. Yeckel, J.J. Derby: On the validity of boundary layer analysis for flow and mass transfer caused by rotation during the solution growth of large single crystals, J. Cryst. Growth **283**, 479–489 (2005)

20.88 E.G. Tsvetkov, V.N. Semenenko, G.G. Khranenko, V.I. Tyurikov: Growth peculiarities and polar structures formation in potassium titanyl phosphate crystals, J. Surf. Invest. X-Ray Synchrotron Neutron Tech. **5**, 65–70 (2002), in Russian

20.89 M. Roth, N. Angert, M. Tseitlin, A. Alexandrovski: On the optical quality of KTP crystals for nonlinear optical and electro-optic applications, Opt. Mater. **16**, 131–136 (2001)

20.90 P.A. Morris, A. Ferretti, J.D. Bierlein, G.M. Loiacono: Reduction of the ionic conductivity of flux grown $KTiOPO_4$ crystals, J. Cryst. Growth **109**, 367–375 (1991)

20.91 M.G. Roelofs: Identification of Ti^{3+} in potassium titanyl phosphate and its possible role in laser damage, J. Appl. Phys. **65**, 4976–4982 (1989)

20.92 V.D. Kugel, G. Rosenman, N. Angert, E. Yaschin, M. Roth: Domain inversion in KTiOPO$_4$ crystal near the Curie point, J. Appl. Phys. **76**, 4823–4826 (1994)

20.93 M.E. Hagerman, V.L. Kozhevnikov, K.R. Poeppelmeier: High-temperature decomposition of KTiOPO$_4$, Chem. Mater. **5**, 1211–1215 (1993)

20.94 D.K.T. Chu, H. Hsiung: Ferroelectric phase transition study in KTiOPO$_4$ – An optical 2nd harmonic generation study, Appl. Phys. Lett. **61**, 1766–1768 (1992)

20.95 K. Zhang, X. Wang: Structure sensitive properties of KTP-type crystals, Chin. Sci. Bull. **46**, 2028–2036 (2001)

20.96 T.F. McGee, G.M. Blom, G. Kostecky: Growth and characterization of doped KTP crystals, J. Cryst. Growth **109**, 361–366 (1991)

20.97 D.K.T. Chu, H. Hsiung, L.K. Cheng, J.D. Bierlein: Curie temperatures and dielectric-properties of doped and undoped KTiOPO$_4$ and isomorphs, IEEE Trans. Ultrason. Ferroelectr. Fraq. Control **40**, 819–824 (1993)

20.98 R.V. Pisarev, R. Farhi, P. Moch, V.I. Voronkova: Temperature-dependence of Raman-scattering and soft modes in TlTiOPO$_4$, J. Phys. Condens. Matter **2**, 1555–1568 (1990)

20.99 G.M. Loiacono, R.A. Stolzenberger, D.N. Loiacono: Modified KTiOPO$_4$ crystals for noncritical phase matching applications, Appl. Phys. Lett. **64**, 16–18 (1994)

20.100 P.F. Bordui, R.G. Norwood, M.M. Fejer: Curie-temperature measurements on KTiOPO$_4$ single-crystals grown by flux and hydrothermal techniques, Ferroelectrics **115**, 7–12 (1991)

20.101 V.K. Yanovskii, V.I. Voronkova, A.P. Leonov, S.Y. Stefanovich: Ferroelectric properties of KTiOPO$_4$ type crystals, Sov. Phys. Solid State **27**, 1506–1516 (1985)

20.102 B.C. Choi, J.B. Kim, B.M. Jin, S.I. Yun, J.N. Kim: Dielectric properties of flux-grown KTiOPO$_4$ single crystals, J. Korean Phys. Soc. **25**, 327–331 (1992)

20.103 Y.V. Shaldin, R. Poprawski: Spontaneous birefringence and pyroelectricity in KTiOPO$_4$ crystals, Ferroelectrics **106**, 399–404 (1990)

20.104 A. Miyamoto, Y. Mori, Y. Okada, T. Sasaki, S. Nakai: Refractive index and lattice constant variation in flux grown KTP crystals, J. Cryst. Growth **156**, 303–306 (1995)

20.105 M.N. Satyanarayan, H.L. Bhat: Influence of growth below and above T_C on the morphology and domain structure in flux-grown KTP crystals, J. Cryst. Growth **181**, 281–289 (1997)

20.106 M. Roth, N. Angert, M. Tseitlin: Growth-dependent properties of KTP crystals and PPKTP structures, J. Mater. Sci. Mater. Electron. **12**, 429–436 (2001)

20.107 N. Angert, M. Tseitlin, L. Kaplun, E. Yashchin, M. Roth: Ferroelectric domain reversal in KTP crystals by high-temperature treatment, Ferroelectrics **157**, 117–122 (1994)

20.108 M.J. Martín, D. Bravo, R. Solé, F. Díaz, F.J. López, C. Zaldo: Thermal reduction of KTiOPO$_4$ crystals, J. Appl. Phys. **76**, 7510–7518 (1994)

20.109 P.A. Morris: Impurities in nonlinear optical oxide crystals, J. Cryst. Growth **106**, 76–88 (1990)

20.110 M. Roth, N. Angert, L. Weizman, A. Gachechiladze, M. Shachman, D. Remennikov, M. Tseitlin, A. Zharov: Development of KTP crystals for laser applications, Ann. Isr. Phys. Soc. **14**, 89–92 (2000)

20.111 L.T. Cheng, L.K. Cheng, R.L. Harlow, J.D. Bierlein: Blue light generation using bulk single crystals of niobium-doped KTiOPO$_4$, Appl. Phys. Lett. **64**, 155–157 (1994)

20.112 V.I. Chani, K. Shimamura, S. Endo, T. Fukuda: Growth of mixed crystals of the KTiOPO$_4$ (KTP) family, J. Cryst. Growth **171**, 472–476 (1997)

20.113 K.T. Stevens, L.E. Halliburton, M. Roth, N. Angert, M. Tseitlin: Identification of Pb-related Ti^{3+} center in flux-grown KTiOPO$_4$, J. Appl. Phys. **88**, 6239–6244 (2000)

20.114 E. Lebiush, R. Lavi, Y. Tsuk, N. Angert, A. Gachechiladze, M. Tseitlin, A. Zharov, M. Roth: RTP as a Q-switch for high repetition rate applications, Proc. Adv. Solid State Lasers, TOPS **34**, 63–65 (2000)

20.115 Y. Stefanovich, L.A. Ivanova, A.V. Astafyev: *Ionic and Superionic Conductivity in Ferroelectrics* (Nitekhim, Moscow 1989) p. 37, in Russian

20.116 F.C. Zumsteg, J.D. Bierlein, T.E. Gier: K$_x$Rb$_{1-x}$TiOPO$_4$: A new nonlinear optical material, J. Appl. Phys. **47**, 4980–4985 (1976)

20.117 W.P. Risk, G.M. Loiacono: Periodic poling and wavegude frequency doubling in RbTiOAsO$_4$, Appl. Phys. Lett. **69**, 311–313 (1996)

20.118 M. Roth, M. Tseitlin, N. Angert: Composition-dependent electro-optic and nonlinear optical properties of KTP-family crystals, Opt. Mater. **28**, 71–76 (2006)

20.119 Y. Jiang, L.E. Halliburton, M. Roth, M. Tseitlin, N. Angert: Hyperfine structure associated with the dominant radiation-induced trapped hole center in RbTiOPO$_4$ crystals, Phys. Status Solidi (b) **242**, 2489–2496 (2005)

20.120 J.D. Bierlein, F. Ahmed: Observation and poling of ferroelectric domains in KTiOPO$_4$, Appl. Phys. Lett. **51**, 1328–1330 (1987)

20.121 G.M. Loiacono, R.A. Stolzenberger: Observation of complex domain walls in KTiOPO$_4$, Appl. Phys. Lett. **53**, 1498–1500 (1988)

20.122 L.P. Shi, J. Chrosch, J.Y. Wang, Y.G. Liu: Twinning in KTiOPO$_4$ crystals, Cryst. Res. Technol. **27**, K76–K78 (1992)

20.123 F. Laurell, M.G. Roelofs, W. Bindloss, H. Hsiung, A. Suna, J.D. Bierlein: Detection of ferroelectric domain reversal in KTiOPO$_4$ wave-guides, J. Appl. Phys. **71**, 4664–4670 (1992)

20.124 R.J. Bolt, W.J.P. Enckevort: Observation of growth steps and growth hillocks on the {100}, {210}, {011} and {101} faces of flux-grown KTiOPO$_4$ (KTP), J. Cryst. Growth **119**, 329–338 (1992)

20.125 A.A. Chernov: Stability of faceted shapes, J. Cryst. Growth **24/25**, 11–31 (1974)

20.126 N.R. Ivanov, N.A. Tikhomirova, A.V. Ginzberg, S.P. Chumakova, E.I. Eknadiosyants, V.Z. Borodin, A.N. Pinskaya, V.A. Babanskikh, V.A. D'yakov: Domain structure of KTiOPO$_4$ crystals, Crystallogr. Rep. **39**, 593–599 (1994)

20.127 W.P. Risk, R.N. Payne, W. Lenth, C. Harder, H. Meier: Noncritically phase-matched frequency doubling using 994 nm dye and diode laser radiation in KTiOPO$_4$, J. Appl. Phys. **55**, 1179–1181 (1989)

20.128 W.P. Risk, W.J. Kozlovsky: Efficient generation of blue-light by doubly resonant sum-frequency mixing in a monolithic KTP resonator, Opt. Lett. **17**, 707–709 (1992)

20.129 L.T. Cheng, L.K. Cheng, R.L. Harlow, J.D. Bierlein: Blue light generation using bulk single crystals of niobium-doped KTiOPO$_4$, Appl. Phys. Lett. **64**, 155–157 (1994)

20.130 H. Mabuchi, E.S. Polzik, H.J. Kimble: Blue-light-induced infrared-absorption in KNbO$_3$, J. Opt. Soc. Am. B **11**, 2023–2029 (1994)

20.131 J.A. Armstrong, N. Bloembergen, J. Ducuing, P.S. Pershan: Interaction between light waves in a nonlinear dielectric, Phys. Rev. **127**, 1918–1939 (1962)

20.132 J.D. Bierlein, D.B. Laubacher, J.B. Brown: Balanced phase matching in segmented KTiOPO$_4$ wave-guides, Appl. Phys. Lett. **56**, 1725–1727 (1990)

20.133 C.J. van der Poel, J.D. Bierlein, J.B. Brown: Efficient type I blue second harmonic generation in periodically segmented KTiOPO$_4$ waveguides, Appl. Phys. Lett. **57**, 2074–2076 (1990)

20.134 M.G. Roelofs, P.A. Morris, J.D. Bierlein: Ion exchange of Rb, Ba, and Sr in KTiOPO$_4$, J. Appl. Phys. **70**, 720–728 (1991)

20.135 M.C. Gupta, W.P. Risk, A.C.G. Nutt, S.D. Lau: Domain inversion in KTiOPO$_4$ using electron beam scanning, Appl. Phys. Lett. **63**, 1167–1169 (1993)

20.136 G. Rosenman, P. Urenski, A. Agronin, A. Arie, Y. Rosenwaks: Nanodomain engineering in RbTiOPO$_4$ ferroelectric crystals, Appl. Phys. Lett. **82**, 3934–3936 (2003)

20.137 M. Yamada, N. Nada, M. Saitoh, K. Watanabe: First-order quasi-phase matched LiNbO$_3$ waveguide periodically poled by applying an external field for efficient blue second harmonic generation, Appl. Phys. Lett. **62**, 435–437 (1993)

20.138 L.E. Myers, R.C. Eckardt, M.M. Fejer, R.L. Byer, W.R. Bosenberg, J.W. Pierce: Quasi-phase-matched optical parametric oscillators in bulk periodically poled LiNbO$_3$, J. Opt. Soc. Am. B **12**, 2102–2116 (1995)

20.139 Q. Chen, W.P. Risk: Periodic poling of KTiOPO$_4$ using an applied electric field, Electron. Lett. **30**, 1516–1517 (1994)

20.140 G.D. Boyd, D.A. Kleinman: Parametric interaction of focused Gaussian light beams, J. Appl. Phys. **39**, 3597–3639 (1968)

20.141 M.M. Fejer, G.A. Magel, D.H. Jundt, R.L. Byer: Quasi-phase-matched second harmonic generation: tuning and tolerances, J. Quantum Electron. **QE-28**, 2631–2654 (1992)

20.142 A. Arie, G. Rosenman, V. Mahal, A. Skliar, M. Oron, M. Katz, D. Eger: Green and ultraviolet quasi-phase-matched second harmonic generation in bulk periodically-poled KTiOPO$_4$, Opt. Commun. **142**, 265–268 (1997)

20.143 M. Peltz, U. Bäder, A. Borsutzky, R. Wallerstein, J. Hellström, H. Karlsson, V. Pasiskevicius, F. Laurell: Optical parametric oscillators for high pulse energy and high average power operation based on large aperture periodically poled KTP and RTA, Appl. Phys. B **73**, 663–670 (2001)

20.144 G. Rosenman, K. Garb, A. Skliar, D. Eger, M. Oron, M. Katz: Domain broadening in quasi-phase-matched nonlinear optical devices, Appl. Phys. Lett. **73**, 865–867 (1998)

20.145 H. Karlsson, F. Laurell: Electric field poling of flux grown KTiOPO$_4$, Appl. Phys. Lett. **71**, 3474–3476 (1997)

20.146 M. Pierrou, F. Laurell, H. Karlsson, T. Kellner, C. Czeranowsky, G. Huber: Generation of 740 mW of blue light by intracavity frequency doubling with a first-order quasi-phase-matched KTiOPO$_4$ crystal, Opt. Lett. **24**, 205–207 (1999)

20.147 G. Rosenman, A. Skliar, D. Eger, M. Oron, M. Katz: Low temperature periodic poling of flux-grown KTiOPO$_4$ and isomorphic crystals, Appl. Phys. Lett. **73**, 3650–3652 (1998)

20.148 A. Garashi, A. Arie, A. Skliar, G. Rosenman: Continuous-wave optical parametric oscillator based on periodically poled KTiOPO$_4$, Opt. Lett. **23**, 1739–1741 (1998)

20.149 C. Canalias, J. Hirohashi, V. Pasiskevicius, F. Laurell: Polarization-switching characteristics of flux-grown KTiOPO$_4$ and RbTiOPO$_4$ at room temperature, J. Appl. Phys. **97**, 124105 (2005)

20.150 P. Urenski, M. Lesnykh, Y. Rosenwaks, G. Rosenman: Anisotropic domain structure of KTiOPO$_4$ crystals, J. Appl. Phys. **90**, 1950–1954 (2001)

20.151 P. Urenski, M. Molotski, G. Rosenman: Bulk ferroelectric domain nucleation in KTiOPO$_4$ crystals, Appl. Phys. Lett. **79**, 2964–2966 (2001)

20.152 C. Canalias, S. Wang, V. Pasiskevicius, F. Laurell: Nucleation and growth of periodic domains during electric field poling in flux-grown KTiOPO$_4$ observed by atomic force microscopy, Appl. Phys. Lett. **88**, 032905 (2006)

20.153 J. Zhang, J. Wang, B. Ge, Y. Liu, X. Hu, G. Zhao, S. Zhu, R.I. Boughton: Growth, conductivity and periodic poled structure of doped KTiOPO$_4$ and its analogue crystals, Opt. Mater. **28**, 355–359 (2006)

20.154 G. Rosenman, P. Urenski, A. Arie, M. Roth, N. Angert, A. Skliar, M. Tseitlin: Polarization reversal and domain grating in flux-grown KTiOPO$_4$ crystals with variable potassium stoichiometry, Appl. Phys. Lett. **76**, 3798–3800 (2000)

20.155 Q. Jiang, P.A. Thomas, D. Walker, K.B. Hutton, R.C.C. Ward, P. Pernot, J. Baruchel: High potassium KTiOPO$_4$ crystals for the fabrication of quasi-phase matched devices, J. Phys. D: Appl. Phys. **36**, 1236–1241 (2003)

20.156 G. Rosenman, A. Skliar, Y. Findling, P. Urenski, A. Englander, P.A. Thomas, Z.W. Hu: Periodically poled KTiOAsO$_4$ crystals for optical parametric oscillation, J. Phys. D: Appl. Phys. **32**, L49–L52 (1999)

20.157 P. Pernot-Rejmánková, P.A. Thomas, P. Cloenets, F. Lorut, J. Baruchel, Z.W. Hu, P. Urenski, G. Rosenman: Periodically poled KTA crystal investigated using coherent X-ray beams, J. Appl. Cryst. **33**, 1149–1153 (2000)

20.158 H. Karlsson, F. Laurell, L.K. Cheng: Periodic poling of RbTiOPO$_4$ for quasi-phase matched blue light generation, Appl. Phys. Lett. **74**, 1519–1521 (1999)

20.159 A. Fragemann, V. Pasiskevicius, J. Nordborg, J. Hellström, H. Karlsson, F. Laurell: Frequency converters from visible to mid-infrared with periodically poled RbTiOPO$_4$, Appl. Phys. Lett. **83**, 3090–3092 (2003)

20.160 H. Karlsson, F. Laurell, P. Hendriksson, G. Arvidsson: Frequency doubling in periodically poled RbTiOAsO$_4$, Electron. Lett. **32**, 556–557 (1996)

20.161 H. Karlsson, M. Olson, G. Arvidsson, F. Laurell, U. Bäder, A. Borsutzky, R. Wallenstein, S. Wikström, M. Gustafsson: Nanosecond optical parametric oscillator based on large-aperture periodically poled RbTiOAsO$_4$, Opt. Lett. **24**, 330–332 (1999)

20.162 G.T. Kennedy, D.T. Reid, A. Miller, A. Ebrahimzadeh, H. Karlsson, G. Arvidsson, F. Laurell: Near- to mid-infrared picosecond optical parametric oscillator based on periodically poled RbTiOAsO$_4$, Opt. Lett. **23**, 503–505 (1998)

20.163 P. Loza-Alvarez, D.T. Reid, M. Ebrahimzadeh, W. Sibbett, H. Karlsson, P. Hendriksson, G. Arvidsson, F. Laurell: Periodically poled RbTiOAsO$_4$ femtosecond optical parametric oscillator tunable from 1.38 to 1.58 µm, Appl. Phys. B **68**, 177–180 (1999)

20.164 W. Chen, G. Mouret, D. Boucher, F.K. Tittel: Mid-infrared trace gas detection using continuous-wave difference frequency generation in periodically poled RbTiOAsO$_4$, Appl. Phys. B **72**, 873–876 (2001)

20.165 I. Yutsis, B. Kirshner, A. Arie: Temperature-dependent dispersion relations for RbTiOPO$_4$ and RbTiOAsO$_4$, Appl. Phys. B **79**, 77–81 (2004)

20.166 S. Moscovich, A. Arie, R. Urenski, A. Agronin, G. Rosenman, Y. Rosenwaks: Noncollinear second-harmonic generation in sub-micrometer-poled RbTiOPO$_4$, Opt. Exp. **12**, 2242 (2004)

20.167 C. Canalias, V. Pasiskevicius, R. Clemens, F. Laurell: Submicrion periodically poled flux-grown KTiOPO$_4$, Appl. Phys. Lett. **82**, 4233–4235 (2003)

20.168 D. Feng, N. Ming, J. Hong, Y. Yang, J. Zhu, Z. Yang, Y. Wang: Enhancement of send-harmonic generation in LiNbO$_3$ crystals with periodic laminar ferroelectric domains, Appl. Phys. Lett. **37**, 607–609 (1980)

20.169 V. Bermudez, E. Callejo, E. Dieguez: Effect of temperature annealing on periodically poled rare-earth doped lithium niobate crystal, J. Optoelectron. Adv. Mater. **5**, 55–59 (2003)

20.170 C.E.M. de Oliveira, G. Orr, N. Axelrod, A.J. Agranat: Controlled composition modulation in potassium lithium tantalate niobate crystals grown by off-centered TSSG method, J. Cryst. Growth **273**, 203–206 (2004)

20.171 J.A. Burton, R.C. Prim, W.P. Slichter: The distribution of solute in crystals grown from the melt. Part I. Theoretical, J. Chem. Phys. **21**, 1987–1991 (1953)

20.172 D. Elwell, H.J. Scheel: *Crystal Growth from High-temperature Solutions* (Academic, New York 1975) p. 294

20.173 F.J. Kumar, D. Jayaraman, C. Subramanian, P. Ramasamy: Nucleation kinetic study of KTiOPO$_4$ crystallizing from high temperature solution, J. Cryst. Growth **137**, 535–537 (1994)

20.174 A.V. Smith: *SNLO nonlinear optics code*, available from: www.sandia.gov./imrl/x1118/xxtab.htm

20.175 J.C. Jacco, D.R. Rockafello, E.A. Teppo: Bulk-darkening threshold of flux-grown KTiOPO$_4$, Opt. Lett. **16**, 1307–1309 (1991)

20.176 R. Blachman, P.F. Bordui, M.M. Fejer: Laser-induced photochromic damage in potassium titanyl phosphate, Appl. Phys. Lett. **64**, 1318–1320 (1994)

20.177 B. Boulanger, M.M. Fejer, R. Blachman, P.F. Bordui: Study of KTiOPO$_4$ gray-tracking at 1064, 532 and 355 nm, Appl. Phys. Lett. **65**, 2401–2403 (1994)

20.178 M.P. Scripsick, D.N. Loiacono, J. Rottenberg, S.H. Goellner, L.E. Halliburton, F.K. Hopkins: Defects responsible for gray tracks in flux-grown KTiOPO$_4$, Appl. Phys. Lett. **66**, 3428–3430 (1995)

20.179 J.P. Feve, B. Boulanger, G. Marnier, H. Albrecht: Repetition rate dependence of gray-tracking in KTiOPO$_4$ during second-harmonic generation at 532 nm, Appl. Phys. Lett. **70**, 1–3 (1997)

20.180 W.R. Bosenberg, D.R. Guyer: Single-frequency optical parametric oscillator, Appl. Phys. Lett. **61**, 387–389 (1992)

20.181 A. Deepthy, M.N. Saryanarayan, K.S.R.K. Rao, H.L. Bhat: Photoluminescence studies on gray tracked KTiOPO$_4$ single crystals, J. Appl. Phys. **85**, 8332–8336 (1999)

20.182 B.V. Andreev, V.A. Maslov, A.A. Mikhailov, S.K. Pak, O.P. Shaunin, I.A. Sherbakov: Exprtimental study of the laser-induced absorption effect and the nature of color centers in potassium titanyl phosphate crystals, Proc. SPIE **1839**, 280–289 (1991)

20.183 M.P. Scripsick, G.J. Edwards, L.E. Halliburton, R.F. Belt, G.M. Loiacono: Effect of crystal growth on Ti^{3+} centers in $KTiOPO_4$, J. Appl. Phys. **76**, 773–776 (1994)

20.184 S.D. Setzler, K.T. Stevens, N.C. Fernelius, M.P. Scripsick, G.J. Edwards, L.E. Halliburton: Electron paramagnetic resonance and electron-nuclear double-resonance study of Ti^{3+} centres in $KTiOPO_4$, J. Phys. Condens. Matter **15**, 3969–3984 (2003)

20.185 P.A. Morris, M.K. Crawford, M.G. Roelofs, J.D. Bierlein, P.K. Gallagher, G. Gashurov, G. Loiacano: Proton effects in $KTiPO_5$, MRS Proc. Opt. Fiber Mater. Process. **172**, 283–289 (1990)

20.186 R.G. Batchko, G.D. Miller, A. Alexandrovski, M.M. Fejer, R.L. Byer: *Limitations of High-Power Visible Wavelength Periodically Poled Lithium Niobate Devices due to Green-Induced Infrared Absorption and Thermal Lensing*, OSA Technical Digest, Vol. 6 (OSA, Washington 1998) pp. 75–76

20.187 Y. Jiang, L.E. Halliburton, M. Roth, M. Tseitlin, N. Angert: EPR and ENDOR study of an oxygen-vacancy-associated Ti^{3+} center in $RbTiOPO_4$ crystals, Physica B **400**, 190–197 (2007)

20.188 J.J. Carvajal, R. Solé, J. Gavaldà, J. Massons, P. Segonds, B. Boulanger, A. Brenier, G. Boulon, J. Zaccaro, M. Aguiló, F. Díaz: Spectroscopic and second harmonic generation properties of a new crystal: Yb-doped $RbTiOPO_4$, Opt. Mater. **26**, 313–317 (2004)

21. High-Temperature Solution Growth: Application to Laser and Nonlinear Optical Crystals

Joan J. Carvajal, Maria Cinta Pujol, Francesc Díaz

Growth methods based on high-temperature solutions, traditionally also known as flux growth methods, and especially the top-seeded solution growth (TSSG) and liquid-phase epitaxy (LPE) methods, are some of the most popular methods by which to grow single crystals. These methods have to be used when the grown materials melt incongruently, melt at very high temperatures, or suffer from polymorphic transitions below the crystallization temperature. In this chapter we review the main advances produced in these crystal growth techniques during recent years, both in bulk and epitaxial films, and for two families of oxide materials, specifically those commonly used for solid-state lasers and nonlinear optical crystals. We intend to focus on the application to the real problems related to crystal growth in solutions with different viscosities, while revisiting some of the main strategies developed to overcome these problems to enable growth of bulk single crystals and single-crystalline films with good optical quality.

21.1	**Basics** 726
	21.1.1 Historical Background and Overview 726
	21.1.2 Most Important Families of Laser and Nonlinear Optical Materials 727
21.2	**High-Temperature Solution Growth** 731
	21.2.1 Top-Seeded Solution Growth (TSSG) 732
	21.2.2 Liquid-Phase Epitaxy (LPE) 734
21.3	**Growth of Bulk Laser and NLO Single Crystals by the TSSG Method** 736
	21.3.1 Crystal Growth from Low-Viscosity Solutions: Fluorides, Tungstates, and Vanadates 736
	21.3.2 Crystal Growth from High-Viscosity Solutions: Phosphates and Borates. 739
21.4	**Liquid-Phase Epitaxy: Growth of Epitaxial Films of Laser and NLO Materials** 746
	21.4.1 Epitaxial Films of Laser Materials: Lanthanide-Doped KLuW on KLuW Substrates 746
	21.4.2 Epitaxies Within the Structural Field of KTP ... 748
References 752	

Interest in crystal growth technology started at the beginning of the 20th century, initially in the jewelery and watch industry, and later on in microelectronics (semiconductors), solid-state devices, and laser technology.

During the period 1900–1939, before World War II, the basis of high-temperature solution growth technology was developed, but industrial applications were limited almost only to the ruby single crystal, so the new crystal growth method, and the crystals grown from it, were limited to the academic level. The widespread technological application of single crystals began with the military application of piezoelectric single crystals for transducers in sonar and radar devices. In 1948, after World War II, the discovery of the transistor effect substantially increased the demand for single crystals. The deficient performance of the first generation of transistors based on germanium, principally related to the existence of micro- and macrodefects in germanium single crystals, induced and impelled the study and improvement of high-temperature solution crystal growth methods.

After 1990, the expansion to large-scale industrial production of transistors and the progress in the field of electronics definitively motivated crystal growth as a new scientific field in the material science area. Semiconductors such as Si, GaAs, InAs, etc., and crystalline materials such as $BaTiO_3$, ferrites, yttrium iron garnet YIG, and yttrium aluminum garnet (YAG), among others, have been obtained with a high enough level of perfection, and have been incorporated as core materials for different solid-state devices.

Now, technology uses thousands of single-crystal materials, especially in microelectronics, optoelectronics, optics, and laser technologies. Research is oriented to develop methods and technologies to control the crystallinity, purity, and homogeneity of the grown crystals. The most extended crystal growth methods, based on producing controlled solidification of a melt of the stoichiometric composition of the material, are related to the production of semiconductor materials such as silicon and GaAs. However, several crystalline materials cannot be obtained from melt methods and have to be grown from a solution with a convenient solvent. In this chapter we discuss the high-temperature solution growth methodologies as some of the most suitable ways to obtain several optical materials, specifically in the fields of the laser and nonlinear optical technologies.

21.1 Basics

21.1.1 Historical Background and Overview

Crystal growth is a central step in the processing of solid-state laser and nonlinear optical (NLO) materials.

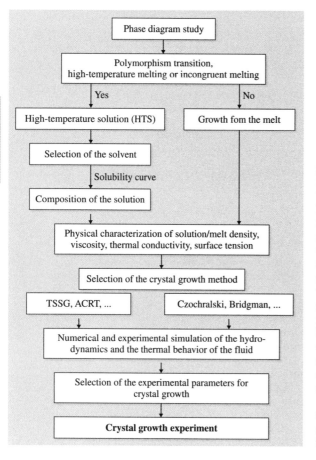

Crystal size, optical losses, and optical uniformity, as well as ultimate crystal cost, tend to be dominated by the crystal growth process. A variety of techniques are practiced, but for most materials, there emerges a preferred crystal growth technique. The process-dependent properties and limitations of a crystal tend to reflect the inherent nature of that growth technique.

In general, crystal growth from the melt is preferable whenever possible. However, other crystal growth methods have to be used when materials melt incongruently, when they suffer from polymorphic transitions below crystallization temperature, or when they melt at very high temperatures. In these cases high-temperature solution methods must be used, in processes analogous to crystal growth from aqueous solutions, but in which the solvent solidifies before reaching room temperature. The main advantage of these methodologies is that crystals are grown below their melting temperature, or in a fluid with lower viscosity than the melt. In high-temperature solution growth, thermal strain is minimized due to the relatively low growth temperature, the much smaller thermal gradients used in these methods compared with classical methods based on growth from melts, and the free growth into a liquid. These factors make crystal growth proceed in physical conditions near thermodynamic equilibrium, allowing crystals to grow in a way that minimizes the superficial energy, developing facets. However, these low growth rates, in general hundreds of times smaller than the growth rates that can be achieved in melt-based growth methods, constitute the main disadvantage of the high-temperature solution growth methods. This, together with the differences in linear growth rates among the different faces of

Fig. 21.1 Flow chart with the criteria used to determine the crystal growth method of a material from the melt or from high-temperature solutions ◄

Table 21.1 Structural, thermal, and optical properties of the most important families of inorganic SSL single crystal

	Space group of symmetry	Optical class	Moh's hardness	Thermal conductivity (W/m K)	Transparency range (nm)	Growth method	References
YAG	$Ia3d$	Isotropic	8.2–8.5	14	200–6000	Czochralski Floating zone μ-Pulling	[21.1, 2]
REVO$_4$[a]	$I4_1/amd$	Uniaxial	5	$c_\parallel = 5.23$ $c_\perp = 5.10$	400–5000	Czochralski HTS EDFF TSSG	[21.3–5]
YAP	$Pbnm$	Biaxial	$\approx$ YAG	11	300–10 000	Czochralski	[21.6]
KREW[b]	$C2/c$	Biaxial	4.5–5	2.2–3.5	300–5000	TSSG INFC TNFC	[21.7–10]
RELF[c]	$I4_1/a$	Uniaxial	4–5	6	120–7500	Bridgman Czochralski VGF	[21.11, 12]
BaREF[d]	$C2/m$	Biaxial	3–3.5	6	125–12 000	Czochralski	[21.13, 14]
LiSAF	$P\bar{3}1c$	Uniaxial		3.3–3.6		Czochralski Bridgman TSSG	[21.15, 16]

[a] Data for YVO$_4$; [b] Data for KGd(WO$_4$)$_2$; [c] Data for YLF; [d] Data for BaYF$_8$

the crystal, which lead to chemical and structural inhomogeneities in the crystal, has hampered the industrial production of crystals grown from high-temperature solution methods.

Figure 21.1 summarizes the criteria used to determine the possibility of use of growth methods based on melt, or when not possible, the use of growth methods based on high-temperature solutions.

21.1.2 Most Important Families of Laser and Nonlinear Optical Materials

Laser Materials

Kaminskii [21.17] mentioned that currently there are over 575 single crystals for solid-state laser (SSL) applications, among them, different structure fluorides, oxides, chlorides, bromides, and compounds with mixed anion pairs such as chlorine–oxygen, fluorine–oxygen, and sulfur–oxygen, which can produce continuous wave (CW) or pulsed stimulated emission (SE) generation over a very wide spectral range from ≈ 0.172 to $\approx 7.24\,\mu$m.

The desirable properties of a good laser material are: feasibility of growth; high capacity to incorporate the active ions without affecting the structure; transparency in the suitable range; favorable spectroscopic properties for the active ions; high thermal conductivity; low and similar thermal expansion coefficients; inertness; low cost; and hardness, which favors polishing. They especially have to favor large absorption and emission cross sections, and have: a high gain coefficient; an optimum distance between the active ions (i.e., if laser transition is based on a scheme where cross-relaxation phenomena favors the population of the emitting level, the distance must be short to increase this populating level; on the other hand, if the laser emitting level could be depopulated by a cross-relaxation or energy transfer among active ions, then the distance must be longer); long lifetime to allow population inversion; and low phonon energy modes of the structure to avoid nonradiative transitions.

Table 21.1 lists the most important families of inorganic SSL single crystals, summarizing some of their main structural, thermal, and optical properties.

Oxides

$Y_3Al_5O_{12}$, Yttrium Aluminum Garnet (YAG). YAG crystallizes in the cubic system. The active ions substitute Y^{3+} in the structure in a D_2 position. It is a very stable compound with high robustness and high thermal conductivity. Optically YAG is isotropic. YAG is one of the most widely produced gain laser hosts. The Nd:YAG laser is by far the most commonly used solid-state laser since its first reported lasing at Bell Telephone Laboratories and RCA Laboratories in 1964. Tm-, Ho-, and Er-doped YAG lasers are also widely studied and used in the 2 μm region.

$REVO_4$, Rare-Earth Vanadate. Rare-earth vanadates crystallize in the tetragonal system. The trivalent active ions substitute the optically inert trivalent cations RE^{3+} at lattice sites with $\bar{4}2m$ symmetry. Optically they are positive uniaxial crystals, with $n_o = n_a = n_b$ and $n_e = n_c$. YVO_4 is a strongly birefringent material ($\Delta n = 0.2225$ at 633 nm at room temperature). Their advantages when compared with YAG are the high absorption and emission cross sections, broad bandwidths, and natural polarization of these materials.

$YAlO_3$, Yttrium Aluminum Perovskite (YAP). Several possible crystalline structures can be found in the literature for $YAlO_3$ [21.6]. The active lanthanide ions substitute the Y^{3+} in the lattice sites in C_{1h}. The physical properties of YAP, such as hardness, thermal conductivity, etc., are quite similar to those of YAG [21.18]. From the optical point of view, YAP is a biaxial, birefringent material (Δn, at 1.06 μm, $n_a = 1.929$, $n_b = 1.943$, and $n_c = 1.952$). The transparency region is from around 300 nm to 10 μm. The advantage of this material is related to its anisotropy, which allows the possibility of short tuning of the wavelength with the variation of the wavevector direction in the crystal, and the generation of linearly polarized output beams.

$KRE(WO_4)_2$, Potassium Double Tungstates (KREW). The monoclinic KREW is another well-known family of laser host compounds. The active lanthanide ions substitute the RE^{3+} cation in the structure at the local symmetry lattice site C_2. The physical properties of KREW materials are governed by their huge anisotropy. Their thermal conductivity is anisotropic, and depends also on the RE^{3+} cation. Optically, monoclinic KREW are biaxial crystals. The N_p orthogonal principal crystallooptic axis is parallel to the twofold symmetry axes. The other two principal axes are in the a–c plane. The principal axis N_g is located at $\kappa = 18.5–21.5°$ (angle value depending on the RE^{3+} of the host) clockwise to the **c**-crystallographic axis when the positive **b**-axis is pointing towards the observer. N_m is rotated at $\beta + \kappa - 90°$ with respect to the **a** crystallographic axis in the clockwise direction. Their anisotropy is responsible for the large absorption and emission cross sections of the active ions. Furthermore, these crystals are also very interesting for their stimulated Raman scattering (SRS) properties, and recently for their high efficiency as laser optical cooling materials [21.19].

Fluorides

$RELiF_4$, Rare-Earth Lithium Fluoride (RELF). This material crystallizes in the tetragonal system. The active ions substitute the RE^{3+} in the S_4 positions of the structure. It is an anisotropic uniaxial crystal, with the optical axis along the **c** crystallographic direction. Its birefringence at 633 nm is $n_o = 1.443$ and $n_e = 1.464$, and at 1064 nm is $n_o = 1.448$ and $n_e = 1.470$. Fluoride crystals are useful for coherent optical sources in the ultraviolet (UV) wavelength region. It can host a high percentage of doping elements, and possesses a low thermal lensing effect, when compared with YAG.

$BaREF_8$, Barium Rare-Earth Fluoride (BaREF). These materials crystallize in the monoclinic system. The rare-earth dopant substitutionally enters the RE^{3+} sites, whose symmetry is S_4. The BaYF refractive index n is 1.5. Optically, $BaREF_8$ is a biaxial crystal. Their monoclinic structure compensates the thermal lens effect under strong pumping better than cubic crystals such as YAG. Additionally, these materials have low-energy phonons: the maximum phonon energy is 400 cm^{-1}.

$LiSrAlF_6$, Lithium Strontium Aluminum Fluoride (LiSAF). LiSAF crystals belong to the colquiriite fluoride family of crystals LiMAF (where M = Sr, Ga, Ca). These materials crystallize in the trigonal system. Optically, these materials are uniaxial. Among them, $LiCaAlF_6$:Cr (LiCAF:Cr) is more robust and has more advantageous thermo-optical properties. It exhibits higher scattering and smaller absorption and emission cross sections when compared with LiSAF. Cr^{3+}-doped single crystals of this family have been described as efficient broadly tunable laser materials [21.20, 21].

Nonlinear Optical Materials

Since its first demonstration in 1961 [21.22], nonlinear frequency conversion has been a field limited by the available materials, with practical advances largely controlled by progress in making improved NLO materials. To date, the most important class of materials used in nonlinear optics has been inorganic single crystals. Organic materials, although promising, have yet to be produced with good enough chemical and mechanical properties to find broad practical application. Relatively few materials find application in nonlinear optics. The physics of the frequency conversion process places severe demands on potential NLO crystals. Beyond the NLO physics are the additional practical requirements of mechanical and chemical stability and the possibility for production in the form of adequately sized and uniform single crystals.

Table 21.2 lists the main inorganic NLO crystals together with their main structural, thermal, and optical properties.

Niobates. LiNbO$_3$ (LN) crystallizes in the trigonal system. Its NLO coefficients are relatively high, and the birefringence is of a magnitude, enabling phase matching for noncritical type I second-harmonic generation (SHG) of a 1064 nm fundamental at room temperature. Its thermal conductivity is very high. LN can suffer from photorefractive damage when illuminated with visible radiation. Certain Mg-doped LN crystals have been measured to have a photorefractive damage threshold more than 10^5 times greater than that of typical

Table 21.2 Structural, thermal, and optical properties of the main inorganic NLO single crystals [21.23]

	Space group of symmetry	Optical class	Thermal conductivity (W/m K)	Transparency range (nm)	NLO coefficients (pm/V)	Growth method
LN	$R3c$	Uniaxial	0.0015	350–5000	$d_{21} = d_{16} = -2.1$ $d_{22} = 2.1$ $d_{14} = d_{25} = d_{36} = 0$ $d_{31} = d_{15} = -4.3$ $d_{32} = d_{24} = -4.3$ $d_{33} = -27$	Czochralski VTE TSSG DC-Czochralski
KN	$Amm2$	Biaxial		400–5500	$d_{21} = d_{16} = 0$ $d_{22} = 0$ $d_{14} = d_{25} = d_{36} = 0$ $d_{31} = d_{15} = -11.3$ $d_{32} = d_{24} = -12.8$ $d_{33} = -19.5$	HTS
KTP	$Pna2_1$	Biaxial	2–3.3	350–4500	$d_{21} = d_{16} = 0$ $d_{22} = 0$ $d_{14} = d_{25} = d_{36} = 0$ $d_{31} = d_{15} = 2.0$ $d_{32} = d_{24} = 3.6$ $d_{33} = 8.3$	HTS Hydrothermal
KDP	$I\bar{4}2d$	Uniaxial	1.86–2.09	180–1800	$d_{21} = d_{16} = 0$ $d_{22} = 0$ $d_{14} = d_{25} = d_{36} = 0.37$ $d_{31} = d_{15} = 2.0$ $d_{32} = d_{24} = 3.6$ $d_{33} = 8.3$	Aqueous solution
β-BBO	$R3c$	Uniaxial	0.001–0.002	198–2600	$d_{21} = d_{16} = -2.3$ $d_{22} = 2.3$ $d_{14} = d_{25} = d_{36} = 0$ $d_{31} = d_{15} = 0.1$ $d_{32} = d_{24} = 0.1$ $d_{33} = 0$	HTS Metastable growth Czochralski

Table 21.2 (cont.)

	Space group of symmetry	Optical class	Thermal conductivity (W/m K)	Transparency range (nm)	NLO coefficients (pm/V)	Growth method
LBO	$Pna2_1$	Biaxial		160–2300	$d_{21} = d_{16} = 0$ $d_{22} = 0$ $d_{14} = d_{25} = d_{36} = 0$ $d_{31} = d_{15} = -0.67$ $d_{32} = d_{24} = 0.85$ $d_{33} = 0.04$	HTS
BIBO	$C2$	Biaxial		286–2500	$d_{222} = 2.53(4)$ $d_{211} = 2.3(2)$ $d_{233} = 1.3(1)$ $d_{231} = 2.3(2)$ $d_{112} = 2.8(2)$ $d_{332} = 0.9(1)$ $d_{312} = 2.4(3)$ $d_{132} = 2.4(3)$	Czochralski
AgGaS$_2$	$I\bar{4}2d$	Uniaxial	0.01	500–13 000	$d_{21} = d_{16} = 0$ $d_{22} = 0$ $d_{14} = d_{25} = d_{36} = 17.5$ $d_{31} = d_{15} = 0$ $d_{32} = d_{24} = 0$ $d_{33} = 0$	Bridgman
LiIO$_3$	$P6_3$	Uniaxial	8×10^{-4}	310–5000	$d_{21} = d_{16} = 0$ $d_{22} = 0$ $d_{14} = d_{25} = d_{36} = 0$ $d_{31} = d_{15} = 4.4$ $d_{32} = d_{24} = 4.4$ $d_{33} = 4.5$	Aqueous solution

undoped LN. Mg doping appears to have little effect on the NLO coefficients.

KNbO$_3$ (KN) is notable for its very large NLO coefficients and birefringence. The transmission range is similar to that of LN, although residual losses are generally higher and much less consistent from crystal to crystal. It suffers from significant processing limitations, due to the material's low Curie temperature and its pyroelectric and ferroelastic character.

Phosphates. KTiOPO$_4$ (KTP) belongs to family of compounds with the general formula ABOXO$_4$ where A = K, Rb, Na, Cs, Tl, NH$_4$; B = Ti, Sn, Sb, Zr, Ge, Al, Cr, Fe, V, Nb, Ta, and X = P, As, Si, crystallizing in the orthorhombic system [21.24]. KTP single crystals are chemically inert. Optically they are biaxial crystals. The KTP transmission range, NLO coefficients, and birefringence are similar to those of LN, although KTP's residual absorption is notably higher. Perhaps most important for KTP are its very large thermal and angular phase-matching bandwidths for SHG of fundamental wavelengths near 1064 nm.

KH$_2$PO$_4$ (KDP) and its homologs are among the most widely used commercial NLO materials. Although members of the family have relatively low NLO coefficients, they feature good UV transmission, high birefringence, and relatively high resistance to laser damage. KDP is widely used to generate second, third, fourth, and fifth harmonics of 1.06 μm radiation.

Borates. Single crystals of the low-temperature phase of barium metaborate, β-BaB$_2$O$_4$ (β-BBO) and LiBO$_3$ (LBO) have found important applications for NLO devices. β-BBO has a wide transparent spectral range, and good mechanical properties. In general, borates combine UV transparency, high-laser damage thresholds, and adequate birefringence for phase-matching a broad range of visible and UV interactions. β-BBO has larger nonlinear susceptibilities than LBO, and can be phase-matched more readily in the UV, but suffers from smaller angular acceptance and lower damage threshold, and is more hygroscopic.

BiB$_3$O$_6$ (BIBO) is another outstanding NLO crystal. It belongs to the monoclinic system. BIBO is an appropriate material for phase-matching condition for near-infrared (NIR) wavelengths. It possesses a large effective nonlinear coefficient and high damage threshold, and is nonhygroscopic. Its nonlinear coefficient is 3.5–4 times higher than that of LBO, and 1.5–2 times higher than that of β-BBO [21.25].

Chalcopyrites. The most important of these compounds are AgGaS$_2$, AgGaSe$_2$, and ZnGeP$_2$. These materials exhibit broad transparency far into the infrared, sufficient birefringence to enable phase-matching over a major portion of the transparency range, and relatively high nonlinear coefficients. Laser damage resistance tends to be very low. The greatest limitation to application of the chalcopyrites has been optical loss.

LiIO$_3$. Because of its large birefringence, lithium iodate, LiIO$_3$ is broadly phase-matchable, but suffers from large walk-off effects that limit conversion efficiencies despite its larger nonlinear susceptibility. Applications taking advantage of the broad tuning range are the most common.

21.2 High-Temperature Solution Growth

When compared with growth from the melt, crystal growth from solution produces remarkable improvements in the quality of the grown crystals, principally due to a much lower crystallization temperature. Using lower temperatures during the growth process means a lower density of structural defects and less contamination in the liquid phase from the crucible or the ambient. Furthermore, the low growth rate developed at lower temperatures enables better control of all the parameters involved in a stable growth process. Finally, careful choice of a suitable solvent can help reduce the viscosity of the solution, minimizing problems related to deficient mass transport that can generate solvent inclusions in the crystals. However, crystal growth by high-temperature solution (HTS) presents some disadvantages when compared with crystal growth techniques from the melt. Substitution, or generation of interstitial defects when the solvent contains ions alien to the grown material, and the relatively low crystal growth rates inherent to HTS techniques constitute the main disadvantages of these techniques, especially when scaling up these processes to industrial production.

Nevertheless, there is no doubt that HTS growth methods have a huge field of application and they constitute the basis of several of the current solid-state technologies, mainly in microelectronics and optoelectronic devices. Theoretically, this method presents no limits – it is always possible to find a suitable solvent and the most convenient crystallization conditions to obtain any material. In practice, however, it is not this easy. Despite this, the list of materials that have been obtained by HTS methods is very long, covering materials from diamond or oxide compounds to metals. Hundreds of single-crystalline materials are grown by HTS-based methods, among them the magnetic ferrites developed in the 1970s, the superconductor materials such as YBa$_2$Cu$_3$O$_{7-x}$ (YBCO) developed during the 1980s and 1990s, many ferroelectric materials such as KTP, etc. Crystal growth from HTS is also one of the most popular growth methods as it does not require sophisticated equipment. At present, several research laboratories around the world are not only growing new and different crystalline materials, but are also studying and modeling the fundamentals of the transport processes involved in crystal growth using these HTS techniques.

The principle of crystal growth by HTS lies in choosing a suitable solvent S for the components A and B of a given α_{AB} phase, all solids at room temperature, with a lower solubility for the α_{AB} phase at a given temperature than for the initial components.

By heating the reagents it will be possible to induce HTS growth

$$A_S + B_S + S_S \rightarrow A_L + B_L + S_L \rightarrow \alpha_{AB} + S_L . \tag{21.1}$$

The driving force for HTS growth is supersaturation of the $\alpha_{AB} + S_L$ solution. This supersaturation can be created by slow cooling of the solution, by evaporation of the solvent or by increasing the solute concentration at constant temperature.

To control spurious nucleation and improve the quality of the grown crystals, modifications to HTS methods have been applied through the years. Methods based on homogeneous nucleation used in the 1960s have been displaced by methods based on induced growth (secondary nucleation), which involve techniques such as rotation and pulling of the growing crystals that enhance mass transport conditions in the solution, avoid supersaturation gradients, and allow growth of larger crystals. Top-seeded solution growth and liquid-phase epitaxy are the most popular methods based on HTS growth technologies. In the next sections the main features of these two methods will be described.

21.2.1 Top-Seeded Solution Growth (TSSG)

To grow a crystal the first step, after having selected a suitable candidate solvent, is the study of the phase diagram of the different compounds that will be mixed to create the growth solution to determine the composition and temperature range in which the desired crystalline phase (solute into the solution) can be grown. As an example the phase diagram of $KGd_{0.5}Nd_{0.5}(PO_3)_4$ in the ternary system $(Nd_2O_3+Gd_2O_3)-K_2O-P_2O_5$ is shown in Fig. 21.2a. Parameters such as the existence of other neighboring crystalline phases, the crystallization temperature, the solute-to-solvent ratio, etc. will influence the selection of the most suitable point of the phase diagram at which to perform the crystal growth experiments. The solubility curve of the crystallizing phase in the chosen solvent must then be determined. Figure 21.2b shows the solubility curve of the KREW–$K_2W_2O_7$ system. When the solubility curve has been determined, the crystallization experiment can be performed by choosing the most suitable solute–solvent composition while accounting for the slope of the solubility curve. If the growth experiment is performed in a solution with a solute–solvent composition in region I of Fig. 21.2b, where the solubility curve shows a pronounced slope, a temperature fluctuation ΔT produces a small change in the concentration of solute Δc_1. However, if the experiment is performed in a solution with a composition in region II, where the solubility curve has a flat slope, the same temperature fluctuation ΔT will produce a larger change in the concentration of solute Δc_2 that can lead to fluctuations in the supersaturation levels created in the solution. In order to obtain crystals with a lower density of defects it is important to avoid fluctuations in the supersaturation levels in the growth solution. Thus, region I is the most suitable region to grow crystals in these solutions, as it provides better control of the solutal fluctuations by small changes in the temperature.

The TSSG method represents an improvement of crystal growth from HTS methodologies as it restricts and forces crystal growth only at a point of the free surface of the solution, i.e., on a crystal seed in contact with the surface of the solution that can be rotated and pulled. Figure 21.3a shows the experimental con-

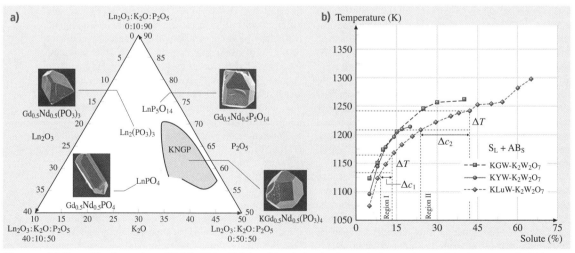

Fig. 21.2 (a) Crystallization region of $KGd_{0.5}Nd_{0.5}(PO_3)_4$ with solution isotherms in the $Ln_2O_3-K_2O-P_2O_5$ system. SEM images of $K(Gd, Nd)(PO_3)_4$ crystals and neighboring phases. (b) Solubility curve of KREW (RE = Gd, Y, and Lu) in $K_2W_2O_7$ solvent

figuration of the TSSG method. Before configuring the values of the TSSG parameters it is necessary to know the physicochemical properties of the solution such as density, viscosity, thermal conductivity, and surface tension [21.26]. This knowledge enables numerical and experimental simulation to determine the best solute–solvent composition and to know the hydrodynamics of the solution needed for a successful crystal growth experiment [21.27]. These experimental simulation studies will provide the most suitable values for parameters such as the aspect ratio (diameter/height) of the crucible that will contain the solution, axial and radial thermal gradients in the solution, rotation rate of the crystal seed, extent of the crucible's rotation, ratio between the size of the growing crystal and the crucible radius, etc. Figure 21.3b shows a nonaxisymmetric horizontal flow pattern of the solution obtained by visualization techniques in an experimental simulation of a growth experiment. The best conditions for real growth should avoid nonaxisymmetric flow patterns. Figure 21.3c shows an axial cross section of a growth solution by visualizing the flow established by a crystal/crucible

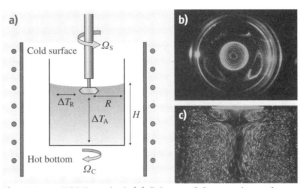

Fig. 21.3a–c TSSG method. (**a**) Scheme of the experimental configuration. (**b**) Nonaxisymmetric horizontal flow pattern obtained by visualization techniques of the solution. (**c**) Axial cross section of a solution of growth, visualizing the flow established by a crystal/crucible counter-rotation configuration

counter-rotation configuration. These experimental simulations have been obtained from liquid simulation under similar thermal and hydrodynamics conditions to

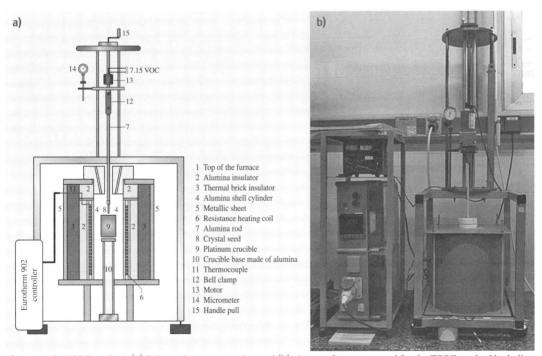

Fig. 21.4a,b TSSG method. (**a**) Schematic representation and (**b**) picture of a system used for the TSSG method including a single-thermal-zone vertical tubular furnace, the crystal seed attach, rotation, and pulling system, the crucible and its support, and temperature controller/programmer

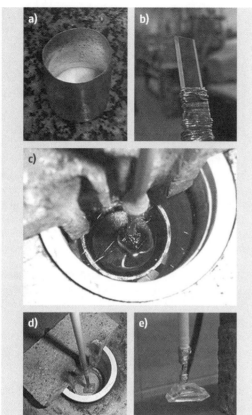

Fig. 21.5a–e Process of growth of a double tungstate single crystal by the TSSG method. (**a**) Platinum crucible containing the flux or solution at room temperature. (**b**) Crystal seed ready to be used for the crystal growth experiment. (**c**) Single crystal growing inside the solution. (**d**) Single crystal after being removed from the solution while cooling the furnace to room temperature. (**e**) Single crystal attached to the crystal seed at the end of the experiment

those of a real growth solution (equal Reynolds and Prandtl numbers). In high-viscosity solutions the use of angular acceleration of the crystal or the crucible, or both, with periodic changes in their direction of rotation, has also been used to improve the homogeneity of the solution. For very-high-viscosity systems a HTS alternative technology such as the accelerated crucible rotation technique (ACRT) must be used [21.28].

Figure 21.4 shows a schematic representation and a picture of the system used for the TSSG method. It includes a single-thermal-zone vertical tubular furnace, the crystal seed attach, rotation, and pulling system, the crucible and its support, and the temperature controller/programmer. The furnace is heated by an electrical resistance, and the thermal insulation and the alumina shell cylinder can also be seen in the cross section of the furnace provided in the schematic representation. The seed attach, rotation, and pulling system consist of a series of high-precision mechanisms attached to two different stepper motors that allow simultaneous rotation and pulling of the crystal seed. Both the speed of rotation and the speed of pulling can be modified according to the voltage provided by the power source. The attach system can be moved up and down to locate the crystal seed precisely in contact with the surface of the solution. This movement can be monitored with a micrometric comparer for better precision. The support of the crucible allows it to be located at the correct height along the axis of the furnace to obtain the desired thermal gradient inside the solution. Finally, the heating or cooling rate and the temperature of the furnace are precisely controlled by the controller/programmer systems, connected to a thyristor.

Figure 21.5 shows a sequence of pictures representing the process of growth of a double tungstate single crystal using the TSSG method. Figure 21.5a shows a platinum crucible containing the flux or solution, which is a solid at room temperature. Figure 21.5b shows a crystal seed attached to an alumina rod and a platinum support ready for the crystal growth experiment. In Fig. 21.5c, the single crystal growing inside the solution while attached to the crystal seed can be seen. Figure 21.5d shows the single crystal after being removed from the solution and standing near the surface of the solution while the furnace is being cooled slowly to room temperature to avoid thermal cracking of the crystal. Finally, Fig. 21.5e shows the single crystal at the end of the crystal growth experiment still attached to the crystal seed.

21.2.2 Liquid-Phase Epitaxy (LPE)

The term epitaxy can be defined as the deposition of a single-crystal layer on a single-crystal substrate such that the crystalline structure of the substrate is continued into the layer. For a given substrate, the lattice parameter mismatch between the layer and the substrate is limited by a critical value, above which continuity of the crystalline construction across the substrate–layer interface is impossible. LPE is the growth of epitaxial layers, hereafter epilayers, from solutions at high temperatures.

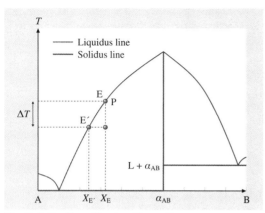

Fig. 21.6 Cooling effect in a binary diagram

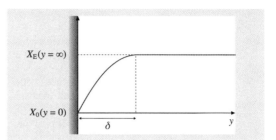

Fig. 21.7 Solutal boundary layer adjacent to substrate surface

The process of LPE growth involves three main steps:

1. Creation of supersaturation
2. Mass transport from solution to solid–liquid interface
3. Nucleation and surface attachment [21.29, 30]

1. *Creation of supersaturation.* Figure 21.6 introduces the fundamentals of LPE as a crystal growth method. A substrate is introduced into a solution, which is cooled by a discrete amount ΔT and the equilibrium liquid-phase composition is shifted from X_E to $X_{E'}$, which has a lower α_{AB} (solute) concentration. The decrease in solubility creates a supersaturated solution and yields deposition of α_{AB} solid phase onto the substrate.
The supersaturation created in the solution

$$\sigma = \frac{X_E(T) - X_{E'}(T)}{X_{E'}(T)} \qquad (21.2)$$

is related to the change of free energy when crystallization occurs $\Delta G \approx RT\sigma$ and to the gradient of chemical potential $\Delta G \approx \Delta \mu$. The chemical potential at point P in Fig. 21.6 of the supersaturated solution is higher than its equilibrium value at the same temperature, point E' in Fig. 21.6. This constitutes the driving force of LPE growth, and induces heterogeneous nucleation, which exhibits a lower activation energy than homogeneous nucleation because the surface of the substrate catalyzes the crystal growth process.

2. *Mass transport from the solution to the solid–liquid interface.* When the epilayer starts to grow, and because of the limitations of mass transport near the substrate surface, a solutal boundary layer is established, as can be seen in Fig. 21.7.
The velocity of mass transport from the solution to the solid–liquid interface in stationary conditions is given by Fick's equation

$$\begin{aligned} R &= \frac{D_v}{\rho} \frac{X_{E'} - X_0}{\delta} = \frac{1}{\rho} \frac{D_v}{\delta}(X_{E'} - X_0) \\ &= \frac{1}{\rho} \frac{1}{\delta/D_v}(X_{E'} - X_0)\,, \end{aligned} \qquad (21.3)$$

where D_v is the volumetric diffusivity of the solute in the solution, ρ is the density, and δ is the thickness of the solutal boundary layer. The quotient δ/D_v is called the diffusional resistivity of mass transport.
In a situation of natural thermal convection under low Rayleigh number (low thermal gradients) and also in growth experiments with forced convection but with low Reynolds number (low rotational velocity of the substrate), the boundary thickness is given by

$$\delta \propto D_v^{1/3} \nu^{1/6} \omega^{-1/2}\,, \qquad (21.4)$$

where ν is the kinematic viscosity and ω is the revolution rate of the substrate.

3. *Heterogeneous nucleation and surface attachment.* The capture of the atoms onto the surface of the

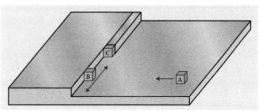

Fig. 21.8 Heterogeneous nucleation and surface attachment process on the substrate surface

substrate takes place in a number of successive steps:
a) The nucleation, or the formation of stable clusters of atoms on the surface of the substrate at sites such as A (Fig. 21.8)
b) The migration of atoms on the surface to the growth step by surface diffusion and their adsorption at sites such as B
c) Migration of the adsorbed atoms in B positions to energetically favorable sites such as C
d) Removing the released heat from the global reaction and transport of the solvent from the lost solute–solvent solvation

The contributions of all surface processes on the kinetics of the crystal growth process can be evaluated by adding a new resistivity $1/\kappa$ to the crystal growth rate formula

$$R = \frac{1}{\rho}\left(\frac{1}{\kappa} + \frac{\delta}{D_v}\right)^{-1}(X_{E'} - X_0). \qquad (21.5)$$

21.3 Growth of Bulk Laser and NLO Single Crystals by the TSSG Method

A favourable viscosity of the growth solution can be defined in the range 1–100 cP, while the maximum practical viscosity that allows development of a crystal growth process would be about 10 P. An ideal solvent was defined by *Elwell* and *Scheel* [21.29] as a solvent with a viscosity in the range 1–10 cP. However, it is not trivial to find compounds that can be used as solvents in high-temperature solutions with such viscosities. In this chapter we define low-viscosity solutions as those with viscosity below 40–50 cP, typical of fluorides, vanadates, and tungstates. Solutions containing phosphates and borates present a much higher viscosity, with values that can reach up to 10 P. Crystal growth using high-temperature solutions with low or high viscosities present different difficulties that we will try to summarize here, together with the strategies proposed to overcome these problems.

21.3.1 Crystal Growth from Low-Viscosity Solutions: Fluorides, Tungstates, and Vanadates

Double Tungstates MREW
(M = Na, K, Rb; RE = Y and Lanthanides)
The crystallization temperature of the required phase, the presence of other phases crystallizing at higher temperature, the nature of possible polymorphic phase transformations, and the melting character of the compound are the factors that determine the most suitable growth method to obtain single crystals of a particular MRE(WO$_4$)$_2$ compound. The Czochralski method is the most used method to obtain double tungstate compounds that melt congruently and do not show polymorphic transformations, such as Na-based double tungstates compounds with RE ions with large

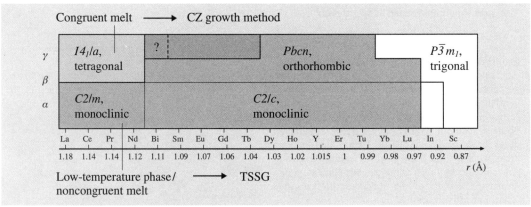

Fig. 21.9 Polymorphism of KREW compounds at room temperature, and recommended methods to grow the different crystal of this family of materials

Table 21.3 Concentration of precursor oxides to form a growth solution of KGdW using $K_2W_2O_7$ as a solvent, with a composition 85 mol % solvent–15 mol % solute

Molar ratio of oxides in the composition of growth: 30.85% K_2O – 2.5% Gd_2O_3 – 66.67% WO_3
Molar ratio of oxides in the solvent: 33.3% K_2O – 66.67% WO_3
Molar ratio of oxides in the solute: 16.6% K_2O – 16.6% Gd_2O_3 – 66.67% WO_3
Solvent: $(K_2O + 2WO_3) \times 0.85$
Solute: $(0.5K_2O + 0.5Gd_2O_3 + 2WO_3) \times 0.15$
Total: $0.925K_2O + 0.075Gd_2O_3 + 2WO_3$ = 30.85% K_2O – 2.5% Gd_2O_3 – 66.67% WO_3

ionic radii, such as $NaGd(WO_4)_2$. However, melting is not congruent when the ionic radius of the lanthanide decreases [21.31]. On the other hand, KREW monoclinic single crystals are usually grown by the TSSG slow-cooling (SC) method, as they present a phase transformation at a lower temperature than the congruent melting temperature. Figure 21.9 summarizes the polymorphism of KREW compounds together with the recommended methods that can be used to grow these crystals.

The appropriate solvent for growing KREW crystals has been chosen among the stable compounds of the K_2O–WO_3 binary system [21.32, 33] to avoid introduction of foreign ions in the solution that can contaminate the crystals. The most used solvents are K_2WO_4 and $K_2W_2O_7$. $K_2W_2O_7$ has a lower melting point when compared with the other stable compounds in the K_2O–WO_3 system, and due to a higher tungsten content, it exhibits a low viscosity during the growth process. K_2WO_4 was used as well [21.34], as a solvent for the growth of KREW, but it did not yield good homogenization of the solution and was more prone to evaporation. The solubility curves of the monoclinic KREW (RE = Gd, Y, and Lu) in $K_2W_2O_7$ are shown in Fig. 21.2b.

The limits of the solubility curve are given by the properties of this binary system. At around 45 mol % solute, the phase that crystallizes first when decreasing the temperature does not belong to the monoclinic system [21.35]. The lower limit at around 5 mol % solute is related to the economical profitability of the growth procedure. Although the monoclinic phase of KREW can be grown in the range 5–42 mol % solute, it is better to use a solution with < 15 mol % solute, where small variations of the temperature lead to small changes in supersaturation. The growth temperatures when using these solutions are around 1090–1230 K and guarantee a low economical cost for the growth procedure.

The thermal gradients in the solutions are usually low, in the range of 1.5 and 11.5 K/cm for axial and radial gradients, respectively, with the bottom and the crucible wall as the hottest spots. Growth of double tungstates is very sensitive to thermal gradients [21.36, 37]. Large thermal gradients may crack the growing crystal due to the high anisotropy of the linear thermal expansion coefficients of monoclinic double tungstates. The solution was prepared by melting the mixed oxide precursors, weighed in stoichiometric ratio. Table 21.3 shows the molar ratio of the precursor oxides for the growth of $K_2W_2O_7$/KGdW with composition 85 mol % solvent–15 mol % solute.

The solution was homogenized by keeping it at ≈ 50 K above the saturation temperature for several hours (up to 24 h in some cases). Then, the saturation temperature was determined by observing the growth/dissolution of an oriented prism crystal of KREW acting as a seed in contact with the surface of the solution.

For KGdW the saturation temperature was in the range 1173–1203 K, for KYbW it was in the range 1170–1195 K, and for KLuW it was in the range 1146–1162 K. The fact that to reach the same solubility value we have to increase the temperature in the sequence $T(\text{sol})_{\text{KGdW}} < T(\text{sol})_{\text{KYbW}} < T(\text{sol})_{\text{KLuW}}$, as observed in Fig. 21.2b, is translated to a lower temperature of crystal growth when decreasing the ionic radius of the lanthanide ion $T_{S,\text{KGdW}} > T_{S,\text{KYbW}} > T_{S,\text{KLuW}}$ for a fixed solute-to-solvent ratio.

Growth experiments could be made using the same growth conditions but not the same crystal seed orientation. Crystal seeds were cut with a parallelepipedal shape and oriented along the b-, c-, and a^*-directions. The use of b-oriented seeds allows us to apply a slightly faster cooling rate and consequently to obtain a higher growth rate than when using crystal seeds oriented along other orientations [21.7, 38], still obtaining defect-free single crystals. Seeding along [$\bar{1}11$] direction has also been reported [21.39]. The seed was rotated at a broad range of rotation velocities from 4.5 to 90 rpm. Once the saturation temperature was determined, the crystal seed was placed in contact with the surface of the solution and slow cooling was applied to cre-

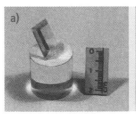

Fig. 21.10a,b Monoclinic undoped KGd(WO$_4$)$_2$. (a) Crystal grown on a *b*-oriented seed. (b) Crystal grown on a *c*-oriented seed

ate supersaturation. The temperature of the solution was usually decreased for ≈ 10 K at a cooling rate of 0.03–0.05 K/h. After that, the crystals were removed slowly from the solution and cooled to room temperature at 15 K/h to avoid thermal shocks. The results obtained clearly showed that the crystal quality depended strongly on the cooling rate. The slower the applied cooling rate, the better the quality of the crystals.

In the growth of KHoW single crystals, *Borowiec* et al. [21.40] used programmed changes of the cooling rate to grow crystals with larger dimensions.

Pulling at rates of 2–5 mm/day has also been used to grow these crystals [21.40, 41]. In this way, crystals grow as cylinders bounded by {110} and {011} prisms and by {100}, {010}, and {001} pinacoids [21.40].

Figure 21.10 shows some examples of KREW crystals grown by the TSSG method.

As the main application of these crystals is to be used as hosts for other active laser lanthanides ions, it is worth mentioning the effects of lanthanide doping on the growth procedure. First, it is important to highlight the feasibility of growing highly doped lanthanide monoclinic tungstates, up to 100% stoichiometric KREW, with RE as the active ion, such as KErW and KYbW [21.42]. The difference of ionic radii between the RE constituting element of the KREW host and the doping Ln^{3+} cation governs the difficulty of the crystal growth process and the attainable limit of substitution. For example, when growing praseodymium-doped KGdW crystals, it was necessary to reduce the cooling rate to 0.02 K/h to obtain high-quality single crystals. The degree of substitution can be evaluated by the distribution coefficient $K_{Ln^{3+}}$, which provides the ratio between the lanthanide concentration in solution and the lanthanide concentration in the crystal. This measures how easy it is to dope the crystal with a particular ion. The distribution coefficient can be calculated by the following expression

$$K_{Ln^{3+}} = \frac{\{[Ln^{3+}]/([Ln^{3+}]+[RE^{3+}])\}_{\text{crystal}}}{\{[Ln^{3+}]/([Ln^{3+}]+[RE^{3+}])\}_{\text{solution}}}. \quad (21.6)$$

Distribution coefficients close to 1 favor homogeneity of doping element inside the crystal. Figure 21.11 shows the distribution coefficient for lanthanide doping in KGdW. Figure 21.11a shows the dependence of the distribution coefficient on the ionic radii of the lanthanide doping ion: the larger the ion, the more difficult it is for it to enter the structure. Figure 21.11b shows that, the smaller the ionic radii difference between the rare-earth host and the lanthanide ion, the larger the distribution coefficient.

Fluorides

LiYF$_4$ (YLF) has been grown by the TSSG method [21.43] based on the phase diagram of the binary system LiF–YF$_3$, which presents a peritectic point at 49 mol% YF$_3$ and 1090 K, and a eutectic point at 19 mol% YF$_3$ and 970 K. Crystals were grown by pulling in a melt containing YF$_3$ and a slight excess of LiF over the stoichiometric amount in a purified helium atmosphere.

KYF$_4$ (KYF) and KLiYF$_5$ (KLYF) have also been grown by the TSSG method [21.44]. Crystals were grown by slow cooling of *b*-oriented crystal seeds, controlling their dimensions by pulling of the crystal from the solution and using weight control feedback. However, the existence of cleavage planes perpendicular to the *b*-axis limits growth to some extent.

Other fluorides such as LiCaAlF$_6$ and LiSrAlF$_6$ can also be grown by TSSG using inert or reactive gas atmospheres to avoid traces of H$_2$O and O$_2$ that can degrade

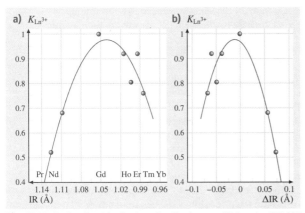

Fig. 21.11a,b Lanthanide doping distribution coefficient in KGdW crystal with RE = Gd, Y, Yb, and Lu. (a) Ionic radii (IR). (b) Difference between RE host and lanthanide ion

their optical quality. Wet chemical methods, ammonium salt methods, or hot hydrofluorination processes have to be used to obtain the high-purity halides to be used in these growth processes, as they are not commercially available [21.45]. The control of the diameter by a weight-sensing feedback system, and pulling of the growing crystal from the solution, have been used to grow these crystals.

Vanadates

REVO$_4$ single crystals are usually grown by the Czochralski method, but these compounds melt congruently at very high temperatures [21.46]. So, the TSSG method has also been reported as a suitable technique to grow these crystals. This technique allows one to avoid the formation of nonpentavalent vanadium oxides. Various fluxes have been used to grow these crystals: NaVO$_3$ [21.47], V$_2$O$_5$ [21.48], Pb$_2$V$_2$O$_7$ [21.4, 47], PbO + PbF$_2$ [21.49], and LiVO$_3$ [21.5]. However, PbO + PbF$_2$ is not very appropriate because of its toxicity, and Pb$_2$V$_2$O$_7$ has a high viscosity and a low solubility for these solutes. Large axial thermal gradients in the growth solution were used to grow crystals with this solvent on c-oriented crystal seeds rotated at 100 rpm [21.50]. LiVO$_3$ has a suitable solubility for YVO$_4$, low viscosity, low evaporation in the working temperature range, and no toxicity [21.51]. LiVO$_3$ is a very hydroscopic soluble in water, which facilitates crystal and crucible cleaning after the growth procedure. Crystals were grown by slow cooling of the solution. Rotation and pulling of nonoriented YVO$_4$ seeds have been used.

21.3.2 Crystal Growth from High-Viscosity Solutions: Phosphates and Borates

Crystal growth in highly viscous solutions is very difficult because the high viscosity limits the hydrodynamics and mass transport in the melted solution. It also creates a thick growth boundary layer and makes both heat extraction and impurity expulsion difficult. All these factors interfere with the growth process, resulting in many crystal macrodefects, such as inclusions. This, coupled with the tendency of the melt to supercool, makes this type of growth especially challenging.

Phosphates

KTP was first synthesized in 1890 by Ouvrard, and was crystallized in a polycrystalline form from a flux by *Masse* and *Grenier* [21.52]. KTP melts incongruently at 1172 °C when heated in air [21.24, 53], and therefore it cannot be grown directly from its melt. Various solvents have been used to grow KTP crystals from high-temperature solutions. Table 21.4 summarizes the chemical reactions used to synthesize KTP and its solvents. The crystallization region of KTP in the K$_2$O–P$_2$O$_5$–TiO$_2$ system was first studied by *Voronkova* and *Yanovskii* [21.54], and later in detail by *Iliev* et al. [21.55]. The solubility of KTP in this system has been shown to increase as the concentration of the anion (P$_2$O$_7$)$^{4-}$ increases [21.56–58]. The viscosity of these solutions range from 50 to 80 cP, increasing as the TiO$_2$ content increases, due to the formation of [Ti(PO$_4$)$_3$]$^{5-}$ complexes that polymerize upon addition of TiO$_2$ [21.55, 59].

KTP single crystals on crystal seeds were first grown from the phosphate system and patented by *Gier* [21.60]. *Alexandrovskii* et al. [21.61] used fused KPO$_3$ with excess K$_2$O for the same purpose. *Loiacono* et al. [21.58] reported that crystals of KTP grown in K$_4$P$_2$O$_7$ solvent resulted in a platy morphology. The growth process has been developed with slow cooling of the saturated solutions [21.54], and with gradient transport at constant ΔT [21.56]. However, the viscosity of phosphate solutions increases to unacceptable levels when decreasing the temperature. This leads to supersaturation gradients in the solution and results in spurious nucleation and solvent inclusion in the crystals. These inclusions can be a source of strain that cracks the crystals on cooling to room temperature. By imposing motion of the growing crystal, inclusion-free single crystals can be obtained, as this rotation improves the mass transport in the solution [21.62, 63]. Crystals can be rotated about the center of the surface of the solution or eccentrically, although the latter method does not use the crucible space as efficiently as the former one [21.57, 64]. The accelerated crucible rotation technique, by which the accelerating motion of the solution relative to the growing crystal surface prevents the adverse effects of constitutional supercooling, has also been used to improve the mass transport in the solution [21.56]. Strictly controlled cooling rates and a specially designed monitoring of crystal weight changes were employed to crystallize the largest KTP inclusion-free crystal reported [21.64]. Figure 21.12 shows some single crystals of the KTP family obtained by the TSSG method.

Dhanaraj et al. [21.65, 66] realized that using cooling rates that are too fast (1 K/h) in these viscous solutions resulted in coarse dendritic structures on the crystal surfaces and flux inclusion in the bulk of the crystal, due to a rapid increase in supersaturation, es-

Table 21.4 Chemical reaction of formation of KTP and several fluxes used for its growth from TSSG method

Formation of KTP
$KH_2PO_4 + TiO_2 \rightarrow KTiOPO_4 + H_2O$
$K_2CO_3 + 2TiO_2 + P_2O_5 \rightarrow 2KTiOPO_4 + CO_2$
$K_2CO_3 + 2TiO_2 + 2NH_4H_2PO_4 \rightarrow 2KTiOPO_4 + CO_2 + 2NH_3 + 3H_2O$
$K_2TiO(C_2O_4)_2 \cdot 2H_2O + KH_2PO_4 \rightarrow KTiOPO_4 + H_2C_2O_4 + K_2C_2O_4 + 2H_2O$
Formation of phosphate solvents
$3K_2CO_3 + 3P_2O_5 \rightarrow 2K_3P_3O_9 + 3CO_2$
$2K_2HPO_4 \rightarrow K_4P_2O_7 + H_2O$
$2K_2CO_3 + P_2O_5 \rightarrow K_4P_2O_7 + 2CO_2$
$KH_2PO_4 + 2K_2HPO_4 \rightarrow K_5P_3O_{10} + 2H_2O$
$3K_4P_2O_7 + K_3P_3O_9 \rightarrow 3K_5P_3O_{10}$
$2K_2HPO_4 + 2KH_2PO_4 \rightarrow K_6P_4O_{13} + 3H_2O$
$2KH_2PO_4 + 2K_2HPO_4 \rightarrow K_6P_4O_{13}$ (or $2KPO_3 \cdot K_4P_2O_7$) $+ 3H_2O$
$4KH_2PO_4 + K_2CO_3 \rightarrow K_6P_4O_{13} + 4H_2O + CO_2$
$3K_4P_2O_7 + 2K_3P_3O_9 \rightarrow 3K_6P_4O_{13}$
$4KH_2PO_4 + 2K_2HPO_4 \rightarrow K_8P_6O_{19} + 5H_2O$
$11KH_2PO_4 + 2K_2HPO_4 \rightarrow K_{15}P_{13}O_{40} + 12H_2O$
Formation of tungstate solvents
$6K_2HPO_4 + 6WO_3 \rightarrow 6K_2WO_4 + 3P_2O_5 + 3H_2O$
Formation of other solvent
$NaF + KH_2PO_4 \rightarrow KNaPO_3F + H_2O$
$5KH_2PO_4 + 2BaCO_3 \rightarrow KBa_2(PO_3)_5 \cdot 2K_2O + 2CO_2 + 5H_2O$

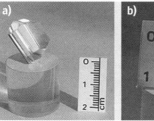 

Fig. 21.12a,b Single crystal of (**a**) KTiOPO$_4$ and (**b**) RbTiOPO$_4$ obtained by the TSSG method in phosphate and tungstate fluxes

pecially during the final stages of growth. When growth is performed on *c*- and *b*-oriented seeds, since the {001} and {010} faces are fast-growing faces which do not occur on KTP naturally, a capping process takes place at the initial stages of growth, until the {201} and {011} faces, or the {110} and {011} faces, have appeared, respectively [21.67]. Crystals grown on [100]-oriented seeds showed inclusions parallel to {011} faces, although this seeding may yield large single-sector KTP crystals [21.68]. [201]-Oriented seeding produced a tilt of the growing crystal to respect the rotation axis that improves the hydrodynamics of the solution, helping in melt mixing and allowing optimal utilization of the available melt volume by placing the long *c*-direction of the crystal along the body diagonal of the crucible [21.69]. KTP crystals have also been grown from phosphate fluxes by the TSSG method with pulling [21.67, 70] to avoid their multifaceted shape that leads to inefficient cutting of optical elements from KTP crystals [21.67]. As crystal growth proceeds on natural faces, they show relatively low dislocation densities, allowing one to obtain single crystals without inclusions or growth striations [21.70].

Carvajal et al. [21.71] developed a crystal growth system comprising a Pt stirrer immersed in the growth solution and two crystal seeds in contact with the solution surface, symmetrically distributed at about 1.5 cm from the rotation axis and 2 cm up the platinum turbine, as can be seen in Fig. 21.13. This system improved the mass transport conditions in the solution, thus minimizing problems associated with nonhomogeneous supersaturation in these viscous solutions. Stirring the solution decreases the frequency of spontaneous nucle-

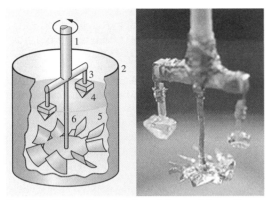

Fig. 21.13 Schematic and picture of the crystal growth system used in TSSG experiments to grow RbTiOPO$_4$ crystal and isostructurals that included a Pt turbine rotating together with the crystal seeds: (1) alumina rod, (2) platinum crucible, (3) crystal seeds, (4) growing crystals, (5) solution, and (6) platinum turbine

ation during the growth process and yields a higher quantity of high-quality inclusion-free single crystals of the KTP family.

Other KTP isostructurals, such as RbTiOPO$_4$ (RTP), which shows almost the same NLO properties of KTP, have been grown from these phosphate solutions. *Oseledchik* et al. [21.72] determined the crystallization region of RTP in the phosphate system and obtained crystals containing some inclusions, indicating the presence of unstable temperature control, high cooling rates, unsatisfactory temperature profile, and unsatisfactory dynamics of the melt flowing around the crystal.

An alternative to the improvement of the mass transport is the reduction of the viscosity of the solution by adding a modifier. *Ballman* et al. [21.73] proposed the low-viscosity, very fluid, non-glass-forming, water-soluble tungstate melts as a solvent, which greatly improves circulation and mixing during the crystallization process and results in a higher yield of inclusion-free material. *Iliev* et al. determined the phase diagram of KTP in the K$_2$O–P$_2$O$_5$–TiO$_2$–WO$_3$ system, as well as the viscosity of these solutions [21.74]. An increase in concentration of WO$_3$ results in a decrease of the solubility of KTP; however, the shallower slope of the solubility curves indicates a low degree of supersaturation for a given temperature drop and allows for a more controllable growth process. The addition of lithium to these solutions further enhances the fluidity of the melt and helps in the crystal growth process [21.69]. However, the use of tungstate fluxes leads to incorporation of tungsten ions in KTP crystals and reduces their optical transparency [21.75]. Striations and growth sectors were observed for all the existing faces on the crystals, which was attributed to changes in tungsten concentration in the crystal caused by variations in growth rates, temperature fluctuations in the furnace, or convective motion of the solution in the crucible [21.75, 76]. *Carvajal* et al. [21.71] determined the variation of the crystallization region of RTP when introducing WO$_3$ in the solution. The crystallization regions became narrower, and were displaced towards Rb$_2$O-rich regions when the concentration of WO$_3$ increased, as can be seen in Fig. 21.14. In these solutions the time of homogenization was shorter, the interval cooling of temperatures could be wider, and the cooling process to obtain high-quality crystals could be made faster than in phosphate fluxes.

Marnier proposed the use of alkaline halide (KF, KCl, and KBr) fluxes as an alternative to tungstate fluxes [21.77]. Alkaline halides also reduce the viscosity of the solution and enable a satisfactory growth rate at low temperature while enhancing the solubility of KTP and isostructurals. Sulfate-containing fluxes also reduce the viscosity of the solution due to the presence of SO$_4^{2-}$ ions in the flux that breaks the titanate chains of the flux. However, a certain amount of incorporation of sulfur into the crystals takes place [21.67]. More recently, *Suma* et al. [21.78, 79] carried out rapid growth of KTP single crystals by using KBa$_2$(PO$_3$)·2K$_2$O and KNaPO$_3$F as fluxes. The steepness of the solubility

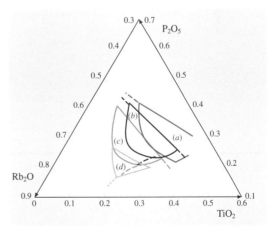

Fig. 21.14 Crystallization region of RbTiOPO$_4$ in the system Rb$_2$O–P$_2$O$_5$–TiO$_2$–WO$_3$ for *(a)* 0 mol % WO$_3$ *(b)* 10 mol % WO$_3$ *(c)* 20 mol % WO$_3$, and *(d)* 30 mol % WO$_3$ in solution

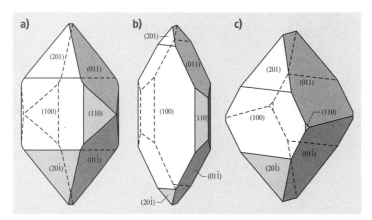

Fig. 21.15a–c Morphology of (**a**) RTP crystal, (**b**) RTP crystal doped with Nb grown on a thin crystal seed in the ***a***-direction, and (**c**) RTP crystal doped with Nb grown on a thin crystal seed in the ***a***-direction. The morphology of these crystals improved, where the {110} form tends to disappear, and the {011} and the {01$\bar{1}$} forms become larger, which provide a larger useful area of crystal in the *a–b* plane with benefits for applications in SHG

curve of KTP in these systems enables rapid cooling of the solution and fast growth of KTP.

In the last decade, much effort has been devoted to doping of KTP and isostructurals with various ions to change some of their physical properties, such as the crystal's optical transmission [21.80, 81], Curie temperature [21.82], ionic conductivity [21.83, 84], refractive indexes [21.85], and NLO properties [21.86, 87]. For this purpose, Nb is one of the most used ions with which KTP and isostructural crystals have been doped. However, this ion increases the difficulty of crystal growth: the growth solution is more viscous, the homogenization time increases, the saturation temperature increases, the solution is more prone to spurious nucleation, the efficiency of the crystal growth process decreases, crystals tend to crack, and they show more solution inclusions. Furthermore, these tendencies increase with increasing Nb concentration in the crystal. The crystal morphology is flat, long, and narrow, and especially small in size in the ***a***-axis direction, suggesting that crystals may grow by a two-dimensional nucleation mechanism. When crystals were grown on ***a***-oriented seeds, they showed poor transparency with many inclusions and twin-crystal flaws along the (100)-plane. In spite of this, this seeding orientation has a good effect on controlling spurious nucleation. When crystals were grown on ***c***-oriented crystal seeds, they had a larger transparent area and fewer inclusions with no twin-crystal flaws. If the crystal seed was rotated, reversing the rotation direction periodically during the growth process, crystals showed even larger transparent areas. Using crystal seeds with the same composition as the crystal to be grown further reduced the number of cracks observed in the crystals [21.88]. By forcing crystal growth in the ***a***-direction using crystal seeds with a larger dimension in this direction than in any other crystallographic direction, crystals with typical dimensions of $5 \times 5 \times 5$ mm^3 in the three crystallographic directions could be obtained [21.87]. Figure 21.15 summarizes these changes in morphology caused by Nb in RTP crystals.

KNd(PO$_3$)$_4$ (KNP) and KGd(PO$_3$)$_4$ (KGdP) single crystals have been grown by the TSSG technique [21.89, 90]. The crystallization regions of KNP, KGdP, and the intermediate KGd$_{0.5}$Nd$_{0.5}$(PO$_3$)$_4$ compound in the system Ln$_2$O$_3$–K$_2$O–P$_2$O$_5$ (where Ln = Nd or Gd) have been studied [21.89–91]. Figure 21.2a shows a comparison of these crystallization regions. Although the viscosity of the solution is very high throughout the crystallization region, when the Ln$_2$O$_3$/K$_2$O molar ratio (Ln = Nd or Gd) is above $\frac{3}{97}$, it is so high that it hinders crystallization.

TSSG experiments were carried out with axial temperature gradients in the solution of 0.75–0.92 K/mm for KNP, and 1.2 K/mm for KGdP. Crystals were grown with constant rotation at 75 rpm using KNP or KGdP parallelepipedal oriented seeds located at the center of the solution surface, by slow cooling at 0.1 K/h for an interval of 15–20 K. To improve the mass transport in the solution, a platinum turbine similar to that described for KTP and isostructural crystals was used, resulting in better quality crystals. Seeding along the ***a********-direction in the crystallographic reciprocal space and the ***c********-crystallographic direction in the reciprocal space was suitable for growing KNP and KGdP single crystals. Seeding along the ***b***-direction in the direct space was unsuitable because crystals tended to crack perpendicularly to this direction when they were too heavy. When solutions contained 65 mol % P$_2$O$_5$ or higher, small crystals with poor quality were obtained, due to the high viscosity of the solution. KGdP:Nd crystals showed some inclusions, especially when the Nd content increased, which could be avoided when the cooling rate was decreased to 0.05 K/h.

Borates

LBO has been known since 1926; in 1958 the Li_2O–B_2O_5 phase diagram was studied [21.93], and in 1978 small LBO single crystals were grown. LBO is a peritectic compound that cannot be grown by congruent melting and solidification. It can be synthesized by the peritectic reaction, a process that is too slow and difficult to carry to completion. Thus, the sole method of obtaining LBO single crystals is the flux growth method [21.94]. However, fluxed melt systems for growing LBO make controlling the temperature gradients difficult, as the main heat transport occurs by means of radiation energy. This leads to the appearance of growth defects, such as inclusions, inhomogeneous distribution of uncontrolled impurities, and nonstoichiometry of composition [21.95].

LBO crystals have been grown in different solution compositions of the Li_2O–B_2O_3 system. Solutions containing B_2O_3 are, in general, too viscous. In this system, the most suitable region of LBO concentrations to grow LBO crystals is 72–82 wt % LBO as the viscosity has a relatively low value and changes slightly with concentration. If the concentration is lower than 72 wt %, the mass transport is rather difficult because of the high viscosity. At concentrations higher than 82 wt %, the initial values of the supersaturation are high and drive the crystallization process too fast and hard for operating [21.96]. Additionally, the steep temperature versus composition slope places tight constraints on the seeding temperature: too high a temperature resulted in melting the seed, whereas too low a temperature resulted in polycrystal forming. The melt also showed a strong tendency to supercool. Furthermore, as LBO is sensitive to thermal shocking it has to be cooled after growth at very low cooling rates (≈ 3 K/h) [21.97]. Use of small thermal gradients in the furnace favors the growth of good-quality crystals [21.94]. Properly thermally insulating the upper part of the growth chamber, reducing temperature gradients above the melt during growth, and lowering the melt from the crystal when growth was terminated instead of lifting the crystal away from the melt helped to reduce cracking in the crystals. This leaves the crystal in the heated position of the chamber, allowing better control of cooling and resulting in smaller thermal gradients in the crystal. However, growth in solution with insufficiently high thermal gradients resulted in the appearance of large inclusions of fluxed melt and oxygen nonstoichiometry in the crystals. This is due to the presence of concentrations of inhomogeneities in the fluxed melt, which cannot be readily removed via the usual diffusion process during the period of their movement across the surface of the growing crystal because of the high viscosity of the fluxed melt. As the axial and radial thermal gradients increase near the crystallization front, the fluxed melts become more homogeneous and the quality of LBO crystals improves.

The surfaces of LBO crystals gradually decompose due to moisture present in the ambient atmosphere, forming a millimeter-thick, optically opaque, white, polycrystalline skin, replicating the original surface of the growing crystal. Upon cooling, severe cracking occurs on the crystal surfaces adjacent to this layer. The formation of this layer, and the subsequent cracking, can be avoided by growing crystals under dry nitrogen atmospheres [21.98].

Increasing rotation rates and the use of accelerated crucible rotation yields larger crystals due to the enhancement of forced convection. Seeding in the direction normal to widely developed faces is thought to enlarge the diameter of the crystal, as happens in LBO when using a seed normal to the (011) face compared with seeding in the [001] direction, while the thickness and the quality of crystal were almost the same, keeping the remaining growth conditions constant. In general, the forced convection in the solution caused by rotation is affected by the ratio of the crystal to crucible diameters. As the diameter of the growing crystal increases, the forced convection in the solution increases, which increases the crystal yield by increasing the mixing and the mass transfer in the solution. Figure 21.16 shows this effect in two LBO single crys-

Fig. 21.16a,b LBO crystals grown by seeding in different directions while maintaining the rest of the conditions of growth [21.92]: (**a**) seeding along [001] and (**b**) seeding along [011]. Each scale is 1 mm

tals grown under the same conditions except for the seeding direction.

As the viscosity of the LBO–B_2O_3 system cannot be decreased by increasing the growth temperature because of the low decomposition temperature of LBO, it is very important to find new solvents which can reduce the viscosity while having sufficient solubility for LBO. B_2O_3 solutions form three-dimensional networks, mainly consisting of randomly oriented boroxol rings interconnected by B–O–B bridges, the structure of which may be altered by the addition of alkali oxides, reducing the viscosity of the solutions as a function of the O-to-B ratio. However, the phase region for the growth of LBO crystals is located near the maximum of the temperature-dependent viscosity curve, so the addition of small amounts of alkali oxides has little effect on the overall viscosity of the solution [21.99]. The addition of halide ions such as Cl^- and F^- can weaken the network in alkali borate melts. Cl^- anions do not take part in the boron–oxygen network, but lie in the space between the network as free ions. Therefore, the large size of the Cl^- anions and the electrical repulsion between the Cl^- ions and the $[BO_4]^{5-}$ ions is effective in reducing the viscosity of the melt. F^- ions terminate the network by forming B–F nonbridging bonds, thereby reducing the viscosity of the solution. In the $Li_2B_4O_7$–B_2O_3–NaCl system, the viscosity of the solution and the seeding temperature were found to decrease as the amount of NaCl in the solution increased, while the metastable zone for crystallization of LBO became wider. This made it possible to increase the growth rate by up to 2–3 times for solutions containing 4 mol% NaCl with no obvious effect on crystal quality. The addition of MoO_3 also reduced the viscosity of the system. However, phases of enriched MoO_3, which can be incorporated into the growing crystal, appear [21.100].

Although BBO melts congruently at 1369 K, the structural reordering that occurs during the phase transition at 1198 K causes cracking of grown crystals during cooling. Of all the growth techniques used to grow β-BBO, only in the TSSG method can large-sized crystals of β-BBO be grown [21.101–105]. However, even when using this technique, the real success in growing β-BBO came with the discovery of a suitable solvent for this material [21.106]. Up to now, B_2O_3, BaF_2, $BaCl_2$, Li_2O, Na_2O and $Na_2B_4O_7$ [21.107], NaF [21.101], NaCl [21.108], Na_2O–NaF [21.109], Na_2O–BO_3 [21.110], $Na_2B_2O_4$ [21.101], Na_2SO_4, and CaF_2 [21.111] have been used as fluxes to grow β-BBO crystals. Using B_2O_3, Li_2O, $Na_2B_2O_4$, and $Na_2B_4O_7$ as fluxes limits the growth of β-BBO crystals due to the high viscosity of the solutions or narrow crystallizing range. Growth of β-BBO using NaCl, Na_2SO_4, $BaCl_2$, BaF_2, and CaF_2 as fluxes is difficult because of the high volatility of the solutions, and in some cases, because of the high-temperature hydrolysis that releases HCl vapor during growth. Na_2O is the flux which produced the best results, in terms of crystal size and defect density, when using the TSSG method with pulling to grow β-BBO crystals [21.104].

In the BBO–Na_2O pseudobinary diagram [21.101], there is a large temperature range from 1198 to 1028 K to grow β-BBO crystals. Normally, after cooling the melts for 70–85 K, an onset of growth interface instability was observed, which could be readily detected since, before its occurrence, melt flow patterns could be clearly seen through the growing crystals, whereas after it happened, the melt convection was obscured. Above a certain diameter of the growing crystal, the free convection flow in the solution around the crystal in the direction from the crystal edge to its center is opposed by a counterflow of forced convection due to the rotation of the crystal. The interaction between these two flows below the crystal leads to a temperature instability, and hence to disturbance of the growth stability [21.110]. Moderate and steeper radial and axial temperature gradients have been used to grow β-BBO crystals with this solvent. A suitable value for the vertical thermal gradient has been estimated to be ≈ 20 K/cm. Nevertheless, large single crystal with optical quality could be grown 3–4 times faster in high thermal gradient furnaces, as it was possible to cool farther before encountering interface instabilities.

Pulling at a rate commensurate with the growth rate has also been applied to the growth of these crystals, which allows the growth of thicker crystals at higher growth rates in smaller crucibles, which prevents seed failure that occurs when crystals touch the crucible wall. However, during pulling, if a steep thermal gradient is used, the quality of the crystals grown will be poor, but if a moderate thermal gradient is used, it is difficult to control the diameter of the crystal. However, if the pulling method is not used, the crystals assumed a shallow lens shape typical of BBO crystals.

Normally, during the growth period, the seed was rotated. However, large rotation rates make the forced convection gradually overcome natural convection, eventually reversing the direction of radial fluid flow along the growth interface, resulting in lower supersaturation at the center of the growth interface and a change in the interface shape from convex to concave. As the highest-quality crystals are usually grown with a flat or slightly convex interface, rotation rates have to be limited to avoid concave interfaces.

A major limitation of using *c*-axis oriented crystal seeds was the tendency toward interface breakdown after 15 mm of growth, and that crystals tended to show a higher defect density in the core region, corresponding to the center of the star-shaped convective flow pattern, where there is stagnation and little mixing due to crystal rotation. Boule cross sections were more elliptical, and faceting on the top surface of the boule was less pronounced, when using *a*- and *b*-oriented crystal seeds, growing in a nearly cylindrical shape when pulled. However, crystals tended to fracture along the (0001) cleavage planes during cooling due to anisotropy in the thermal expansion.

Inclusions are generally linked to unstable growth conditions, and a standard method of improving the growth stability is by mechanical stirring of the solution. However, none of the conventional stirring techniques, including convective stirring, uniform crucible rotation, accelerated crucible rotation, and uniform boule rotations have produced any substantial change in the quality of β-BBO crystals. The change of the symmetry and the rotation of the thermal field that create inhomogeneous stationary and cyclically varying external thermal fields to control the convective heat and mass transfer, and the application of a vibrational temperature mode in the growth zone, has been proved to be very successful in the growth of β-BBO crystals. An appropriate commutation of the heating elements around the crucible creates a rotating thermal field that considerably intensifies the stirring both in the flux bulk and in the vicinity of the crystallization front, as can be seen in Fig. 21.17. Thus, the thickness of the diffuse layer at the crystallization front decreases and the concentration supercooling is delayed, making it possible to grow large β-BBO single crystals with circular cross section and high structural quality [21.112].

Another method used to improve the quality of the β-BBO crystals grown in BBO–Na$_2$O solution is continuous feeding during growth by adding pure BBO. This provides the possibility of isothermal growth at a suitable temperature. Gradual depletion of the solution during growth was constantly compensated by manual addition of pure BBO through an alumina tube that transferred the feed material to a small platinum crucible partially immersed in the melt whose bottom was punched, allowing for slow dissolution of the feed [21.103]. The main drawback of using this solvent is that β-BBO crystals contained up to 200 ppm of Na, which affects the optical transmission of the crystals, especially in the UV region [21.113].

Adding Nd$_2$O$_3$ to the BBO–Na$_2$O system with the aim of obtaining a new self-frequency-doubling material resulted in a strong stabilization effect of the β-BBO phase (Fig. 21.18). However, as the Nd$_2$O$_3$ concentration and the saturation temperature in the system become higher, the creeping of the solution along the crucible wall was stronger, which meant that relatively good β-BBO single crystals, doped with up to 2 mol % of Nd^{3+}, were only obtained from solutions with low Nd$_2$O$_3$ concentration (12 mol %) [21.114].

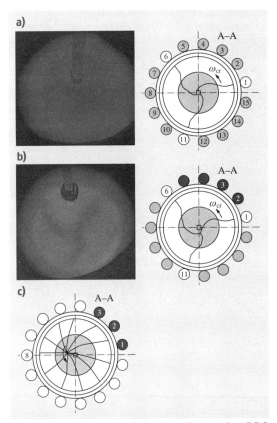

Fig. 21.17a–c Schematic of the setup for growing BBO crystal with the change of the symmetry and rotation of the thermal field and convective patterns observed on the free surface and through a growing BBO crystal [21.112]: (1–15) heating elements (*dark color* indicates switched-on elements, and *white color* indicates switches-off elements), (**a**) and (**b**) creation of thermal fields with a threefold and a quasi-threefold symmetry; (**c**) rotating thermal fields

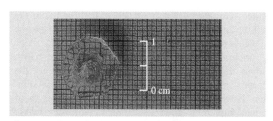

Fig. 21.18 Crystal of β-BaB$_2$O$_4$ growth in a solution of BBO–Na$_2$O

Roth and *Perlov* [21.104] determined the BBO–NaF pseudobinary diagram. Compared with the BBO–Na$_2$O system, in the BBO–NaF system the slope of the liquid curve is smoother and the maximal range of the BBO composition change wider, which allows higher yields of β-BBO crystals. The BBO–NaF pseudobinary phase diagram has a single eutectic point at a composition of 61 mol % NaF and a temperature of 1027 K. The solution viscosities are lower by about 15% on average over the entire temperature range of interest when compared with BBO–Na$_2$O solutions. Although the volatility of BBO–NaF solutions is almost an order of magnitude higher than that of BBO–Na$_2$O, the average escape of 6 mg of solution per hour (mainly NaF) does not affect the crystal growth process considerably and can be corrected by adjusting the cooling rate during the growth process. A breakdown of the solid–liquid interface into a cellular structure also occurs at a certain stage of growth. However, the crystal yield before breakdown is larger than in the case of Na$_2$O-containing solutions. Crystals grown from NaF solutions had smaller diameters and larger thickness than those grown from Na$_2$O solution, which could be attributed to the low viscosity of the solution, which would make natural convection dominant over forced convection, and which resulted in vertical growth dominating radial growth [21.115]. Adding Na$_2$O to the BBO–NaF system reduces the volatility of the solutions to less than 1% during the entire growth process [21.107].

21.4 Liquid-Phase Epitaxy: Growth of Epitaxial Films of Laser and NLO Materials

The thin-disk laser approach was introduced by *Giesen* et al. in 1994 [21.116] to ameliorate the quality of the beam and the thermal loading using efficient longitudinal cooling; moreover the thermal lensing effect is also minimized, especially in the high-power regime [21.117]. The thickness of the active layer minimizes reabsorption phenomena in quasi-three-level systems such as ytterbium or 2 μm thulium emissions. Some examples of thin-disk lasers have been realized [21.118, 119], including those which employ monoclinic KREW active layers [21.120, 121]. KREW materials are good hosts for thin-disk lasers due to the high absorption and emission cross sections of lanthanide-doped KREW. This has propelled, in recent years, the growth of single-crystalline thin films of these materials by LPE techniques.

Growth of thin films of NLO materials, specifically phosphate materials of the structural field of KTiOPO$_4$ (KTP), attracts attention as a means to fabricate optical waveguides. The confinement of light in a micrometer-sized waveguide and its propagation without appreciable diffraction greatly increases the optical fields, and the efficiencies of NLO processes [21.122]. Waveguides of KTP and related materials could be used to control or convert high-intensity optical beams with input wavelengths extending from the visible to the infrared (IR) with thermal and mechanical stability [21.123].

21.4.1 Epitaxial Films of Laser Materials: Lanthanide-Doped KLuW on KLuW Substrates

For the thin-disk approach, as well as for waveguide lasers, homoepitaxial growth of lanthanide-doped KREW (RE = Y, Gd, Lu) by LPE has been recently reported for Yb:KYW films grown on KYW substrates [21.124, 125] and for Yb:KLuW films grown on KLuW substrates [21.126].

Successful growth of these epitaxial layers was demonstrated in a special vertical furnace with a wide zone of uniform temperature to achieve a zero temperature gradient in the solution. The most common solvent for this epitaxial growth is K$_2$W$_2$O$_7$ in a solution with a 5 mol % solute and 85 mol % solvent composition. Taking into account the solubility curves of KREW in K$_2$W$_2$O$_7$ (Fig. 21.2b), the average degree of supersaturation at 5 mol % solute is 0.16×10^{-2} g/K (g of solution). This low level of supersaturation allows the growth to occur near equilibrium. Also Yb:KYW

Table 21.5 Mismatch for Ln^{3+}:KLuW/KLuW (Ln = Yb and Tm) on different faces

Epitaxial layer	$f(010)$	$f(110)$	$f(310)$	$f(\bar{1}11)$
$KLu_{0.88}Yb_{0.12}W$/KLuW	−0.074	−0.081	−0.085	−0.089
$KLu_{0.78}Yb_{0.22}W$/KLuW	−0.107	−0130	−0.142	−0.135
$KLu_{0.48}Yb_{0.52}W$/KLuW	−0.143	−0.220	−0.259	−0.211
KYbW/KLuW	−0.215	−0.354	−0.425	−0.324
$KLu_{0.97}Tm_{0.03}W$/KLuW*	−0.069	−0.064	−0.062	−0.073
$KLu_{0.95}Tm_{0.05}W$/KLuW*	−0.086	−0.087	−0.087	−0.090
$KLu_{0.925}Tm_{0.075}W$/KLuW*	−0.085	−0.099	−0.106	−0.103
$KLu_{0.90}Tm_{0.10}W$/KLuW*	−0.116	−0.136	−0.147	−0.140
$KLu_{0.80}Tm_{0.20}W$/KLuW*	−0.198	−0.261	−0.293	−0.248

* The epitaxial layer stoichiometry is expressed by the initial solution composition (not corrected by distribution coefficient)

thin films on KYW substrates have been grown using a NaCl–KCl–CsCl solvent. Although this leads to precipitation of yttrium and tungsten oxides on the bottom of the crucible, this problem can be solved by following the methodology of *Kawaguchi* et al. [21.126] using LPE from a solid–liquid coexisting solution.

The step of homogenization of the flux is to increase the temperature 50 K above the expected saturation temperature for 24 h.

Before being placed in the furnace, the substrates must be carefully cleaned in $HNO_3 : H_2O$ (1 : 1 in volume), distilled water, acetone, and ethanol in 5 min steps each. They are then slowly introduced into the furnace to prevent thermal stress and kept at a constant temperature for about 1 h above the surface of the solution. The temperature of the solution is then reduced to 1 K above the saturation temperature and the substrate is introduced into the solution and kept at this temperature for 5 min, so that the outer layer of the substrate dissolves, without introducing defects into the subsequent epitaxial growth.

The growth of the epilayers is achieved by creating a ≈ 5 mol % supersaturation with a decrease of the saturation temperature by 2–6 K. The substrate is rotated at 15–60 rpm. In some experiments, a cooling rate of 0.67 K/h is used.

Wiping off the solution is not difficult due to its low viscosity. When the substrate is removed from the solution, it is still rotating while the whole system cools to room temperature.

Aznar et al. mentioned that on the (010) faces the epitaxial growth mainly exhibits a flat surface, which indicates a layer-by-layer growth mechanism. However *Romanuyk* et al. reported a three-dimensional (3-D) nucleation, also known as the Volmer–Weber growth mode [21.127].

The substrates were oriented perpendicular to the *b*-crystallographic direction, as preliminary studies of thin-film growth had demonstrated that epitaxial growth on the (010) face resulted in high-quality films at the fastest growth rate. Generally, higher density of growth steps is observed on epilayers grown on (310) and ($\bar{1}$11). A possible reason is the higher growth rate of this faces when compared with their neighboring faces. The layer–substrate mismatches on the ($\bar{1}$11) and (010) faces are rather similar. The lattice mismatch for Yb-doped KLuW thin films grown on KLuW substrates and Yb-doped KYW thin films grown on KYW substrates are listed in Table 21.5. No thermal mismatch data are available for these epitaxial layers.

Growth hillocks have been observed on Yb:KLuW films grown on KLuW substrates, which is a typically observed micromorphology that develops during the growth process.

Figure 21.19 shows the comparison of the thickness on different faces.

The chemical composition of the layer is usually measured by x-ray spectrometry or electron probe microanalysis to quantify the concentration and to calculate the distribution coefficient of the doping element in the epilayer, and also to determine if diffusion of the doping element into the substrate occurs. The distribution coefficient of the different lanthanide doping ions in the epilayers is near unity. No diffusion into the substrate has been observed. Higher doping levels of the epitaxial film will lead to an increase of the lattice mismatch and, consequently, difficulties in the growth of the epitaxial layers. When doping KYW with Yb the doping level obtained in the epitaxial films with high crystalline quality was 10 at. % in the growth solution. In the case of KLuW, it is possible to increase the dop-

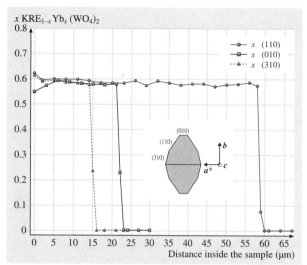

Fig. 21.19 Ytterbium concentration profile for the KLu$_{0.48}$Yb$_{0.052}$W layer on KLuW substrates on three different faces: (310), (110), and (010) (after [21.128])

ing concentration to 50 at. % while maintaining high quality of the epitaxial layer.

Recently, the growth of thulium-doped epitaxial layers on KYW and KLuW substrates has been reported. For the KLuW host the doping level can reach 7.5 at. % substitution of lutetium by thulium while maintaining high film quality.

21.4.2 Epitaxies Within the Structural Field of KTP

Optical waveguides in KTP have been produced by ion exchange of K$^+$ by Rb$^+$, Cs$^+$ or Tl$^+$ on the surface of a KTP substrate immersed in molten salts of Rb, Cs, and Tl [21.129]. A metal mask (e.g., Al, Au, Ti) has been used to fabricate channel waveguides by this procedure. These waveguides appear to be optically uniform over the width of the channel and show no evidence of lateral ion diffusion [21.130], which enables fabrication of high-density waveguide arrays and modulated index waveguides, while at the same time optimizing the electric-field overlap for modulators, switches, and nonlinear waveguide devices [21.123]. When Ba(NO$_3$)$_2$ is added to the molten RbNO$_3$, the ion-exchange process changes not only the optical indices of the crystal but also its polarity, enabling the fabrication of not only Rb-exchanged quasi-phase-matched waveguides of KTP [21.131].

However, ion-exchange processes have their drawbacks. Due to the inherently diffusive and strongly anisotropic nature of the process, it is the difficult to control the waveguide depth, producing guides with a broad, poorly defined index profile along the c-axis [21.24, 131]. This index profile, although satisfactory for many applications, is less effective for confinement of optical fields, especially in waveguiding second-harmonic generation [21.132] where variations of ionic conductivity due to crystal growth methods and with impurities make the device fabrication process difficult and results in poor yields.

Optical waveguides in KTP have been produced by proton or ammonium exchange [21.122], yielding a more step-like index profile by sol–gel chemistry [21.133] producing continuous KTP polycrystalline films with a grain size of 0.3 μm, and by pulsed excimer laser ablation [21.134], allowing growth of KTP films on foreign substrates such as sapphire, silicon [21.135], and quartz.

Unlike other techniques used to fabricate waveguides, the LPE technique enables growth of a homogeneous single-crystal film and allows control of the thickness by adjusting the supersaturation of the solution and the growth time. Films with well-defined step-like refractive-index profile can be grown directly by LPE [21.131]. Good-quality single-crystal epitaxial layers of KTP suitable for producing optical waveguides can be obtained by LPE on substrates of the same family. This can be done by controlling the mismatch between the cell parameters of the substrate and those of the film using the solid solutions offered by this family of crystals. Thin films of KTiOP$_x$As$_{1-x}$O$_4$ were grown by *Cheng* et al. on KTP substrates using both tungstate and the pure phosphate–arsenate self-fluxes [21.132]. The KTA–KTP system was chosen, since the substitution of arsenic for phosphorus provides the desired refractive-index difference without compromising the nonlinearity of the material. KTiOP$_x$As$_{1-x}$O$_4$ films with a thickness of 5–50 μm were grown on polished KTP substrates. These substrates were polished with diamond-based polishing powder and finished with a 30 s chemical–mechanical polish in colloidal silica prior to epitaxial growth. A small (≈ 0.75 mm) hole, drilled at one corner of the substrate, allows it to be tied vertically onto a crystal rotation–pulling head with a thin platinum wire to assist flux drainage after dipping. Slight etching of the substrate in warm dilute hydrochloric acid prior to the dipping improved the quality of the epitaxial layer. {100}, {011}, {110}, and {201} oriented plates cut from a single crystal grown by the flux method were chosen as substrates. Although {001} films would be preferred in device applications, since the largest nonlinear optical coefficients of KTP lie in this plane, growth of high-quality {001} films has not been successful due to the fast growth and c-capping habit of KTP. The same problem was observed when growing thin epitaxial layers of KTP on {010} substrates. It should be noted that the use of substrates cut from hydrothermally grown crystals leads to optical degradation with the formation of fine white filaments in the substrate due to the precipitation of fine water-based inclusions [21.131].

The dipping setup [21.131] consisted of a 250 ml crucible placed at the bottom of a short-zone top-loading crucible furnace. The melt (≈ 200 ml) was homogenized overnight at ≈ 50 K above its liquidus temperature, which was determined accurately by repeated seeding. The substrate was introduced into the growth furnace slowly ($\approx 5-25$ mm/min) to avoid cracks due to thermal stress, and the flux was cooled to $\approx 1.5-3$ K below the saturation point and allowed to equilibrate for 30 min prior to dipping the substrate into the melt. The substrate was spun unidirectionally at 10 rpm. The dipping time varied depending on the desired film thickness, the degree of supersaturation used, the choice of flux, and the growth temperature. Experimentally it was found that back-etching of the substrate in the same LPE solution prior to growth resulted in significantly better quality films. This was accomplished by taking advantage of the thermal inertia of the system and submerging the substrate before the melt reached the growth temperature. Upon completion of the dipping, the substrate–epitaxy was removed from the flux and washed with warm diluted hydrochloric acid.

Optical-quality films could be readily obtained using a 3–4 K supersaturation and a growth temperature of ≈ 1173 K. The observed growth rate was found to be weakly dependent on the orientation of the substrates. An abrupt increase in the arsenic concentration at the film–substrate interface was revealed, and resulted in an abrupt, step-like refractive index profile. It was concluded that the maximum lattice mismatch between the film and the substrate that still yields high-quality films was $\approx 1\%$, which corresponds to a $\approx 35\%$ increase in arsenic content in the $KTiOAs_xP_{1-x}O_4$ film. Film cracking was observed for films grown on substrates with larger lattice mismatch.

Significantly different growth properties were observed for tungstate and $K_6P_4O_{13}$ fluxes. First, a longer soak time was needed when using the $K_6P_4O_{13}$ flux. Second, under the same growth conditions (temperature and supercooling), the growth rate normal to the natural face used as the substrate was substantially lower in the $K_6P_4O_{13}$ flux, necessitating supercooling roughly twice that used in tungstate flux to achieve a comparable growth rate. Third, films grown from $K_6P_4O_{13}$ flux tend to show film–substrate interfaces of poorer quality due to the slow dissolution kinetics of this flux, which makes the implementation of pregrowth etching difficult.

Appropriate replacement of the titanyl group via solid solution formation (e.g., Sn or Ge) or impurity doping could also generate epitaxial films with well-defined refractive index boundaries [21.24]. However, the growth of $KTi_{1-x}Sn_xOPO_4$ films proved to be difficult due to the anomalously slow dissolution properties of $KSnOPO_4$. As an alternative, growth of $KTi_{1-x}Ge_xOPO_4$ thin films on KTP substrates using a 20% Ge solution has been tested [21.131]. Films (10 μm thick) of $KTi_{0.96}Ge_{0.04}OPO_4$ on {011} KTP substrates were grown. Discouragingly, even with a low 4.3% Ge incorporation, numerous cracks perpendicular to the c-crystallographic direction were observed in thicker (30 μm) films because the $KTi_{0.96}Ge_{0.04}OPO_4$ films grew under tensile stress.

Cheng et al. [21.131] suggested that this situation can be improved by reversing the film–substrate configuration, such as growing KTP films on $KTi_{1-x}Ge_xOPO_4$ substrates. In this way, *Solé* et al. [21.85, 136] produced KTP thin films grown on $KTi_{1-x}Ge_xOPO_4$ substrates also using both tungstate and pure phosphate self-fluxes. These LPE experiments were performed in a special vertical furnace built to provide a wide enough region in which there was practically no axial gradient so that the epitaxial film thickness did not depend on the solution depth. Platinum cylindrical crucibles, 30 mm in diameter and 40 mm high, filled with about 50 g of solution with composition of $K_2O : P_2O_5 : TiO_2 = 49.8 : 33.2 : 17$ (mol %) in self-flux and $K_2O : P_2O_5 : TiO_2 : WO_3 = 42 : 14 : 14 : 30$ in tungstate flux were used. After homogenization of the solution, special attention was paid to determining the saturation temperature of the solution using a c-oriented seed rotating at 16 rpm. A new approach was used to study the quality of the films simultaneously on different substrates with different orientation. The epitaxial growth was made directly on the natural faces of as-grown single crystals of $KTi_{1-x}Ge_xOPO_4$. Thus, the growth conditions were forced to be the same for all the different orientations of the crystal faces used as substrates. The substrates were fixed using their own growth crystal seeds at the end of an alumina rod and cleaned for 5 min in $HNO_3 : H_2O = 1 : 1$ by volume,

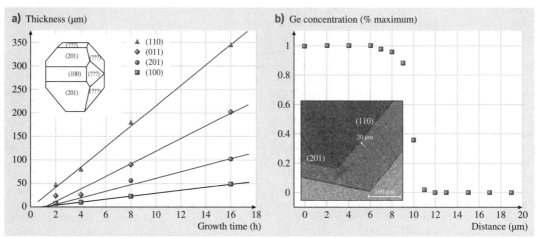

Fig. 21.20 (a) Epitaxial film thickness on different faces of a $KTi_{0.988}Ge_{0.012}OPO_4$ crystal as a function of the time of growth. The thin films were grown at 2 K below the saturation temperature of the solution. A morphology sketch with the difference faces of the $KTi_{0.988}Ge_{0.012}OPO_4$ crystal is included. (b) Normalized germanium concentration around the substrate–film interface determined by electron probe microanalysis. A scanning electron microscopy image of a cross section of the substrate–film interface taken with a secondary-electron detector is included. The difference in contrast indicates where the substrate–film interface is located, suggesting also a sharp interface (after [21.85])

for 5 min in distilled water, and for 5 min in ethanol. The whole cleaning process was carried out with crystal rotation. The crystal was then introduced into the furnace slowly to avoid cracks caused by thermal stress and heated for 30 min above the solution surface. The substrate was then dipped into the solution at a temperature 3–5 K above the saturation temperature for 1 h to dissolve the substrate surface before the beginning of growth. In tungsten solutions, the dissolution of the substrate was performed at 1 K above the saturation temperature for 10 min since this solution is significantly less viscous than the solution without WO_3, and all these processes are much quicker. The temperature of the solution was maintained at 2 K below the saturation temperature for several hours to induce epitaxial growth when self-flux was used, or for several minutes when tungstate flux was used. The growth was performed at 1 K below the saturation temperature only when the growth time was long (16 h) and the degree of substitution was high. In all cases, the crystal rotation was maintained at 16 rpm. After the epitaxial growth, the crystal was removed from the flux and extracted from the furnace slowly to avoid thermal stresses in the crystal.

In general, the surface morphology and quality of the epitaxial films obtained depends on the crystal face on which they grew. Other parameters such as Ge content in the substrate and time of growth have significantly less influence. The films on (201) and (100) faces were found to be of high quality and smooth. Films grown on (100) faces tended to show small macrosteps that could be a reproduction of the steps already existing on the substrates. The films on (011) and (110) faces tended to show small hillocks when the time of growth and the Ge in the substrate increased and a tendency for films grown on (011) face to be of best quality compared with those grown on (110) face. This difference in quality could be related to the faster velocity of film growth on these faces. The worst quality films grew on (101) face. This face, although it exists in $KTi_{1-x}Ge_xOPO_4$ crystals, is not a usual face in pure KTP. The defects in films grown on the (101) face are due to the faster growth velocity in pure KTP relative to the growth velocity on the other faces. In general, films grown on $KTi_{0.918}Ge_{0.082}OPO_4$ substrates showed some cracks. The quality of epitaxial growth in tungstate flux on (100), (201), and (011) faces was found to be good up to the highest concentration of Ge studied and the longest time studied. The epitaxy on (011) face, however, seemed to show a greater tendency to have macrosteps and growth hillocks than the other two faces. Similarly, when the Ge substitution for Ti in the substrate, and/or the growth time increased, the epitaxial film on (110) face showed a slight tendency to exhibit more defects.

A linear correlation between the epitaxial thickness and the time of growth was observed for all films grown on the faces of KTi$_{0.988}$Ge$_{0.012}$OPO$_4$. Figure 21.20a shows the epitaxial film thickness on the different faces as a function of the growth time for films grown at 2 K below the saturation temperature of the solution. The epitaxy on the (100) face showed the lowest rate of growth, followed by the film grown on the (201) face. The quality of these epitaxial films remained good even for long growth times and high concentrations of Ge in the substrate. The epitaxial films on (011) and (110) faces showed a faster rate of growth. Thus, when the time of growth increased, the density of defects also increased. Because of the poor quality of the epitaxial film on (101) face it was difficult to measure the epitaxial thickness accurately. For the case of growth in tungstate solutions, the epitaxy on (100) face showed the lowest rate of growth, followed by the films grown on (201), (011), and (110) faces, which showed similar rates of film growth. A sharp change in the germanium concentration at the substrate–film interface was observed. Figure 21.20b shows the normalized germanium concentration around the substrate–film interface.

The mismatch between the substrate and the film, defined as $f_{s(hkl)} = (S_{s(hkl)} - S_{0(hkl)})/S_{0(hkl)}$, where $S_{s(hkl)}$ and $S_{0(hkl)}$ are the areas obtained from the periodicity vectors of the substrate and the film, respectively, is listed for each face in Table 21.6. For {100} and {201} faces, these mismatches were positive, meaning the substrate was larger than the film. For {110} and {011} faces the mismatches were always negative, which is believed to produce films of lower quality. The different quality of the epitaxial films grown on the different crystal faces could also be explained by considering the position of titanium in the KTP structure with respect to the different planes that constitute the external morphology of the crystals used as the substrate on which the epitaxial films are grown. The Ti planes parallel to (100), (201), and (011) faces are regularly located in the structure and equidistant, while planes parallel to the (110) and (101) faces are not equidistant. Thus, the structural distortion produced by the Ge substitution in these planes was not distributed as homogeneously as in the case of the (100), (201), and (011) faces. This could lead to a larger tendency to generate defects in the film.

Epitaxial Rb$_x$K$_{1-x}$TiOPO$_4$ films were grown from a 20% Rb solution on a KTP substrate [21.133]. Although the grown layer was measured to be 50 μm, the diffusion of Rb into the substrate at the growth temperature (≈ 1125 K) significantly broadened the waveguiding layer up to ≈ 125 μm. It is therefore more appropriate to view the liquid-phase epitaxy of Rb$_x$K$_{1-x}$TiOPO$_4$ on KTP as an ion-exchange waveguide fabrication process using molten tungstate instead of nitrate.

Liquid-phase epitaxy was also used as a tool to test the morphological stability of the (00$\bar{1}$) face, which is a natural face in KTiOAsO$_4$ crystals and on KTP crystals. For this purpose, a c-cut KTP plate was submerged into a slightly supercooled (≈ 4 K) K$_6$P$_4$O$_{13}$ solution of KTP at 920 °C for ≈ 35 min. Smooth planar growth steps at the center of the KTP plate suggest that the (00$\bar{1}$) face is indeed singular in KTA. Stable {01$\bar{1}$} and {20$\bar{1}$} lamellae extend rapidly along the ⟨00$\bar{1}$⟩ and ⟨20$\bar{1}$⟩ directions to form overhangs which eventually enclose the (001) and (00$\bar{1}$) faces and thus eliminate them from the final growth form [21.137].

A completely new combination of both top-down and bottom-up approaches has been reported recently to grow two-dimensional (2-D) photonic crystals of KTP involving liquid-phase epitaxial techniques [21.138]. KTP rods grew inside the air holes of an ordered silicon matrix closely bound to a c-oriented KTP substrate and following the orientation of the substrate. The 2-D KTP patterning implemented can be summarized in a four-step procedure, shown in Fig. 21.21a, that involves preparation of a high-quality ordered macroporous silicon template, epitaxial growth of the KTP rods into the silicon template, polishing of the top or bottom surface of the KTP columns, and finally partial selective etching of the silicon matrix. A platinum wire was used to bind the silicon template to a KTP substrate. The template–substrate was dipped for 5 min into a high-temperature solution with a mol % composition K$_2$O : P$_2$O$_5$: TiO$_2$: WO$_3$ = 42 : 14 : 14 : 30 without any additional thermal gradient. We initiated the epitaxial growth of the KTP 2-D photonic structure

Table 21.6 Mismatches between KTi$_{1-x}$Ge$_x$OPO$_4$ substrates and KTP films

Epitaxial layer	$f(100)$	$f(201)$	$f(101)$	$f(110)$	$f(011)$
KTi$_{0.988}$Ge$_{0.012}$OPO$_4$/KTP	−0.003	0.006	0.018	−0.017	−0.007
KTi$_{0.955}$Ge$_{0.045}$OPO$_4$/KTP	0.009	0.032	0.060	−0.035	−0.009
KTi$_{0.918}$Ge$_{0.082}$OPO$_4$/KTP	−0.071	−0.029	0.023	−0.156	−0.107

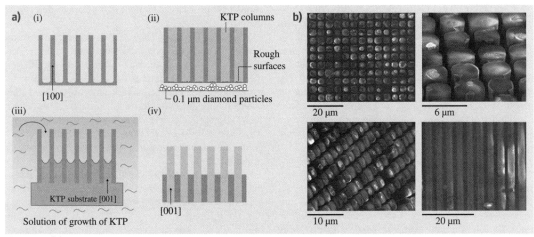

Fig. 21.21 (a) Schematic view of the four stages of growth of KTP 2-D photonic crystals: (i) preparation of the oriented 2-D macroporous silicon membrane, (ii) the silicon template is attached to an oriented KTP substrate and then immersed in the growth solution of KTP. The direction of growth of the KTP rods is the same as oriented of the KTP substrate, [001]. (iii) After growth, the top of the rods is polished with diamond particles in order to obtain an optical-quality surface. (iv) Silicon is partially removed by selective chemical etching. (b) Scanning electron microscopy images of (i) the top of a 2-D KTO phonic crystal after polishing and partially etching of the macroporous silicon template, (ii) side view of the KTP rods after removing the silicon template, (iii) a detailed view of a 2-D KTP phonic crystal with a period of 4.5 μm, and (iv) side view of a plane of rods of a 2-D KTP photonic crystal lattice

2 K below the saturation temperature, which provided a supersaturation in the solution of about 2%. The template–substrate–epitaxy composite was then removed from the solution, but kept inside the furnace above the surface of the solution while the furnace was cooled to room temperature at a rate of 15 K/h to avoid thermal stress that could result in cracks either in the 2-D photonic structures or in the substrate. The final 2-D photonic structures are formed from independent rods of KTP with square cross section, perfectly aligned with the orientation of the KTP substrate. Figure 21.21b shows several scanning electron microscopy (SEM) images of these photonic structures.

References

21.1 M.M. Kuklja: Defects in yttrium aluminium perovskite and garnet crystals: atomistic study, J. Phys. Condens. Matter **12**, 2953–2967 (2000)

21.2 V. Lupei: RE^{3+} emission in garnets: multisites, energy transfer and quantum efficiency, Opt. Mater. **19**, 95–107 (2002)

21.3 J. Petit, B. Viana, P. Goldner, D. Vivien, P. Louiseau, B. Ferrand: Laser oscillation with low quantum defect in Yb:$GdVO_4$, a crystal with high thermal conductivity, Opt. Lett. **29**, 833–835 (2004)

21.4 S.H. Smith, G. Garton, B.K. Tanner: Top-seeded flux growth of rare-earth vanadates, J. Cryst. Growth **23**, 335–340 (1974)

21.5 S. Erdei: Growth of oxygen deficiency-free YVO_4 single crystal by top-seeded solution growth technique, J. Cryst. Growth **134**, 1–13 (1993)

21.6 D.I. Savytskii, L.O. Vasylechko, A.O. Matkovskii, I.M. Solskii, A. Suchocki, D.Y. Sugak, F. Wallrafen: Growth and properties of $YAlO_3$:Nd single crystals, J. Cryst. Growth **209**, 874–882 (2000)

21.7 R. Solé, V. Nikolov, X. Ruíz, J. Gavaldà, X. Solans, M. Aguiló, F. Díaz: Growth of β-$KGd_{1-x}Nd_x(WO_4)_2$ single crystals in $K_2W_2O_7$ solvents, J. Cryst. Growth **169**, 600–603 (1996)

21.8 M.C. Pujol, M. Aguiló, F. Díaz, C. Zaldo: Growth and characterisation of monoclinic $KGd_{1-x}RE_x(WO_4)_2$ single crystals, Opt. Mater. **13**, 33–40 (1999)

21.9 G. Métrat, N. Muhlstein, A. Brenier, G. Boulon: Growth by the induced nucleated floating crystal (INFC) method and spectroscopic properties of $KY_{1-x}Nd_x(WO_4)_2$ laser materials, Opt. Mater. **8**, 75–82 (1997)

21.10 A. Brenier, F. Bourgeois, G. Métrat, N. Muhlstein, M. Boudelle, G. Boulon: Spectroscopic characterization of Nd^{3+}-doped $KY(WO_4)_2$ single crystal, J. Lumin. **81**, 135–141 (1999)

21.11 D. Gabbe, A.L. Harmer: Scheelite structure fluorides: The growth of pure and rare earth doped $LiYF_4$, J. Cryst. Growth **3**, 544 (1968)

21.12 K. Shimamura, H. Sato, A. Bensalah, V. Sudesh, H. Machida, N. Sarukura, T. Fukuda: Crystal growth of fluorides for optical applications, Cryst. Res. Technol. **36**, 801–813 (2001)

21.13 A. Agnesi, G. Carraro, A. Guandalini, G. Reali, E. Sani, A. Toncelli, M. Tonelli: 1-mJ Q-switched diode-pumped $Nd:BaY_2F_8$ laser, Opt. Exp. **12**, 3766–3769 (2004)

21.14 S. Bigotta, D. Parisi, L. Bonelli, A. Toncelli, M. Tonelli, A. Di Lieto: Spectroscopic and laser cooling results on Yb^{3+}-doped BaY_2F_8 single crystal, J. Appl. Phys. **100**, 013109 (2006)

21.15 R.F. Belt, R. Uhrin: Top seeded solution growth of $Cr^{3+}:LiCaAlF_6$ in HF atmosphere, J. Cryst. Growth **109**, 340–344 (1991)

21.16 D. Klimm, P. Reiche: Ternary colquiriite type fluorides as laser hosts, Cryst. Res. Technol. **34**, 145–152 (1999)

21.17 A.A. Kaminskii: Modern developments in the physics of crystalline laser materials, Phys. Status Solidi (a) **200**, 215–296 (2003)

21.18 A.A. Kaminskii: *Laser Crystals. Their Physics and Properties* (Springer, Berlin 1981)

21.19 C.E. Mungan, S.R. Bowman, T.R. Gosnell: Solid-state laser cooling of ytterbium-doped tungstate crystals, Proc. Lasers 2000 (2001) pp. 819–826

21.20 A.N. Medina, A.C. Bento, M.L. Baesso, F.G. Gandra, T. Catunda, A. Cassanho: Temperature dependence of the Cr^{3+} site axial distortion in $LiSrAlF_6$ and $LiSrGaF_6$ single crystals, J. Phys. Condens. Matter **13**, 8435–8443 (2001)

21.21 S.A. Payne, L.L. Chase, H.W. Newkirk, L.K. Smith, W.F. Krupke: $LiCaAlF_6:Cr^{3+}$: A promising new solid-state laser material, IEEE J. Quantum Electron. **QE-24**, 2243–2252 (1988)

21.22 J.A. Armstrong, N. Bloembergen, J. Ducuing, P.S. Pershan: Interactions between light waves in a nonlinear dielectric, Phys. Rev. **127**, 1918–1939 (1962)

21.23 P.F. Bordui, M.M. Fejer: Inorganic crystals for nonlinear optical frequency conversion, Annu. Rev. Mater. Sci. **23**, 321–379 (1993)

21.24 M.E. Hagerman, K.R. Pöppelmeier: Review of the structure and processing defect property relationships of potassium titanyl phosphate: A strategy for novel thin film photonic devices, Chem. Mater. **7**, 602–621 (1995)

21.25 H. Hellwig, J. Liebertz, L. Bohaty: Exceptional large nonlinear optical coefficients in the monoclinic bismuth borate BiB_3O_6 (BIBO), Solid State Commun. **190**, 249–251 (1998)

21.26 R. Solé, X. Ruíz, R. Cabré, M. Aguiló, F. Díaz, V. Nikolov, P. Peshev: High temperature solutions of the $(0.4Na_2O-0.6\,B_2O_3)-BaO-((6-x)+xCoSnO_3)$ system. Physical properties, Mater. Res. Bull. **30**, 779–788 (1995)

21.27 X. Ruíz, M. Aguiló, J. Massons, F. Díaz: Numerical and experimental study of the forced convection inside a rotating disk-cylinder configuration, Exp. Fluids **14**, 333–340 (1993)

21.28 H.J. Scheel, E.O. Schulz-Dubois: Flux growth of large crystals by accelerated crucible-rotation technique, J. Cryst. Growth **8**, 304–306 (1971)

21.29 D. Elwell, H.J. Scheel: *Crystal Growth from High-Temperature Solutions* (Academic, London 1975)

21.30 M.G. Astles: *Liquid-Phase Epitaxial Growth of III–V Compound Semiconductor Materials and Their Device Applications* (Adam Hilger, New York 1990)

21.31 A.A. Maier, M.V. Provotorov, V.A. Balashov: Double molybdates and tungstates of the rare earth and alcali metals, Russ. Chem. Rev. **42**, 822–833 (1973)

21.32 E. Gallucci, C. Goutadier, G. Boulon, M.T. Cohen-Adad: Growth of $KY(WO_4)_2$ single crystal: Investigation of the rich WO_3 region in the $K_2O-Y_2O_3-WO_3$ ternary system. 1. The K_2O-WO_3 binary system, Eur. J. Solid State Inorg. Chem. **34**, 1107–1117 (1997)

21.33 R. Guérin, P. Caillet: Sur les phases du système $K_2WO_4-WO_3$ et leur reduction par la vapeur de potassium, C. R. Acad. Sci. Ser. C **271**, 815–817 (1970), in French

21.34 G. Wang, Z.D. Luo: Crystal growth of $KY(WO_4)_2$: Er^{3+},Yb^{3+}, J. Cryst. Growth **116**, 505–506 (1992)

21.35 P.V. Kletsov, L.P. Kozeeva, L.Y. Kharchenko: Study on the crystallization and polymorphism of double potassium and trivalent metal tungstates, $KR(WO_4)_2$, Sov. Phys. Crystallogr. **20**, 732–735 (1976)

21.36 A. Majchrowski: *Supermaterials* (Kluwer, Dordrecht 2000)

21.37 A. Majchrowski, M.T. Borowiec, E. Michalski: Top seeded solution growth of $KHo(WO_4)_2$ single crystals, J. Cryst. Growth **264**, 201–207 (2003)

21.38 M.C. Pujol, R. Solé, J. Gavaldà, J. Massons, M. Aguiló, F. Díaz: Growth and ultraviolet optical properties of $KGd_{1-x}RE_x(WO_4)_2$ single crystals, J. Mater. Res. **14**, 3739–3745 (1999)

21.39 K. Wang, J. Zhang, J. Wang, W. Yu, H. Zhang, Z. Wang, X. Wang, M. Ba: Predicted and real habits of flux grown potassium lutetium tungstate single crystals, Cryst. Growth Des. **5**, 1555–1558 (2006)

21.40 M.T. Borowiec, A. Majchrowski, V. Domuchowski, V.P. Dyakonov, E. Michalski, T. Zayarniuk, J. Żmija, H. Szymczak: Crystal growth and x-ray structure investigation of the $KHo(WO_4)_2$, Proc. SPIE **5136**, 20–25 (2003)

21.41 A.A. Kaminskii, J.B. Gruber, S.N. Bagaev, K. Ueda, U. Hömmerich, J.T. Seo, D. Temple, B. Zandi, A.A. Kornienko, E.B. Dunina, A.A. Pavlyuk, R.F. Klevtsova, F.A. Kuznetsov: Optical spectroscopy and visible stimulated emission of Dy^{3+} ions in monoclinic α-$KY(WO_4)_2$ and α-$KGd(WO_4)_2$ crystals, Phys. Rev. B **65**, 125108 (2002)

21.42 M.C. Pujol, M. Bursukova, F. Güell, X. Mateos, R. Solé, J. Gavaldà, M. Aguiló, J. Massons, F. Díaz, P. Klopp, U. Griebner, V. Petrov: Growth, optical characterization, and laser operation of the stoichiometric crystal $KYb(WO_4)_2$, Phys. Rev. B **65**, 165121 (2002)

21.43 H.P. Jenseen, A. Linz: Analysis of the optical spectrum of Tm^{3+} in $LiYF_4$, Phys. Rev. B **11**, 92–101 (1975)

21.44 B. Chai, J. Lefaucheur, A. Pham, G. Lutts, J. Nicholls: Growth of high-quality single crystals of KYF_4 by TSSG method, Proc. SPIE **1863**, 131–135 (1993)

21.45 P. Hagenmüller: *Inorganic Solid Fluorides* (Academic, New York 1985)

21.46 E.M. Levin: The system Y_2O_3-V_2O_5, J. Am. Ceram. Soc. **50**, 381–382 (1967)

21.47 L.G. Van Uitert, R.C. Linares, R.R. Soden, A.A. Ballman: Role of f-orbital electron wave function mixing in the concentration quenching of Eu^{3+}, J. Chem. Phys. **36**, 702–705 (1962)

21.48 W. Hintzmann, G. Müller-Vogt: Crystal growth and lattice parameters of rare-earth doped yttrium phosphate, arsenate and vanadate prepared by the oscillating temperature flux technique, J. Cryst. Growth **5**, 274–278 (1969)

21.49 S.H. Smith, B.M. Wanklyn: Flux growth of rare earth vanadates and phosphates, J. Cryst. Growth **21**, 23–28 (1974)

21.50 B.M. Wanklyn: Use of a crystalline seal in flux growth – Rare-earth borates, vanadates and garnets, $KNiF_3$ and $CsNiMF_6$ (M = Fe, Cr), J. Cryst. Growth **54**, 610–614 (1981)

21.51 V.A. Timofeeva: Physicochemical aspects of flux crystallization of the oxide materials. In: *Growth of Crystals* (Consultants Bureau, New York 1988)

21.52 R. Masse, J.C. Grenier: Étude des monophosphates du type $M'TiOPO_4$ avec M' = K, Rb et Tl, Bull. Soc. Fr. Mineral. Cristallogr. **94**, 437–439 (1971), in French

21.53 M.N. Satyanarayan, A. Deepthy, H.L. Bhat: Potassium titanyl phosphate and its isomorphs. Growth, properties, and applications, Crit. Rev. Solid State Mater. Sci. **24**, 103–191 (1999)

21.54 V.I. Voronkova, V.K. Yanovskii: Growth of $KTiOPO_4$-group crystals from a solution in a melt and their properties, Inorg. Mater. **24**, 273–277 (1988)

21.55 K. Iliev, P. Peshev, V. Nikolov, I. Koseva: Physicochemical properties of high-temperature solutions of the K_2O-P_2O_5-TiO_2 system suitable for the growth of $KTiOPO_4$ (KTP) single crystals, J. Cryst. Growth **100**, 219–224 (1990)

21.56 J.C. Jacco, G.M. Loiacono, M. Jaso, G. Mizell, B. Greenberg: Flux growth and properties of $KTiOPO_4$, J. Cryst. Growth **70**, 484–488 (1984)

21.57 P.F. Bordui, J.C. Jacco, G.M. Loiacono, R.A. Stolzenberger, J.J. Zola: Growth of large single crystals of $KTiOPO_4$ (KTP) from high-temperature solution using heat pipe based furnace system, J. Cryst. Growth **84**, 403–408 (1987)

21.58 G.M. Loiacono, T.F. McGee, G. Kostecky: Solubility and crystal growth of $KTiOPO_4$ in polyphosphate solvents, J. Cryst. Growth **104**, 389–391 (1990)

21.59 P.F. Bordui, J.C. Jacco: Viscosity and density of solutions used in high-temperature solution growth of $KTiOPO_4$, J. Cryst. Growth **82**, 351–355 (1987)

21.60 T.E. Gier: Method for flux growth of $KTiOPO_4$ and its analogues, US Patent 4231838 (1980)

21.61 A.L. Aleksandrovskii, S.A. Akhmanov, V.A. D'yakov, N.I. Zheludev, V.I. Pryalkin: Efficient nonlinear optical converters made of potassium titanyl phosphate crystals, Sov. J. Quantum Electron. **15**, 885–886 (1985)

21.62 P.F. Bordui, S. Motakef: Hydrodynamic control of solution inclusion during crystal growth of $KTiOPO_4$ (KTP) from high-temperature solution, J. Cryst. Growth **96**, 405–412 (1989)

21.63 B. Vartak, Y.I. Kwon, A. Yeckel, J.J. Derby: An analysis of flow and mass transfer during the solution growth of potassium titanyl phosphate, J. Cryst. Growth **210**, 704–718 (2000)

21.64 T. Sasaki, A. Miyamoto, A. Yokotani, S. Nakai: Growth and optical characterization of large potassium titanyl phosphate crystals, J. Cryst. Growth **128**, 950–955 (1993)

21.65 G. Dhanaraj, T. Shripathi, H.L. Bhat: Defect characterization of KTP single-crystals, Bull. Mater. Sci. **15**, 219–227 (1992)

21.66 G. Dhanaraj, H.L. Bhat: Dendritic structures on habit faces of potassium titanyl phosphate crystals grown from flux, Mater. Lett. **10**, 283–287 (1990)

21.67 R.J. Bolt, M.H. van der Mooren, H. de Haas: Growth of $KTiOPO_4$ (KTP) single crystals by means of phosphate and phosphate/sulphate fluxes out of a three-zone furnace, J. Cryst. Growth **114**, 141–152 (1991)

21.68 M. Roth, N. Angert, M. Tseitlin, A. Alexandrovski: On the quality of KTP crystals for nonlinear optical and electro-optic applications, Opt. Mater. **16**, 131–136 (2001)

21.69 L.K. Cheng, J.D. Bierlein, A.A. Ballman: Crystal growth of $KTiOPO_4$ isomorphs from tungstate and molybdate fluxes, J. Cryst. Growth **110**, 697–703 (1991)

21.70 N. Angert, L. Kaplun, M. Tseitlin, E. Yashchin, M. Roth: Growth and domain structure of potassium titanyl phosphate crystals pulled from high-temperature solutions, J. Cryst. Growth **137**, 116–122 (1994)

21.71 J.J. Carvajal, V. Nikolov, R. Solé, J. Gavaldà, J. Massons, M. Rico, C. Zaldo, M. Aguiló, F. Díaz: Enhancement of the erbium concentration in RbTiOPO$_4$ by co-doping with niobium, Chem. Mater. **12**, 3171–3180 (2000)

21.72 Y.S. Oseledchik, S.P. Belokrys, V.V. Osadchuk, A.L. Prosvirnin, A.F. Selevich, V.V. Starshenko, K.V. Kuzemchenko: Growth of RbTiOPO$_4$ single crystals from phosphate systems, J. Cryst. Growth **125**, 639–643 (1992)

21.73 A.A. Ballman, H. Brown, D.H. Olson, C.E. Rice: Growth of potassium titanyl phosphate (KTP) from molten tungstate melts, J. Cryst. Growth **75**, 390–394 (1986)

21.74 K. Iliev, P. Peshev, V. Nikolov, I. Koseva: Physicochemical properties of high-temperature solution of the K$_2$O-P$_2$O$_5$-TiO$_2$-WO$_3$ system, suitable for the growth of KTiOPO$_4$ (KTP) single crystals, J. Cryst. Growth **100**, 225–232 (1990)

21.75 D.P. Shumov, M.P. Tarassov, V.S. Nikolov: Investigation of optical inhomogeneities in KTiOPO$_4$ (KTP) single crystals grown from high-temperature tungsten-containing solutions, J. Cryst. Growth **129**, 635–639 (1993)

21.76 A. Yokotani, A. Miyamoto, T. Sasaki, S. Nakai: Observation of optical inhomogeneities in flux grown KTP crystals, J. Cryst. Growth **110**, 963–967 (1991)

21.77 G. Marnier: Process for the flux synthesis of crystals of the KTiOPO$_4$ potassium titanyl monophosphate type, US Patent 4746396 (1988)

21.78 S. Suma, N. Santha, M.T. Sebastián: Growth of KTP crystals from potassium sodium fluoride phosphate solution, Mater. Lett. **34**, 322–325 (1998)

21.79 S. Suma, N. Santha, M.T. Sebastián: A new flux for the fase growth of potassium titanyl phosphate (KTP) single crystals, J. Mater. Sci. Mater. Electron. **9**, 39–42 (1998)

21.80 A. Miyamoto, Y. Mori, T. Sasaki, S. Nakai: Improvement of optical transmission of KTiOPO$_4$ crystals by growth in nitrogent ambient, Appl. Phys. Lett. **69**, 1032–1034 (1996)

21.81 J. Zhang, J. Wang, B. Ge, Y. Liu, X. Hu, R.I. Boughton: Growth, conductivity and generation of blue coherent laser of cesium doped KTiOPO$_4$ crystals, J. Cryst. Growth **267**, 517–521 (2004)

21.82 N. Angert, M. Tseitlin, E. Yashchin, M. Roth: Ferroelectric phase transition temperatures of KTiOPO$_4$ crystals grown from self-fluxes, Appl. Phys. Lett. **67**, 1941–1943 (1995)

21.83 P.A. Morris, A. Ferretti, J.D. Bierlein, G.M. Loiacono: Reduction of the ionic conductivity of flux grown KTiOPO$_4$ crystals, J. Cryst. Growth **109**, 361–366 (1991)

21.84 J. Zhang, J. Wang, B. Ge, Y. Liu, X. Hu, G. Zhao, S. Zhu, R.I. Boughton: Growth, conductivity and periodic poled structure of doped KTiOPO$_4$ and its analogue crystals, Opt. Mater. **28**, 355–359 (2006)

21.85 R. Solé, V. Nikolov, A. Vilalta, J.J. Carvajal, J. Massons, J. Gavaldà, M. Aguiló, F. Díaz: Growth of KTiOPO$_4$ films on KTi$_{1-x}$Ge$_x$OPO$_4$ substrates by liquid-phase epitaxy, J. Mater. Res. **17**, 563–569 (2002)

21.86 J.Y. Wang, Y.G. Liu, J.Q. Wei, L.P. Shi, M. Wang: Crystal growth and properties of rubidium titanium oxide phosphate, RbTiOPO$_4$, Z. Kristallogr. **191**, 231–238 (1990)

21.87 S. Ganesa Moorthy, F.J. Kumar, C. Subramanian, G. Bocelli, P. Ramasamy: Structure refinement of nonlinear optical material K$_{0.97}$Ti$_{0.97}$Nb$_{0.03}$OPO$_4$, Mater. Lett. **36**, 266–270 (1998)

21.88 D.Y. Zhang, H.Y. Shen, W. Liu, W.Z. Chen, G.F. Zhang, G. Zhang, R.R. Zeng, C.H. Huang, W.X. Lin, J.K. Liang: Crystal growth, x-ray diffraction and nonlinear optical properties of Nb:KTiOPO$_4$ crystal, J. Cryst. Growth **218**, 98–102 (2000)

21.89 I. Parreu, R. Solé, J. Gavaldà, J. Massons, F. Díaz, M. Aguiló: Crystallization region, crystal growth, and phase transitions of KNd(PO$_3$)$_4$, Chem. Mater. **15**, 5059–5064 (2003)

21.90 I. Parreu, R. Solé, J. Gavaldà, J. Massons, F. Díaz, M. Aguiló: Crystal growth, structural characterization, and linear thermal evolution of KGd(PO$_3$)$_4$, Chem. Mater. **17**, 822–828 (2005)

21.91 I. Parreu, J.J. Carvajal, X. Solans, F. Díaz, M. Aguiló: Crystal structure and optical characterization of pure and Nd-substituted type III KGd(PO$_4$)$_3$: A new material for laser and nonlinear optical applications, Chem. Mater. **18**, 221–228 (2006)

21.92 H.G. Kim, J.K. Kang, S.H. Lee, S.J. Chung: Growth of lithium triborate crystals by the TSSG technique, J. Cryst. Growth **187**, 455–462 (1998)

21.93 B.S.R. Sastry, F.A. Hummel: Studies in lithium oxide systems: I, Li$_2$O$\cdot$B$_2$O$_3$-B$_2$O$_3$, J. Am. Ceram. Soc. **41**, 7–17 (1958)

21.94 Z. Shuqing, H. Chooen, Z. Hongwu: Growth of lithium triborate (LBO) single crystal fiber by the laser-heated pedestal growth method, J. Cryst. Growth **112**, 283–286 (1990)

21.95 S.A. Guretskii, A.P. Ges, D.I. Zhigunov, A.A. Ignatenko, N.A. Kalanda, L.A. Kurnevich, A.M. Luginets, A.S. Milanov, P.V. Molchan: Growth of lithium triborate single crystals from molten salt solution under various temperature gradients, J. Cryst. Growth **156**, 410–412 (1995)

21.96 D.P. Shumov, V.S. Nikolov, A.T. Nenov: Growth of LiB$_3$O$_5$ single crystals in the Li$_2$O-B$_2$O$_3$ system, J. Cryst. Growth **144**, 218–222 (1994)

21.97 T. Ukachi, R.J. Lane, W.R. Bosenberg, C.L. Tang: Phase-matched second-harmonic generation and growth of a LiB$_3$O$_5$ crystal, J. Opt. Soc. Am. B **9**, 1128–1133 (1992)

21.98 E. Bruck, R.J. Raymakers, R.K. Route, R.S. Feigelson: Surface stability of lithium triborate crystals grown

21.99 J.W. Kim, C.S. Yoon, H.G. Gallagher: The effect of NaCl melt-additive on the growth and morphology of LiB_3O_5 (LBO) crystals, J. Cryst. Growth **222**, 760–766 (2001)

21.100 C. Parfeniuk, I.V. Samarasekera, F. Weinberg: Growth of lithium triborate crystals. I. Mathematical model, J. Cryst. Growth **158**, 514–522 (1996)

21.101 W. Chen, A. Jiang, G. Wang: Growth of high-quality and large-sized β-BaB_2O_4 crystal, J. Cryst. Growth **256**, 383–386 (2003)

21.102 W.R. Bosenberg, R.J. Lane, C.L. Tang: Growth of large, high-quality beta-barium metaborate crystals, J. Cryst. Growth **108**, 394–398 (1991)

21.103 D. Perlov, M. Roth: Isothermal growth of β-barium metaborate single crystals by continuous feeding in the top-seeded solution growth configuration, J. Cryst. Growth **137**, 123–127 (1994)

21.104 M. Roth, D. Perlov: Growth of barium borate crystals from sodium fluoride solutions, J. Cryst. Growth **169**, 734–740 (1996)

21.105 A. Liang, F. Cheng, Q. Lin, Z. Cheng, Y. Zheng: Flux growth of large single crystals of low temperature phase barium metaborate, J. Cryst. Growth **79**, 963–969 (1986)

21.106 R.S. Feigelson, R.J. Raymakers, R.K. Route: Solution growth of barium metaborate crystals by top seeding, J. Cryst. Growth **97**, 352–366 (1989)

21.107 L.H. Brixner, K. Babcock: Inorganic single crystals from reactions in fused salts, Mater. Res. Bull. **3**, 817–824 (1968)

21.108 P.F. Bordui, G.D. Calvert, R. Blachman: Immersion-seeded growth of large barium borate crystals from sodium chloride, J. Cryst. Growth **129**, 371–374 (1993)

21.109 Y.S. Oseledchik, V.V. Osadchuk, A.L. Prosvirnin, A.F. Selevich: Growth of high-quality barium metaborate crystals from Na_2O-NaF solution, J. Cryst. Growth **131**, 199–203 (1993)

21.110 V. Nikolov, P. Peshev, K. Khubanov: On the growth of β-BaB_2O_4 (BBO) single crystals from high-temperature solutions: II. Physicochemical properties of barium borate solutions and estimation of the conditions of stable growth of BBO crystals from them, J. Solid State Chem. **97**, 36–40 (1992)

21.111 Q. Huang, Z. Liang: Studies on flux systems for the single crystal growth of β-BaB_2O_4, J. Cryst. Growth **97**, 720–724 (1989)

21.112 A.E. Kokh, N.G. Kononova, T.B. Bekker, V.A. Vlezko, P.V. Mokrushnikov, V.N. Popov: Change of symmetry and rotation of thermal field as a new method of control of heat and mass transfer in crystal growth (by example of β-BaB_2O_4), Crystallogr. Rep. **50**, 160–166 (2005)

21.113 S.C. Sabharwal, S.M. Goswami, S.K. Kulkarni, B.D. Padalia: Growth, optical transmission and x-ray photoemission studies of BaB_2O_4 single crystals, J. Mater. Sci. Mater. Electron. **11**, 325–329 (2000)

21.114 R. Solé, V. Nikolov, M.C. Pujol, J. Gavaldà, X. Ruíz, J. Massons, M. Aguiló, F. Díaz: Stabilization of β-BaB_2O_4 in the system BaB_2O_4-Na_2O-Nd_2O_3, J. Cryst. Growth **207**, 104–111 (1999)

21.115 H.G. Kim, J.K. Kang, S.J. Park, S.J. Chung: Growth of the nonlinear crystals of lithium triborate and beta barium borate, Opt. Mater. **9**, 356–360 (1998)

21.116 A. Giesen, H. Hügel, A. Voss, K. Wittig, U. Brauch, H. Opower: Scalable concept for diode-pumped high-power solid-state lasers, Appl. Phys. B **58**, 363–372 (1994)

21.117 S. Chénais, F. Balembois, F. Druon, G. Lucas-Leclin, P. Georges: Thermal lensing in diode-pumped ytterbium lasers – Part I: theoretical analysis and wavefront measurements, IEEE J. Quantum Electron. **QE-40**, 1217–1233 (2004)

21.118 C. Stewen, K. Contag, M. Larionov, A. Giesen, H. Hügel: A 1-kW CW thin disc laser, IEEE J. Sel. Top. Quantum Electron. **6**, 650–657 (2000)

21.119 R. Paschotta, J. Aus der Au, G.J. Spühler, S. Erhard, A. Giesen, U. Keller: Passive mode locking of thin-disk lasers: effects of spatial hole burning, Appl. Phys. B **72**, 267–278 (2001)

21.120 U. Griebner, J. Liu, S. Rivier, A. Aznar, R. Grunwald, R.M. Solé, M. Aguiló, F. Díaz, V. Petrov: Laser operation of epitaxially grown Yb:KLu$(WO_4)_2$-KLu$(WO_4)_2$ composites with monoclinic crystalline structure, IEEE J. Quantum Electron. **QE-41**, 408–414 (2005)

21.121 Y.E. Romanyuk, C.N. Borca, M. Pollnau, S. Rivier, V. Petrov, U. Griebner: Yb-doped KY$(WO_4)_2$ planar waveguide laser, Opt. Lett. **31**, 53–55 (2006)

21.122 M.G. Roelofs, A. Ferretti, J.D. Bierlein: Proton-exchanged and ammonium-exchanged waveguides in $KTiOPO_4$, J. Appl. Phys. **73**, 3608–3613 (1993)

21.123 J.D. Bierlein, H. Vanherzeele: Potassium titanyl phosphate-properties and new applications, J. Opt. Soc. Am. B. **6**, 622–633 (1989)

21.124 A. Aznar, R. Solé, M. Aguiló, F. Díaz, U. Griebner, R. Grunwald, V. Petrov: Growth, optical characterization, and laser operation of epitaxial Yb:KY$(WO_4)_2$/KY$(WO_4)_2$ composites with monoclinic structure, Appl. Phys. Lett. **85**, 4313–4315 (2004)

21.125 Y.E. Romanyuk, I. Utke, D. Ehrentraut, V. Apostolopoulos, M. Pollnau, S. Garcia-Revilla, R. Valiente: Low-temperature liquid-phase epitaxy and optical waveguiding of rare-earth-ion-doped KY$(WO_4)_2$ thin layers, J. Cryst. Growth **269**, 377–384 (2004)

21.126 T. Kawaguchi, D.H. Yoon, M. Minakata, Y. Okada, M. Imaeda, T. Fukuda: Growth of high crystalline quality $LiNbO_3$ thin films by a new liquid phase epitaxial technique from a solid-liquid coexisting melt, J. Cryst. Growth **152**, 87–93 (1995)

21.127 E. Bauer: Phänomenologische Theorie der Kristallabscheidung an Oberflächen II, Z. Kristallogr. **110**, 395–431 (1958), in German

21.128 A. Aznar, O. Silvestre, M.C. Pujol, R. Solé, M. Aguiló, F. Díaz: Liquid-phase epitaxy crystal growth of monoclinic $KLu_{1-x}Yb_x(WO_4)_2/KLu(WO_4)_2$ layers, Cryst. Growth Des. **6**, 1781–1787 (2006)

21.129 J.D. Bierlein, A. Ferretti, L.H. Brixner, W.Y. Hsu: Fabrication and characterization of optical waveguides in $KTiOPO_4$, Appl. Phys. Lett. **50**, 1216–1218 (1987)

21.130 A. Raizman, D. Eger, M. Oron: X-ray characterization of Rb exchanged KTP, J. Cryst. Growth **187**, 259–267 (1998)

21.131 L.K. Cheng, J.D. Bierlein, C.M. Foris, A.A. Ballman: Growth of epitaxial thin-films in the $KTiOPO_4$ family of crystals, J. Cryst. Growth **112**, 309–315 (1991)

21.132 L.K. Cheng, J.D. Bierlein, A.A. Ballman: $KTiOP_xAs_{1-x}O_4$ optical wave-guides grown by liquid-phase epitaxy, Appl. Phys. Lett. **58**, 1937–1939 (1991)

21.133 M.A. Harmer, M.G. Roelofs: Sol-gel synthesis of thin-films of potassium titanyl phosphate for nonlinear optical applications, J. Mater. Sci. Lett. **12**, 489–491 (1993)

21.134 F. Xiong, R.P.H. Chang, M.E. Hagerman, V.L. Kozhevnikov, K.R. Pöppelmeier, H. Zhou, G.K. Wong, J.R. Ketterson, C.W. White: Pulsed excimer-laser deposition of potassium titanyl phosphate films, Appl. Phys. Lett. **64**, 161–163 (1994)

21.135 P.M. Lundquist, H. Zhou, D.N. Hahn, J.B. Ketterson, G.K. Wong, M.E. Hagerman, K.R. Pöppelmeier, H.C. Ong, F. Xiong, R.P.H. Chang: Potassium titanyl phosphate thin-films on fused quartz for optical waveguide applications, Appl. Phys. Lett. **66**, 2469–2471 (1995)

21.136 R. Solé, V. Nikolov, A. Vilalta, J.J. Carvajal, J. Massons, J. Gavaldà, M. Aguiló, F. Díaz: Liquid phase epitaxy of $KTiOPO_4$ on $KTi_{1-x}Ge_xOPO_4$ substrates, J. Cryst. Growth **237–239**, 602–607 (2002)

21.137 L.K. Cheng, L.T. Cheng, J. Galperin, P. Morris-Hotsepiller, J.D. Bierlein: Crystal-growth and characterization of $KTiOPO_4$ isomorphs from the self-fluxes, J. Cryst. Growth **137**, 107–115 (1994)

21.138 A. Peña, S. Di Finizio, T. Trifonov, J.J. Carvajal, M. Aguiló, J. Pallares, A. Rodriguez, R. Alcubilla, L.F. Marsal, F. Díaz, J. Martorell: Two dimensional KTP photonic crystal grown using a macroporous silicon template, Adv. Mater. **18**, 2220–2225 (2006)

758

22. Growth and Characterization of KDP and Its Analogs

Sheng-Lai Wang, Xun Sun, Xu-Tang Tao

Crystals of potassium dihydrogen phosphate (KDP, KH_2PO_4) and its deuterated analogs (DKDP, $K(D_xH_{1-x})_2PO_4$) have been studied for their interesting electrical and optical properties, structural phase transitions, and ease of crystallization. They are the only nonlinear crystals currently applied in inertial confinement fusion (ICF), which has made them a hot topic of research for decades. To yield enough large crystals exceeding 50 cm in all three dimensions, the point-seed technique was recently developed. This method can grow crystals one order of magnitude faster than conventional methods. Recent developments in both the techniques and science of growth phenomena and defect formation under various conditions are described in this chapter, which also reviews significant advances in understanding of the fundamentals of KDP crystal growth, other growth methods to yield large high-quality crystals, growth defects and optical performance, and evaluations of crystal quality.

22.1	Background	759
22.2	Mechanism and Kinetics of Crystallization	761
	22.2.1 Studies of KDP Crystal Surfaces	761
	22.2.2 Nucleation Studies in Supersaturated Solution	763
	22.2.3 Dislocation Mechanism	765
	22.2.4 Growth on Two-Dimensional Nuclei	767
	22.2.5 Growth from Crystal Edges	767
22.3	Growth Techniques for Single Crystals	769
	22.3.1 Parameters Affecting Growth Rate	769
	22.3.2 Stability of Solution	770
	22.3.3 Conventional Methods	771
	22.3.4 Rapid Growth from a Point Seed	773
22.4	Effect of Growth Conditions on Defects of Crystals	776
	22.4.1 Impurity Effect	776
	22.4.2 Supersaturation	779
	22.4.3 Filtration	780
	22.4.4 Hydrodynamic Effects	781
22.5	Investigations on Crystal Quality	783
	22.5.1 Spectroscopic Studies	783
	22.5.2 Homogeneity	784
	22.5.3 Laser Damage Threshold	787
References		789

22.1 Background

Crystals of potassium dihydrogen phosphate (KDP) and its isomorphs have been the subject of a wide variety of investigations for over half a century owing to their interesting electrical and optical properties, structural phase transitions, and ease of crystallization [22.1–4]. Today, these crystals are widely used in both laboratory and industrial settings to control the parameters of laser light, such as pulse width, polarization, and frequency, through first- and second-order electrooptic effects [22.4, 5]. Their application in inertial confinement fusion (ICF) research [22.6] has made them a hot topic of research for decades.

The very high-energy Nd-glass lasers used for ICF research need large plates of nonlinear crystals for electrooptic switches and frequency converters (Fig. 22.1). The lasers under construction in the USA and France with about $40 \times 40\,\mathrm{cm}^2$ aperture require single-crystal boules with linear dimensions in the 50–100 cm range. KDP (KH_2PO_4) and its deuterated analogs DKDP ($K(D_xH_{1-x})_2PO_4$) are the only nonlinear crystals currently used for these applications due to their unique physical properties, which include transparency over a wide region of the optical spectrum, resistance to damage by laser radiation, and relatively high nonlin-

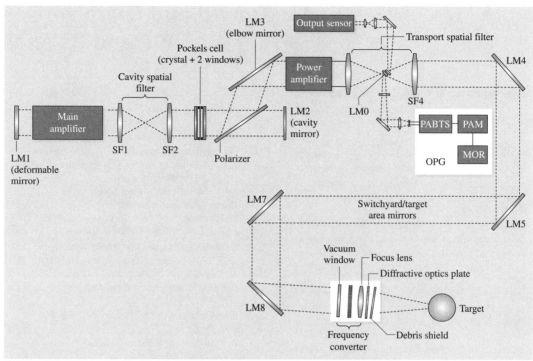

Fig. 22.1 Schematics of single NIF beamline showing positions of major optical components (after [22.7], with permission of ASM) (LM – laser mirror, SF – spatial filter, OPG – optical pulse generation)

ear efficiency, in combination with reproducible growth to large size and perfection.

The main limitation in growth of such large crystals by traditional techniques is the growth rates of only 0.5–1 mm/day typical for low-temperature solution growth, which leads to growth cycles exceeding 1–2 years. Work on rapid growth of KDP crystals started in the early 1980s when the world's biggest laser of that time, Nova, was being built at the Lawrence Livermore National Laboratory (LLNL). New rapid growth techniques [22.9–11] explored at that time involved radical modification of standard crystallization equipment. However, the complicated designs proposed could not solve the problems such as hydrodynamic conditions [22.9] and control of high supersaturation needed to avoid defect formation and spontaneous nucleation [22.10]. As a result, the $27 \times 27\,\text{cm}^2$ plates of KDP crystals for the frequency-conversion arrays on Nova were grown by traditional techniques.

Also, a rapid crystal growth method called the point-seed technique, based on the traditional solution growth process, was initiated at Moscow State University [22.12]. This technique demonstrated that KDP and DKDP crystals could be grown in standard Holden-

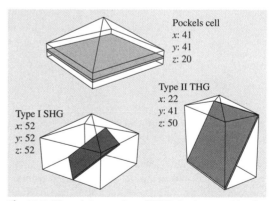

Fig. 22.2 The minimum size of KDP single-crystal boules (in cm) needed for obtaining Pockels cell and harmonic-generation plates of different types for the NIF project (after [22.8], with permission of Elsevier) (SHG – second harmonic-generation, THG – third harmonic-generation)

type rotary crystallizers without spontaneous nucleation and visible defects, one to two orders of magnitude faster than by conventional methods. Later this method was further developed and scaled for production of crystals for Nova's successor – the National Ignition Facility (NIF), under construction at LLNL [22.8]. To achieve economically useful yield, crystals grown for NIF should exceed 50 cm in all three dimensions (Fig. 22.2).

The development of new techniques required both technical solutions and scientific knowledge about growth phenomena, defect formation at varying degrees of supersaturation, and growth rate. The purpose of this chapter is to describe recent developments in growth and characterization of KDP and its analogs. The chapter also includes significant advances in understanding the fundamentals of KDP crystal growth, developing growth methods to yield large-dimension high-quality crystals, and the relation between growth defects and optical performance. To this end, the chapter is divided into four sections encompassing the mechanism and kinetics of crystallization, growth techniques for single crystals, the effect of growth conditions on crystal defects, and problems with crystal quality.

22.2 Mechanism and Kinetics of Crystallization

As we know, since the growth of crystals takes place at the crystal–solution interface, structural information on the immersed solid is essential for understanding crystal growth from solution or melt. This is also a very important topic for KDP-type crystals, which can help us to understand the relaxations and possible reconstructions of the top layers, the influence of impurities on morphology, and even the possible reason for change in optical properties of crystals.

22.2.1 Studies of KDP Crystal Surfaces

The crystallographic theory of Hartman and Perdok aims to predict the morphology of growing crystals dominated by the so-called F (flat) faces, referred to by the Miller indices (*hkl*) [22.14, 15]. However, the theory has its limitations. Often more than one surface termination is possible for a given orientation (*hkl*) and it is impossible to predict which of the alternatives will control the crystal growth.

The theory predicts that the pyramidal faces {101} and the prismatic faces {100} of KDP are flat in solution [22.16, 17], in agreement with the observed habit of these crystals (Fig. 22.3a). For the prismatic {100} faces exactly one surface termination is predicted (Fig. 22.4). For the pyramidal {101} faces, however, two alternative terminations are theoretically possible. One has the negative $H_2PO_4^-$ groups on the outside, and the other the positive K^+ ions (Fig. 22.3b). The difference in polarity of the layers and, especially, the differences in size and polarizability of the ions will result in a different surface free energy. From the surface morphology observed with interference-contrast reflection microscopy

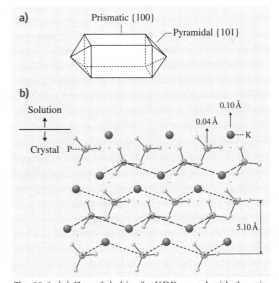

Fig. 22.3 (a) Growth habit of a KDP crystal with the prismatic and pyramidal faces indicated. (b) Schematic side view of the pyramidal face, KDP {101}, projected on the {111} plane. The *big circles* are the potassium atoms while the PO_4 groups are depicted as a *circle* for the phosphor atom connected by sticks to the four neighboring oxygen atoms, shown as *small white circles*. The *dots* give the positions of the hydrogen atoms between two oxygen atoms. The layers with the K^+ ions on top are schematically indicated by the *dashed lines* and the layers with the $H_2PO_4^-$ groups on top by the *discontinuous lines*. *Arrows* indicate the relaxations in the topmost layer as determined from fitting the experimental data (after [22.13])

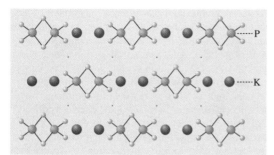

Fig. 22.4 The prismatic KD P{100} face projected on the {010} plane. Here only one termination is possible (after [22.13])

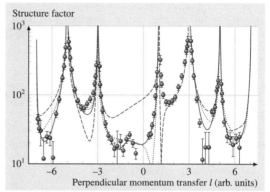

Fig. 22.5 Structure-factor amplitudes along the $(hk) = (10)$ crystal truncation rod for KDP {101} as a function of the diffraction index, which is expressed in reciprocal lattice units. The *dotted line* is a calculation for a bulk K^+-terminated surface, the *dashed curve* for a $H_2PO_4^-$-terminated one. The *solid line* is the best fit starting from a K^+-terminated surface and allowing the K^+ ions and the $H_2PO_4^-$ groups in the top layer to relax (after [22.13])

and considering the symmetry of the crystal [22.17, 18] it can be concluded that the surface is bounded by only one of the polar layers. This is confirmed by atomic force microscopy measurements where the height of the steps on the {101} face is always found to correspond to 0.5 nm, the thickness of double layers [22.19].

De Vries and coworkers used the technique of surface x-ray diffraction at a third-generation synchrotron radiation source to determine the structure of KDP crystal surfaces in air, vacuum, and solution [22.13, 20]. They measured the distribution of diffracted intensities along so-called crystal truncation rods (CTRs) [22.21], which is hardly influenced by the solution and mainly depends on the crystal surface atomic structure. All data show that the pyramidal {101} faces are terminated with K^+ atoms rather than with $H_2PO_4^-$ groups, while the prismatic {100} faces terminate in alternating rows of K^+ and $H_2PO_4^-$ ions (Fig. 22.5).

From the atomic structure of both faces, it is easy to be understood why small traces of trivalent metal ion impurities such as Fe^{3+} or Cr^{3+} block the growth of the prismatic faces, but affect the growth of the pyramidal faces to a much lesser extent [22.17, 22, 23]. With only K^+ ions on the surface of the crystal, metal impurities such as Fe^{3+} and Cr^{3+} ions will experience a large barrier to adsorption onto the positively charged face. The impurity content is limited by adsorption kinetics on the terraces rather than incorporation kinetics at the steps. On the prismatic faces, however, these ions can adsorb easily, and small amounts of Fe^{3+} or Cr^{3+} will already block the growth.

Furthermore, De Vries et al. present the results of experiments about the solid–liquid interface during growth, with and without the addition of Fe^{3+} impurities. The surface was roughened, as seen from the decrease in intensity in between the Bragg peaks. The overall shape of the CTR is unchanged, so no ordered layer of Fe^{3+} is formed on the surface. Apparently, the amounts of adsorbed Fe^{3+} are very small. These impurities locally pin the moving steps [22.24–26], which causes an increased meandering of the steps and thus leads to a rougher surface.

The molecular structure of interface boundary layers in the crystal growth of KDP and DKDP has been an interesting topic for a long period. The Raman spectra of saturated KDP solution show that only 40% of the phosphates exist as monomers and that anion–anion association via hydrogen bonds does cease at the dimer [22.27]. *Lu* et al. [22.28] studied the growth units of KDP crystal from the growth solution by using space group theoretical analyses and Raman spectra; the result proved that the growth unit is dimers of $[H_2PO_4^-]$ anions. *Yu* et al. [22.29] used holography to measure the thickness of boundary layers to be a few hundred micrometers during KDP crystal growth in free convection. Their real-time Raman microprobe experiments show that the smectic ordering structure of the anions–cations called the crystallization unit is formed within the boundary layer of KDP and DKDP growth solutions [22.30].

An in situ x-ray diffraction study of growth surface by *Reedijk* et al. [22.31] revealed interface-induced ordering in the first four layers of water molecules. The first two layers behave ice-like and are strongly bound

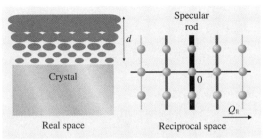

Fig. 22.6 Schematic drawing of a crystal covered with a partly ordered liquid layer of thickness d. The contribution of the (partially) ordered liquid to the substrate diffraction rods is strongest in the specular rod and diminishes at higher parallel momentum transfer ($Q_\parallel$) in reciprocal space (after [22.31])

to the surface. The next two layers are more diffuse and show only minor lateral and perpendicular ordering (Fig. 22.6). The highly ordered *liquid* at the interface is expected to slow down the incorporation and diffusion of the growth units.

Further research [22.33] on the solid–liquid interface structure of the {101} and {100} faces of KDP crystals in contact with growth solutions of different pH values showed in all cases several liquid layers with varying degrees of lateral and perpendicular order. The structural changes are large for the {101} face and small for the {100} face. The changes at the {101} face are likely due to the pH-dependent competition between K^+ and H_3O^+ bonding.

22.2.2 Nucleation Studies in Supersaturated Solution

Nucleation theory has developed since the early 18th century. *Mullin* [22.34] classified nucleation into two major group: primary nucleation and secondary nucleation, where nuclei are generated in the vicinity of crystals already present in a supersaturated system. Primary nucleation consists of homogeneous nucleation and heterogeneous nucleation. The study on the stability of solution by *Zaitseva* et al. [22.32] provided the basis to develop fast growth techniques for KDP and DKDP crystals.

Primary Nucleation

Cooling a solution with concentration C to a temperature T under its equilibrium temperature leads to a supersaturated solution. The supersaturation ΔC is $C - C_0$. In classic nucleation theory, the homogeneous nucleation rate of the supersaturated solutions, i.e., the frequency of formation of a particle with critical nucleus in a unit volume, is expressed as

$$J = B_1 \exp\left(-\frac{16\pi\alpha^3\omega^3}{3k_B^3 T^3 \left(\ln\frac{C}{C_0}\right)^2}\right). \tag{22.1}$$

The induction period is inversely proportional to the nucleation rate, which leads to the expression

$$\text{int}_i = B + \frac{16\pi\alpha^3\omega^3}{3k_B^3 T^3 \left(\ln\frac{C}{C_0}\right)^2}. \tag{22.2}$$

Equations (22.1) and (22.2) express the width of the metastable zone in terms of the supersaturation and induction time for homogeneous nucleation. However, the quantitative estimation of these parameters for real crystallization systems using these equations is practically impossible because of the many unknown parameters, such as α, the specific free energy of the interface, that form the constants. It is not of much help in practical work because a real growth system obviously does not deal with homogenous nucleation. The presence of crystallizer walls, parts of equipment, and extraneous particles in the supersaturated solution makes heterogeneous nucleation more probable. Some theoretical calculations and experiments in pure containerless systems show that homogeneous nucleation occurs at extremely high concentrations, exceeding the solubility limit by several factors [22.35–37].

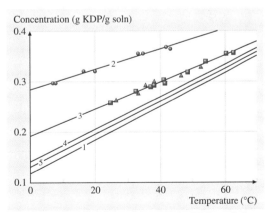

Fig. 22.7 Stability of supersaturated KDP solutions: (1) solubility curve; (2) and (3) metastable boundaries of solutions without (●) and with (■) a growing crystal, respectively; (▲) experiments with the empty platform; (4) and (5) traditional level of stability (after [22.32])

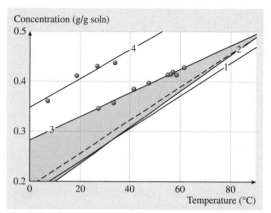

Fig. 22.8 Metastable zone of DKDP solutions ($X = 98\%$). (1) Monoclinic phase solubility. (2) Tetragonal phase solubility. (3) Metastable boundary in the presence of a growing crystal. The region of tetragonal crystals growth is shaded. For comparison the data of [22.38] are shown (*dashed line*). (4) Metastable boundary without crystals (after [22.32])

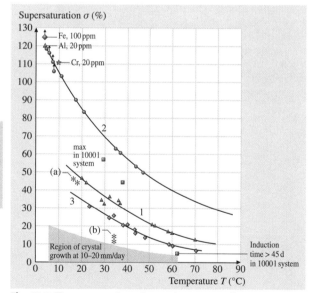

Fig. 22.9 Temperature dependence of supersaturation σ_{max} reached in KDP solutions without spontaneous nucleation: (○) filtered and (△) unfiltered solutions without crystals overheated at 80 °C; (◇) filtered solution in the presence of a crystal overheated at 80 °C; (∗) filtered (a) and unfiltered (b) solutions without crystals overheated at 54 °C (after [22.42])

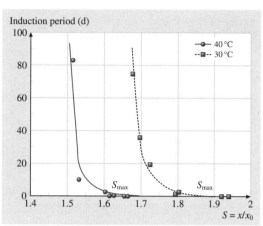

Fig. 22.10 Effect of supersaturation on induction period in KDP solutions (after [22.32])

In order to understand the nature of nucleation in a real crystallization system there was no other way but to perform experiments in the system. *Zaitseva* et al. measured the metastable zone of KDP solution by the polythermal method in standard Holden-type crystallizers with volumes of 5–20 l and 1000 l [22.32]. Figures 22.7–22.9 present the results, which show the maximum concentration and supersaturation reached without spontaneous nucleation in stirred KDP and DKDP solutions without growing crystals. The level of supersaturation that can be reached is much higher than that expected from previous studies [22.10, 38]. The experiments show a much narrower metastable zone when the solution is not filtered. The drop of stability is due to the effect of typical heterogeneous particles present in the salts generated during their commercial production. Insufficient overheating can cause a similar drop in stability as in the absence of filtration. Solution overheating of 80 °C can result in reproducible stability, independent of the saturation temperature. They found no correlation between spontaneous nucleation and the chemical purity of the solutions, which is different from other investigations [22.34, 39–41].

The induction period was also measured for various supersaturated solutions by the isothermal method in the same crystallizers [22.32]. The results (presented in Fig. 22.10) show a sharp increase of induction period at supersaturation of less than some certain values (35% for 40 °C, 50% for 30 °C). Experimentally this means that no precipitation was observed for 3 months with supercooling more than 30 °C below the saturation point.

Secondary Nucleation

Secondary nucleation is more difficult to explain in detail as there are at least three categories of secondary nucleation: apparent (small fragments washed from the surface of crystalline seeds), true (when the current level of supersaturation is higher than the supersaturation level or the solute particles present in solution), and contact nucleation (when a growing particle contacts the walls of the baffles, stirrer or other objects, thus leaving behind residual solute particles that have been broken off from primary crystals) [22.43].

Randolph and *Larsen* stated that, in either continuously stirred crystallizers or seeded batch crystallizers, the main source of secondary nuclei is the crystal suspension itself [22.44]. *Boistele* also agrees that a major source is the crystal surface; secondary nuclei form whenever the tiny embryos, or crystallites, that are removed from the surface and dispersed into a supersaturated solution exceed the critical size [22.45].

Secondary nucleation initiated by a well-faceted crystal is often treated as aggregates appearing and being held in a stagnant supersaturated (transitional boundary) layer under the influence of the force field of the crystal [22.46]. These aggregates are thought to be stripped off the crystal surface by fluid motion, providing secondary nuclei [22.46, 47]. However, there is still a certain contradiction, as supersaturation in the immediate vicinity of the growing crystal is lower than elsewhere in the solution.

Most empirical expressions based on the results obtained in real crystallization systems predict very narrow metastable regions with a width of a few degrees [22.38, 48–50]. However, the metastable zone reported by *Zaitseva* et al. is much larger than those listed above [22.32]. The experiments show that a defectless crystal does not produce any secondary nucleation and does not influence spontaneous nucleation from solution. The initial results shown in Figs. 22.7–22.9 show that the presence of a growing crystal reduces the solution stability, which was supposed to not be caused by secondary nucleation but rather by introduction of the platform into the otherwise closed system.

22.2.3 Dislocation Mechanism

The bulk of a faceted crystal is built up by deposition of surface layers parallel to the crystallographic faces. Recent experimental studies by means of optical interferometry [22.1] and atomic force microscopy (AFM) [22.25, 51] present clear evidence that the dislocation growth mechanism remains dominant during the growth of KDP-type crystals. Dislocation bunches which give rise to growth hillocks on crystal faces form during seed regeneration. During further growth, dislocations arise from mutual displacement of the layers overlapping a solution inclusion or an extraneous particle. Two or more growth sources of the same activity can exist on a face simultaneously. The growth sources initiated during regeneration can be replaced by new leading hillocks that become more active under the changing conditions. The change of the leading hillock is inevitably accompanied by the formation of growth bands due to the appearance of new growth steps of various orientations [22.52]. When very strong and dense dislocation bunches form, cracking may occur due to great internal stress in the crystal during the growth process [22.53].

Dislocation hillocks on the pyramidal {101} and prismatic {100} faces of KDP and DKDP crystals have the different shapes shown in Fig. 22.11. The shapes of the hillocks reflect the crystallography of the face such that each hillock is comprised of a set of sectors with unique slopes and step. Growth occurs on monomolecular steps, and the hillocks on the {101} faces are shaped as trihedral pyramids with three neighboring slopes (vicinal sectors) of different steepness. The hillocks on the {100} faces have an elliptical shape that changes from almost round to nearly a parallelogram with four vicinal sectors at increasing supersaturation [22.1]. The intersection of the vicinal sectors of the hillocks form the edges of a hillock, called vicinal-sectoral boundaries, which are typically straight lines on both sets of faces. Intervicinal boundaries are the geometrical

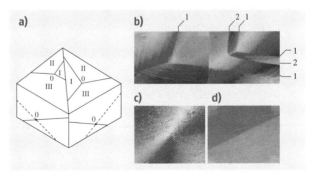

Fig. 22.11a–d Geometry and shape of the dislocation hillocks of KDP: (**a**) position on the adjacent faces; I, II, and III are vicinal sectors on {101} faces. (**b**) Micrograph of the hillocks on {101} face; 1 – vicinal boundaries; 2 – intervicinal boundary; as grown surface at $t = 30\,°C$ and $\sigma = 0.06$; (**c,d**) growth hillocks on {100} face of KDP at low and high supersaturation, respectively (after [22.42])

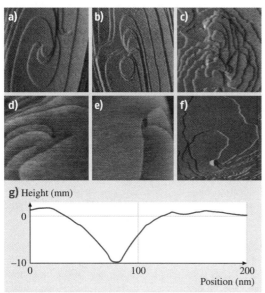

Fig. 22.12a–g AFM images of typical growth hillocks on {100} face of KDP showing complex structure of dislocation sources (**a–c**) and {101} face showing hollow cores at dislocation sources (**d–f**): cross-section of typical hollow core is shown in (**g**) (after [22.7]). (**a**) $1\times 1\,\mu m^2$; (**b**) $1.5\times 1.5\,\mu m^2$; (**c**) $2\times 2\,\mu m^2$; (**d**) $3\times 3\,\mu m^2$; (**e**) $2.6\times 2.6\,\mu m^2$; (**f**) $8.4\times 8.4\,\mu m^2$

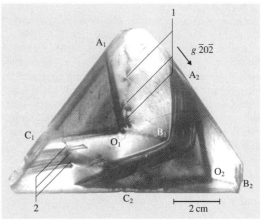

Fig. 22.13 Projection x-ray topograph of a plate cut parallel to the {101} dipyramid face of KDP crystal. O_1 and O_2, are tips of vicinal hillocks formed by dislocations, O_1A_1, O_1B_1, O_1C_1 and O_2A_2, O_2B_2, O_2C_2, are vicinal-sectorial boundaries, $A_2B_1C_2$ is an intervicinal boundary; 1, 2 are dislocations which do not form vicinal hillocks (after [22.42])

points where the steps from the neighboring vicinal hillocks meet [22.42]. The position of the vicinal-sectorial boundaries is determined by the location of the dislocation outcrop on a crystal face. Their orientation, strictly connected with the crystal symmetry, can vary only within several degrees, depending on growth conditions [22.1, 54].

The AFM studies revealed that, for Burgers vectors in excess of one unit step, even simple sources exhibit hollow cores at the dislocation source on {101} faces (Fig. 22.12) [22.7, 19]. The shape of the cores demonstrates that the step edge energy is isotropic. X-ray projection topograph of a {101} KDP plate shows that the adjacent vicinal sectors are each slightly disoriented relative to the others, and the average difference of the lattice parameters is of the order of 10^{-6} of the average lattice parameter value (Fig. 22.13) [22.18, 55]. As a rule, the lattice deformations are maximal at these boundaries that form the vicinal sector I on {101} faces, whereas the steps of the vicinal sectors II and III smoothly convert into each other without clearly discernible deformations. Different impurities selectively incorporate into the different vicinal sectors depending on their nature [22.55]. For example, Fe^{3+} and other metal impurities preferentially incorporate into sector I [22.56]. In DKDP crystals this defect can be

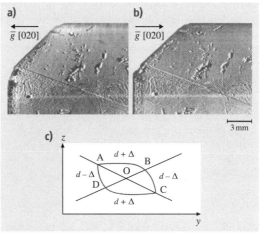

Fig. 22.14a–c X-ray topographic images of a vicinal hillock on the prismatic face of a KDP crystal (**a,b**) and diagram with notations for analyzing the contrast (**c**). d – lattice parameter, Δ – the difference of the parameter in adjacent vicinal sectors (after [22.42])

more pronounced because of the different composition of deuterium and hydrogen in the three vicinal sectors [22.42].

On the {100} face, the dislocation sources are often complex (Fig. 22.14). Vicinal hillocks on the face are shown in Fig. 22.11. Growth steps change their orientation on the edge AC abruptly, while on the edge BD the orientation changes more smoothly. The x-ray topograph reveals that the lattice parameters are greater in vicinal sectors AOB and COD than in sectors BOC and AOD [22.42]. Formation of these vicinal sectors can be also explained by the different incorporation of impurities into the differently oriented steps.

22.2.4 Growth on Two-Dimensional Nuclei

The main mechanism of growth was clear because the dislocation growth sources could be easily seen on the growing faces. At the same time, the growth of KDP {101} surfaces occurs also on islands formed by two-dimensional (2-D) nucleation (Fig. 22.16) at moderate supersaturations (10%) [22.19]. Furthermore, on the terraces of the vicinal hillocks formed by 2-D nucleation, island growth competes with step flow when the interisland spacing is comparable to the terrace width. For example, 2-D nucleation [22.57, 58] was found to contribute to the rapid growth of crystal on the *negative* pyramidal faces {10$\bar{1}$} just above the platform (Fig. 22.15). All the dislocations formed during regeneration very quickly vanish from these negative pyramidal faces because of their orientation at some angle to the platform surface. The dislocation-free growth rate measured in the range of 63–60 °C at $\sigma = 0.06$ is about one order of magnitude (1–1.5 mm/day) slower than for the case of dislocation growth (about 15–17 mm/day under the same conditions).

De Yoreo et al. [22.19] estimated that, at supersaturation below 5%, layer growth on {101} faces is dominated by the dislocation mechanism, and that growth on 2-D nuclei begins to compete at 5–10% supersaturation. Results from *Alexandru* [22.59] showed that growth by the dislocation mechanism on {100} surface is severely retarded by the stopper action of impurities, particularly towards smaller supersaturation. At supersaturation less than 8–10% the dislocation mechanism of growth appears to compete with 2-D nucleation mechanism at lower impurity concentration. The two-dimensional nucleation mechanism of growth becomes dominant at higher supersaturation.

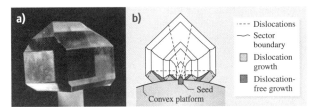

Fig. 22.15 (a) KDP crystal with clearly pronounced {101} low pyramids growing periodically by dislocation and dislocation-free mechanism; (b) schematic of dislocation geometry on convex-shaped platform (after [22.42])

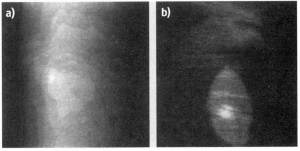

Fig. 22.16 (a) AFM image (5.4×5.4 μm²) of typical hillock on KDP {101} at which no dislocations are observed. (b) Higher-resolution (715×715 nm²) image of one such hillock showing the topmost island, for which the radius is 42 nm (after [22.19])

22.2.5 Growth from Crystal Edges

Existing models of dislocation growth typically do not take into account mutual effects of growing crystal faces and participation of the edges in the growth process. The edges of faceted crystals are often considered as passive places where the steps produced on adjacent faces meet, although evidence of growth-step generation from the edges has been reported in many experimental works [22.60, 61].

Zaitseva et al. proposed a possible mechanism of growth-step generation from the edges of faceted crystals obtained from experimental observations with KDP crystals [22.62]. It suggests that growth from the crystal edges is initiated by the deviation of the edges from their crystallographic orientation and formation of incomplete shapes of singular facets. An incomplete crystallographic shape of a singular face is determined by the existence of concave angles formed by the edges in the plane of this face. These concave angles are sources of growth steps in the surface layer. The surface layer generated from a concave angle on a singular crystal face completes the crystallographic shape of the face. Growth of this surface layer does not require the existence of a preceding layer of the same orientation.

Formation of the layers on the crystal edges, such as the z-cut seed regeneration process, can be observed during growth of KDP crystals. The thin surface layers can grow from the edges in crystallographic planes without pre-existing layers under them. A typical explanation for this phenomenon is attributed to a supersaturation gradient and better hydrodynamic conditions on the edges [22.63].

Experiments [22.62] including regeneration of singular faces with incomplete crystallographic shape (Fig. 22.17) and formation of the thin surface layers during joining of two equally oriented crystals were performed to clarify what effect such deviations produce during growth and formation of the surface structure. These experiments suggested that the process of step generation from the edges was connected with the deviation of faces from crystallographic orientations and shape, resulting from vicinal hillocks on the crystal surface.

The growth velocity of these layers has the same order of magnitude as the tangential growth rate V of a dislocation step. Estimations give values for the velocity of the thin-film advancement of 1×10^{-4}–4×10^{-4} cm/s at temperature of $30\,°\mathrm{C}$ and supersaturation of about 0.09, while the value of V from [22.64] is 7×10^{-4} cm/s for about the same growth conditions. The measurements done at $60\,°\mathrm{C}$ gave higher values of about 1×10^{-3} cm/s at approximately the same supersaturation [22.62].

According to this mechanism, the crystal surface should be considered as an entire system. The structure of one separate face is connected with the phenomena that take place on the other faces through the crystal edges, which play an important role during crystal growth. During the growth of faceted crystals, this mechanism may work in combination with the dislo-

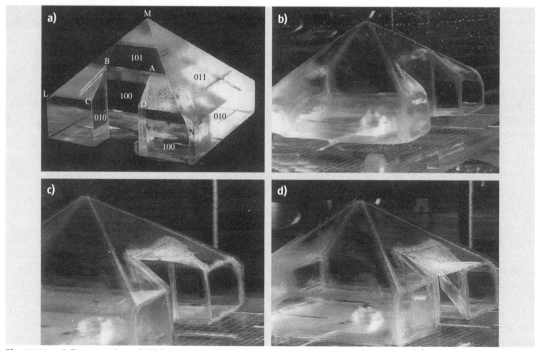

Fig. 22.17a–d Regeneration of KDP crystal surfaces of incomplete crystallographic shape: (**a**) initial crystal with a removed part; (**b**) first stage of the regeneration process; (**c**) formation of a thin surface layer from concave angles A and B on the edge of the {101} face; (**d**) formation of the {100} surface by thin layers; $t_0 = 65\,°\mathrm{C}$; $\sigma = 0.08$; crystal cross section about $10 \times 10\,\mathrm{cm}^2$ (after [22.62])

cation mechanism of growth. Dislocation sources of steps form vicinal slopes which lead to a deviation of the crystal faces and edges from their crystallographic orientation. The reconstruction of the complete crystallographic shape of adjacent faces occurs by the incorporation of building units into the crystal edges. This process results in the generation of growth steps to the deviated faces. The deviation from singularity caused by the presence of vicinal slopes of a crystal face is compensated for and corrected by growth steps generated from the edges.

However, the question that still exists is why these crystals grow not by close packing of the volume but by creating thin surface layers and hollow spaces under them. Formation of the thin layers creates at least double the surface area. In this regard, close packing should be more beneficial because of minimization of the surface energy.

22.3 Growth Techniques for Single Crystals

There have been more than 80 years of research on the growth of KDP/DKDP crystals. The need for large single-crystal plates for use as Q-switches and laser radiation converters [22.6] has stimulated the development of growth techniques for KDP crystals, but the key problems to be solved are centered on the rapidly growing large-aperture crystals and improving the optical quality of the crystal.

22.3.1 Parameters Affecting Growth Rate

The growth rate of KDP crystals depends on many parameters, such as growth temperature, supersaturation of growth solution, impurities, and hydrodynamic conditions, as described in detail by *Zaitseva* et al. [22.55] and other authors [22.7, 66–68]. KDP/DKDP crystals typically grow by the screw dislocation mechanism. The growth rate R of a crystal face in the direction normal to it is given by the geometrical relation

$$R = pV. \tag{22.3}$$

Here p is the slope of the dislocation hillock and V is the tangential speed of the elementary steps. As follows from (22.3), the hillock structure and the step velocity affect the growth rate. In the kinetic limit, the step speed V is linearly connected with the supersaturation σ through

$$V = b\beta\sigma, \tag{22.4}$$
$$\beta = \beta_0 \exp(-E_A/k_B T), \tag{22.5}$$

where b and β_0 are constants, β is the kinetic coefficient, E_A is the activation barrier of the slowest stage of growth, T is the temperature, and k_B is the Boltzmann constant. According to (22.4) and (22.5), the step speed V can be increased by increasing either σ or T.

Increasing supersaturation has always been known as an obvious way to accelerate crystal growth. The shift of the growth process to higher temperature can also increase the growth rate by increasing the value of the kinetic coefficient [22.65, 69]. In agreement with the above analysis, the growth rates empirically measured by *Zaitseva* et al. [22.42] show very close relationship with temperature T and supersaturation σ, as shown in Fig. 22.18.

The slope p in the expression (22.3) is determined by the structure of a dislocation hillock. It depends on the amount of dislocations in the growth source m; the distance between them, defined by the length $2L$; the el-

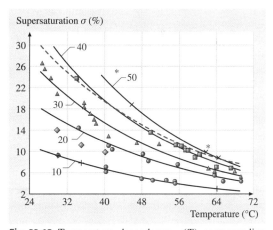

Fig. 22.18 Temperature dependence $\sigma(T)$ corresponding to approximately constant growth rates of KDP crystals Rz: 10 mm/day (○), 20 (●), 30 (▲), 40 (□), 50 (×), and 60 (∗); + and ◆ are data from [22.65] for 10 and 20 mm/day, respectively. The *dashed line* is the boundary of the metastable zone (after [22.42], with permission of Elsevier)

ementary step height h; and the critical radius r_c, given by

$$p = mh/(19r_c + 2L), \quad (22.6)$$
$$r_c = \omega\alpha/k_B T\sigma, \quad (22.7)$$

where ω is the volume of a molecule of the crystallizing matter and α is the specific free energy of a step edge. Expression (22.7) is obtained from the Gibbs–Thomson formula. According to (22.6) and (22.7), p (and hence R) can be increased by increasing σ or changing the number of dislocations in the growth source. Direct control of the growth rate by changing the dislocation structure is difficult because of the complicated relationship between the structure of a dislocation source, σ, and T [22.71]. However, as recently shown [22.68], for $\sigma > 5\%$ the activity of a growth hillock is dominated by the presence of strain-induced dislocation cores and is nearly independent of σ.

There are two more important parameters affecting growth rate: impurities and mass transfer, or hydrodynamic conditions. The metallic cations, especially those with high valency, were considered to greatly affect the growth rate. It is well known that metallic cations (such as Sn^{4+}, Fe^{3+}, Cr^{3+}, and Al^{3+}) affect the growth rate of prismatic faces of KDP crystals much more than the pyramidal ones [22.72]. Fe^{3+} and Al^{3+} ions can decelerate the growth rate of {100} faces [22.66]. With increasing impurity concentration of Fe^{3+} or Cr^{3+} ions, the crystal growth rate of {101} faces first increases and then decreases [22.67]. Sn^{4+} ions can decrease the growth rate of both {100} and {101} faces [22.73].

Anions with strong H-bond affinity such as oligophosphate, are easily adsorbed on the pyramidal faces and inhibit their growth, resulting in extended prismatic faces [22.74–76]. Organic materials such as glycol and ethylene diamine tetraacetic acid (EDTA) show growth-promoting effects on both pyramidal and prismatic faces at very low concentration [22.77], but the growth rate decreases with continuous increase of additive concentration.

Thus, the growth rate can be increased by purification of the raw material, as well as by shifting the growth process into the kinetic regime by increasing the velocity of the solution flow relative to the surface of the growing crystal [22.1, 34]. Unfortunately, wide variations in impurity levels and hydrodynamic conditions are limited because of their undesirable influence on the optical quality of growing crystals. Greater acceleration of the growth rate can obviously be obtained by increasing the supersaturation.

22.3.2 Stability of Solution

Conventional methods could tolerate the presence of some spontaneous crystals during the growth process, but spontaneous nucleation means termination of the growth run for rapid crystal growth. Zaitseva et al. [22.32] performed experiments to show that KDP solution has a wide metastable zone, which provided the basis to develop fast growth techniques for KDP crystals.

A number of methods are reportedly used to improve the stability of the growth solution. These include superheating of the solution to dissolve microcrystals, filtration to remove insoluble particles, reduction of rough surfaces in the crystallizer, and prevention of cavitations by solution stirring. Nakatsuka et al. [22.70] found that the maximum supersaturation reached 120% with overheating above the saturation temperature (by 25 K) at pH 3.2. They also found that lower pH was effective to yield higher supersaturation. Strong acoustic energy can be used to achieve greater solution supersaturation, replacing overheating (Fig. 22.19). The applied acoustic energy may greatly accelerate cluster disconnection. Zaitseva found that a continuous filtration system (CFS) (Fig. 22.20) could improve the stability and minimize the potential for spontaneous nucleation. They designed a continuous filtration system for 1000 l crystallizers to grow large KDP crystals [22.78]. Degassing the solution is also a useful means to prevent nucleation [22.79].

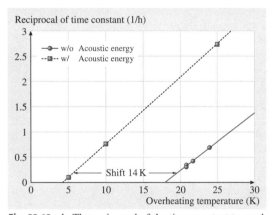

Fig. 22.19a,b The reciprocal of the time constant to reach a steady state of the supersaturation (a) without acoustic energy, and (b) with acoustic energy, as a function of overheating temperature (after [22.70], with permission of Elsevier)

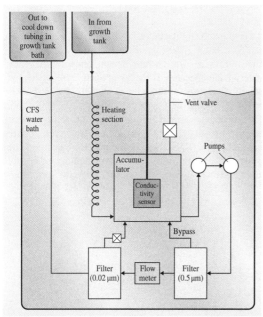

Fig. 22.20 Schematic of the continuous filtration system (CFS) (after [22.55], with permission of Elsevier)

There are some factors associated with crystallization systems that may cause spontaneous nucleation, which lowers the stability of solution during growth [22.42]. Insufficient sealing of the system leads to the possibility of nucleation on the liquid–vapor interface. Nucleation can occur from embryos retained in small cracks and cavities in the surface or joined parts of the equipment immersed in the solution. Nucleation can happen easily at spots of intensive evaporation or dry spots on the parts of the system with an interfacial boundary between solution and vapor that appear as a result of the absence of condensation of the solute on these parts. A classic example of such nucleation is crystallization on the walls of the crystallizer. Secondary nucleation occurs from the main crystal as a result of defect formation or friction between the crystal and supporting parts during the process of stirring.

The impurity composition of the raw material is typically considered an important factor affecting the stability of supersaturated solutions during growth. Some researchers have described the effect, both positive and negative, of different impurities on the stability of solutions [22.40, 73]. However, experiments performed by *Zaitseva* et al. [22.32] show no correlation between spontaneous nucleation and the chemical purity of the solutions. Solutions intentionally doped with Fe, Cr, and Al were measured to have the same stability as regular ones up to very high supersaturation (Fig. 22.9).

22.3.3 Conventional Methods

The temperature-reduction method (TRM) and solution circulating method were two major conventional techniques used to grow large KDP/DKDP crystals before the development of rapid growth from point seed; there are many other methods for KDP/DKDP crystals growth, such as the gel growth and electrodialysis growth methods [22.81].

KDP/DKDP crystals are grown from aqueous solutions by using TRM [22.82, 83] in standard Holden-type crystallizers due to their positive temperature coefficient of solubility [22.32]. The growth equipment is simple, and supersaturation during growth is only controlled by the rate of temperature reduction (Fig. 22.21). Using this method it is difficult to maintain stabilization of supersaturation and the growth rate is generally slow, $1-2$ mm/day. *Sasaki* [22.83] grew large KDP crystal of size, $400 \times 400 \times 600$ mm^3 in 10 months by this method (Fig. 22.22).

Polymorphism of DKDP crystal gives rise to difficulties in the growth of useful tetragonal crystal by TRM. Due to the isotope effect, DKDP crystal may exist in two polymorphs: the tetragonal form (point group $\bar{4}2m$) and monoclinic form (point group 2). The tetragonal–monoclinic phase transition temperature decreases with increasing deuterium concentration in the crystal, which depends on that of the solu-

Fig. 22.21 A 1.5 m^3 growth vessel for large KDP crystals (TRM) (after [22.80])

Fig. 22.22 KDP crystals grown at 2 mm/day on average by TRM (size $40 \times 40 \times 60$ cm^3, growth period 10 months) (after [22.83], with permission of Elsevier)

tion in which the crystal grows. The phase equilibrium transition point of 99.6 mol% deuterated solution is $21 \pm 0.5\,°C$, whereas that of 90 mol% deuterated solution is $49 \pm 1\,°C$. The growth of tetragonal DKDP crystal is very difficult once the tetragonal–monoclinic phase transition takes place or monoclinic phase spurious crystals appear. Thus the degree of deuteration of DKDP imposes strict temperature limits on the growth process; for example, tetragonal DKDP is difficult to grow in a solution of 99.3% deuterium concentra-

tion from a starting temperature higher than $43\,°C$ by TRM [22.38].

In the solution circulating method the supersaturation for crystal growth is maintained by circulating the supersaturated solution at constant temperature. Raw material can be added during growth. The system is usually composed of three tanks, namely, the saturation tank, buffer tank, and growth tank (Fig. 22.23) [22.83]. The supersaturation is provided by the difference in temperature between the saturation tank and growth tank. The three tanks are connected by tubes and the growth solution is circulated by using a pump, so crystal can be grown at constant temperature and supersaturation. This system is more complex and the tube is susceptible to blockage due to crystallization [22.82].

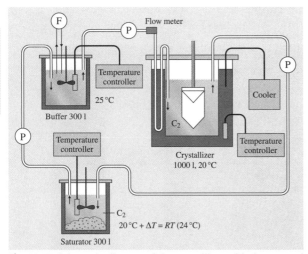

Fig. 22.23 Schematic diagram of the crystallizer with three-vessel solution circulating method (TVM) (after [22.83], with permission of Elsevier) (F – filter, P – pump)

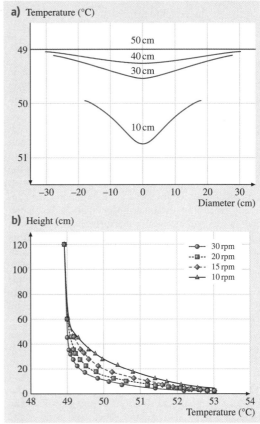

Fig. 22.24 (a) Radial temperature distribution at different height from the bottom of the crystallizer; (b) relation between crystal rotation rate and temperature distribution in axial direction (after [22.84])

In order to resolve the problem of blocking, *Sasaki* made the temperature of the growth tank lower than room temperature and the others higher [22.83]. *Lu* et al. [22.84] further modified the solution circulating system to resolve all of those problems. The stability of the growth solution and the growth temperature were improved by overheating the growth solution and adjusting the flux of solution and the temperature difference between the growth tank and saturation tank. Also, a suitable temperature gradient inside the growth tank was obtained along the radial and axial directions with respect to the crystal rotation rate (Fig. 22.24). The crystal growth rate was thereby increased 3–5 times over that of conventional TRM. Figure 22.25 shows the bulk KDP crystal grown by them.

Besides, *Vladimin* et al. [22.85] designed a method to grow KDP crystal at constant temperature and constant supersaturation. In this method, supersaturation was provided by adding solution. This method, which is very simple, has the advantage of solution circulating at constant temperature. It has only a tank without tubes, and can decrease the volume of the growth tank. Especially for growth of DKDP crystal, this method can effectively decrease the dosage of heavy water to reduce the cost of crystal growth.

22.3.4 Rapid Growth from a Point Seed

With the development of ICF, more and larger KDP/DKDP crystals with higher quality were needed.

Fig. 22.25 Bulk KDP crystal grown by solution circulating method (weight: 110 kg; size: $260 \times 250 \times 770 \, \text{mm}^3$)

However, the problem of low growth rate was particularly daunting. The main limitation on growth of large crystals by conventional techniques is the growth rates of only 1–2 mm/day for KDP and not faster than 1 mm/day in the case of DKDP crystal. In addition, the growth of each crystal required the production of seed material of equal cross-section, adding significantly to the production time. Difficulties in providing reliable equipment, high risk of failure, and defect formation during such long periods resulted in low yield and high cost of final crystals. These reasons stimulated the development of new techniques to accelerate the growth rate without the sacrificing optical quality of large crystals.

Work on rapid growth of KDP/DKDP crystals started in the early 1980s, aimed at conducting the growth process in the kinetic regime [22.9]. It improved the growth rate efficiently by increasing the velocity of the solution flow relative to the crystal surface. A high z-direction rate of 5–25 mm/day was achieved using a novel turbine to enhance mass transport [22.11]. Up to $150 \times 150 \times 80 \, \text{mm}^3$ KDP and DKDP crystals were grown at a rate of 0.5–1 mm/h by the technique of properly feeding the growing surfaces [22.86]. The quality of crystals was reported to remain the same as those grown by conventional technique, but the results proved that this was not the ultimate solution.

Zaitseva et al. [22.8, 32] realized that the primary factor controlling the growth rate was supersaturation, while a number of other researchers focused on increasing mass transport to the crystal surface. Results of nucleation and kinetic experiments made it possible to achieve stable growth rates that were one order of magnitude faster than conventional techniques. Rapid growth on both {100} and {101} faces of a point seed was carried out in Holden-type crystallizers (Fig. 22.26) at high supersaturation by using extensively purified raw material. The process is depicted in references [22.7, 42].

Growth solution was prepared by dissolving salt in water with saturation temperature depending on the required mass of the crystal. The solubility data in the temperature range $T = 10\text{–}80\,°\text{C}$ for KDP, and DKDP with 90% deuteration, are expressed as [22.42]

$$C_0 = 0.1165 + 3.0017 \times 10^{-3}T + 8.5768 \times 10^{-6}T^2, \tag{22.8}$$

$$C_0 = 0.1701 + 3.4817 \times 10^{-3}T + 2.8062 \times 10^{-6}T^2, \tag{22.9}$$

respectively.

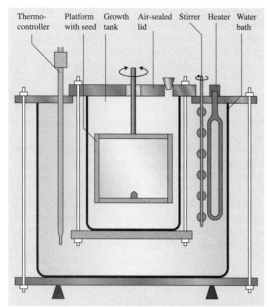

Fig. 22.26 Scheme of the crystallizer used for fast growth (after [22.32], with permission of Elsevier)

Solutions were filtered through submicrometer filters in two ways: prior to growth, or continuously during the whole growth process. After the filtration was stopped, all solutions were overheated for not less than 24 h. Both filtration and overheating were carried out in air-sealed crystallizers with continuous stirring at 80 °C.

Square seeds with T-shaped cross-section were cut from previously grown crystals without visible defects. A water-polished seed was glued into a square hole in the middle of the platform. After the glue had dried, the platform with the seed was assembled in an empty crystallizer and slowly heated to 80 °C. Growth solution that was overheated at the same temperature in a separate vessel was carefully introduced into the crystallizer without splashing. Alternatively, for small-scale growth, the platform with the seed was heated to 80 °C in hot water vapor in a separate crystallizer and then quickly introduced into the overheated solution.

After the seed had been introduction, the crystallizer was tightly sealed and the solution was maintained at 80 °C for 10–15 min. During this time, the seed was slightly dissolved to eliminate any surface crystallites. Additionally, all dry surfaces that could produce spontaneous nucleation were wetted by condensation of the hot solvent vapor. This overheating time was limited by seed dissolution. Typically, the solution was cooled to the saturation temperature before the upper part of the seed had dissolved to the edge of the hole.

Continuous cooling was stopped when the temperature corresponded to the supersaturation for regeneration. Under these conditions, the regeneration process was completed in 1–2 h. During regeneration, the solution was stirred by reversible rotation of the platform at 40–60 rpm with a period of about 30 s in one direction and a 2–3 s pause. This platform rotation was also used during the crystal growth procedure that followed.

After the regeneration process was completed, the temperature reduction continued according to a program that provided stable growth at the desired growth rate. The process could be done completely automatically by using concentration sensors [22.87, 88]. Crystals could be grown at nearly constant growth rate using this procedure. The growth rate could be monotonically decreased toward the end of growth to avoid increasing supersaturation at lower temperature. When the growth was completed, the solution was removed from the crystallizer. If the process finished below or above room temperature T_r, the system was slowly heated or cooled to T_r before the crystal was taken out.

To meet the demands of the National Ignition Facility (NIF) for a large number of 0.5 m high-quality KDP/DKDP crystals, many scientific developments and engineering design changes were incorporated into the rapid growth crystallizer system (Figs. 22.26 and 22.27). The crystal is rotated alternately in one direction and then in the other using a symmetrically programmed schedule with controlled acceleration, de-

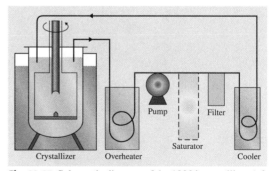

Fig. 22.27 Schematic diagram of the 1000 l crystallizer (after [22.7], with permission of ASM Int.)

Fig. 22.28 Photograph of the large 1000 l crystallizer (after [22.7], with permission of ASM Int.)

Fig. 22.29 First KDP crystal of the NIF SHG size ($53 \times 54 \times 55$ cm^3) grown on the convex platform with continuous filtration (after [22.42], with permission of Elsevier)

celeration, and rotation rates. A special device, called a seed protector, is inserted within the platform shaft in order to cover the point seed during the filtration or overheating procedures [22.89]. The continuous filtration system shown in Fig. 22.20 contains three temperature-controlled sections: a superheater and filter (operating at 80 °C), and the third section where the filtered solution is cooled to the growth temperature [22.78]. The details of this classical loop are designed such that the solution can be continuously filtered during the entire growth process at supersaturations up to 20% without generating spontaneous nucleation. Other important changes include a rigid and streamlined platform compatible with hot KDP solution that could rotate at sufficiently high rates without flexing, oscillating or creating a large wave at the solution–air interface [22.7], and a semi-automated system for temperature reduction [22.87]. Successful incorporation of these design elements led to a reliable method. By using this method, a KDP crystal of 317.97 kg was grown in 52 days in LLNL (ICF Monthly Highlights, January 2000 UCRL-TB-128550-00-04) (Fig. 22.29).

The combination of the regeneration conditions with a convex-shaped platform and continuous filtration makes it possible to control the KDP crystal habit for production. The ratio of dimensions along the crystallographic axes was controlled by two methods: (1) creation of special dislocation structures during the seed regeneration process and (2) change in the orientation of the seed [22.58]. These methods enable the growth of large crystals with specified habit without the deleterious effects of intentionally introduced chemical impurities. KDP and DKDP crystals of various habits were grown at rates of 10–20 mm/day to linear sizes near 90 cm (Fig. 22.30).

In addition, for DKDP crystal, growth by the point-seed rapid growth method not only decreases the growth period but also avoids the emergence of monoclinic DKDP crystal. This is very important for the growth of highly deuterated DKDP crystals [22.32, 53, 90].

Fig. 22.30 DKDP crystal grown for THG on a horizontally oriented seed; $X : Y : Z = 1 : 2.2 : 3$; length $Z = 86$ cm; convex platform (after [22.58], with permission of Elsevier)

22.4 Effect of Growth Conditions on Defects of Crystals

Growth conditions as well as the after-growth treatment [22.91, 92] affect the optical quality of crystals. Discussion of the mechanisms of growth and their influence on the defect formation process showed that the major factors that determine the defect structure of KDP crystals are the impurities content and the dislocation structure of the crystals. Vicinal sectorality, striations, and strain are the main parameters affecting the optical quality of crystals. Solution inclusions and cracks, which make crystals completely inappropriate for optical applications, are also largely influenced by the same factors.

22.4.1 Impurity Effect

Impurities, which originate from the precursors and from dissolution of the crystal growth tank, can be divided into three main kinds: metal cations such as Al^{3+}, Fe^{3+}, Cr^{3+}, Ca^{2+}, Mg^{2+}, Na^+, etc., anions such as SO_4^{2-}, NO_3^-, Cl^-, etc., and organic impurities such as EDTA, organic dyes, etc. The impurities, as well as the crystal growth rate, affect the defect structure of KDP crystal, which in turn affects the optical quality of KDP crystals.

The metallic cations, particularly those with high valency were considered to strongly affect the growth habit and optical properties of crystals. Trivalent metal cations such as Fe^{3+}, Cr^{3+}, and Al^{3+} have a deleterious effect on the growth of {100} faces of KDP [22.1]. Their introduction into the growth solution poisons the {100} faces, producing a region of no growth at low supersaturation commonly referred to as the *dead zone*. They also promote macrostep formation as the supersaturation is increased and growth begins. At impurity concentrations in the range 1–10 ppm, this phenomenon often results in the formation of solution inclusions [22.1, 93]. In addition, these impurities are strongly incorporated into the {100} face with effective segregation coefficient of the order of 10. In contrast, these impurities have little or no effect on the growth of the {101} faces and are strongly rejected with segregation coefficient of 1.0 [22.92]. This nearly hundredfold difference in impurity incorporation results in strong optical inhomogeneities which are manifested in the refractive indices, the linear and two-photon absorption coefficients, and the anomalous birefringence near the {100}/{101} boundaries of the crystal [22.94, 95]. It was found that a correlation exists between *foggy* inclusions (Fig. 22.31) at the corners of the prismatic sectors and aluminum concentrations in the growth solution when a certain concentration is exceeded [22.96].

Table 22.1 presents the relationship between the concentration of the most common impurities in initial solutions and their distribution in pyramidal and prismatic sectors of KDP crystals. From these results one can see that most impurities, especially trivalent metal cations, go preferably into the prismatic growth sectors. However, some impurities, such as Rb, have approximately the same low concentration in both prismatic and pyramidal parts of the crystals. Some impurities (B, Na, Mg, and Ca) seem to be completely rejected by the crystals.

Anions were traditionally considered to have slight effect on the quality and growth of KDP crystals. However, recent research gives us new insights into this hypothesis [22.74–76, 97, 98]. Some anionic impuri-

Fig. 22.31 Photograph of foggy inclusions along prism–prism boundary of NIF-size KDP boule grown in solution having high Al content during middle of growth run (after [22.7], with permission of ASM Int.)

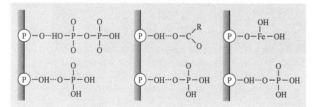

Fig. 22.32 Scheme of adsorption of anionic and cationic species (after [22.75], with permission of Elsevier)

Table 22.1 Distribution of impurity concentration C_i in raw material and pyramidal and prismatic sectors of KDP crystals

Impurity	MDL (ng/g KDP)	C_i raw material (ng/g KDP)	C_i pyramid (ng/g KDP)	C_i prism (ng/g KDP)
B	1000	14000	ND	ND
Na	300	86000	ND	ND
Mg	5	100	ND	D
Al	100	900	200	4400
Si	100	12000	D	390
Ca	100	3600	ND	ND
V	0.5	< 80	0.1	6.0
Cr	100	2000	490	11000
Fe	50	5300	110	12000
As	200	< 200	ND	D
Rb	10	10000	2600	2100
Sr	5	320	870	1800
Mo	10	790	66	1100
Sb	4	70	30	280
Ba	5	52	120	600

MDL – method detection limit, D – detected, ND – not detected

ties (sodium metaphosphate, potassium pyrophosphate, formic acid, acetic acid, oxalic acid, and tyrosine) were found to have an inhibiting effect on the growth of KDP crystals. Complete stoppage of pyramidal face growth was observed in the presence of high concentration of both $[H_{x-n}(PO_3)_x]^{n-}$ and $[H_2P_2O_7]^{2-}$. In these cases, the crystal habit was changed into a closed octahedron and the whole crystal stopped growing. The inhibition is assumed to be caused by the adsorption of these anions at the growing crystal surface, especially on pyramidal surface, through the formation of H-bonds between anions and H_2PO_4 groups (Fig. 22.32). When the doped metaphosphate concentration is high enough, mother-liquor macroinclusions occur along the sector boundaries and in pyramidal sectors (Fig. 22.33). In observations using laser tomography, the density of scatter increases with increasing concentration of doped metaphosphate. The scatters are suggested to be microsolution inclusions caused by the inhibition effect of metaphosphate on the growing faces. These anions have a harmful effect on the laser damage threshold due to possible incorporation (Table 22.2). Sulfate

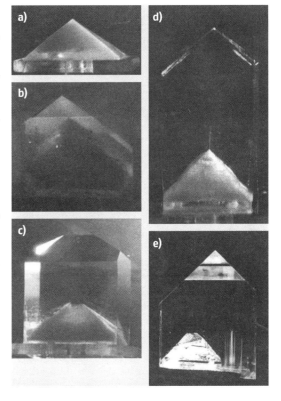

Fig. 22.33a–e Photographs of KDP crystals grown on Z-cut plate seed from metaphosphate-doped solutions with addition of (**a**) 50 ppm, (**b**) 30 ppm, (**c**) 15 ppm, (**d**) 10 ppm, and (**e**) 1 ppm (after [22.76], with permission of Elsevier) ▶

Table 22.2 The effect of anionic species on the optical properties of KDP crystal

Dopants	Mass ratio (ppm)	A (%/cm)	F_{th} (J/cm²)
Potassium pyrophosphate	100	5.1	34.3
EDTA	300	5.5	52.0
Tyrosine	20	4.7	40.4
Tyrosine	50	4.8	40.4
Tyrosine	300	4.9	35.1
Formic acid	300	6.1	42.7
Oxalic acid	300	4.5	45.8
Pure KDP material		5.2	52.0

A – absorption coefficient; F_{th} – laser damage threshold, 1-on-1, 1.06 μm, 10 ns

(SO_4^{2-}) anions, which can join into the KDP crystal lattices through H-bonding and electron attraction, have a great effect on the growth habit of KDP crystals at high dopant concentration. KDP crystals showed many defects such as mother liquor inclusions, parasite crystals, and cracks. Besides, SO_4^{2-} content also adversely affects crystal transparency. At high dopant concentration, KDP crystal transmittance decreases significantly in the ultraviolet region [22.98].

Organic additives such as glycerol, ethylene glycol, polyethylene glycol, and EDTA show growth-promoting effects on both pyramidal and prismatic faces at very low concentration [22.75–77, 99]. In KDP crystals grown in the presence of optimal amounts of organic additives, there is a decrease in the concentration of inorganic background impurities of the cations, an increase in the transparency in the (100) growth sector in a certain range, and an increase in optical uniformity and laser damage threshold.

The distribution of impurities in the KDP crystal is nonuniform, as is evident in crystals grown from solutions doped with impurities that can produce visible color. KH_2PO_4 shows a strong tendency to incorporate anionic dyes only on the {101} faces, presumably because these faces are terminated with K^+ ions [22.93]. However, impurities may inhomogeneously deposit not only between growth sectors, but also within a single growth sector depending on the crystal surface topography. Surfaces of crystals grown in the regime of lower supersaturation often propagate through dislocations that produce growth spirals or hillocks, and shallow-stepped pyramids with single or multiple dislocations at the apex [22.100, 101]. Polygonization of hillocks partitions faces into vicinal regions, each having slightly different inclinations. Impurity partitioning among vicinal slopes and intrasectoral zoning result from selective interactions of impurities with particular stepped hillock slopes.

Zaitseva and coworkers [22.1] perfected KH_2PO_4 crystal growth conditions as a prerequisite to the development of the National Ignition Facility. Amaranth, which displayed exclusive affinity for the {101} surfaces of KH_2PO_4 [22.93, 102], was both inter- and intrasectorally zoned [22.92]. This observation required introduction of the dye during late growth, thereby coloring only a thin surface layer so that patterns of color were not confounded by moving dislocation cores. Figure 22.34 highlights the {101} faces of KH_2PO_4/amaranth crystals. The heterogeneities resulted from amaranth having distinguished among the A, B, and C slopes of the polygonized hillocks prevalent on the pyramid faces. Incorporation followed the trend B > A > C. On the other hand, Chicago sky blue preferred C [22.103]. At low Chicago sky blue concentration, B remained colorless. The incorporation was associated with a critical temperature above which the dye was not captured. In situ interferometry [22.104] was used to show the influence of the dye on the surface morphology at different concentrations and KH_2PO_4 supersaturations.

The effect of impurities on the recovery of surfaces out of the dead zone was traditionally explained by the classic theory of Cabrera and Vermilyea in terms of pinning of elementary step motion by impurities [22.105]. In this model, impurities adsorbed on the terraces create a field of *impurity stoppers* that act to block the motion of elementary steps. When the average impurity spacing is less than a critical distance, whose magnitude is approximately given by the Gibbs–

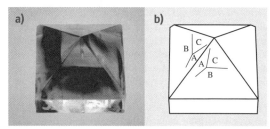

Fig. 22.34a,b Intrasectoral zoning in KH_2PO_4/amaranth. (a) Photograph of KH_2PO_4/amaranth crystal grown by Zaitseva et al. [22.55]. (b) Idealized representation of crystal in (a), illustrating the hillocks observed on the surface

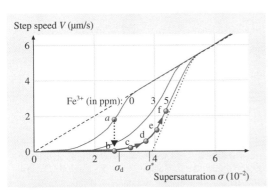

Fig. 22.35 Dependence of step velocity on supersaturation in the presence of Fe^{3+} (after [22.26], with permission of Macmillan)

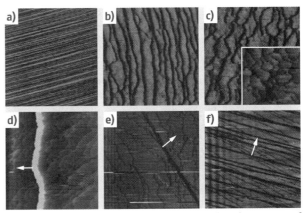

Fig. 22.36a–f Series of AFM images illustrating the process of impurity poisoning and recovery (after [22.7], with permission of ASM Int.): **(a–c)** $15 \times 15\,\mu m^2$ scans collected at $\sigma = 0.04$ along heavy *dashed line* trajectory from point a to point b in Fig. 22.35 showing poisoning of surface by introduction of 12 ppm Fe^{3+} (per mole of KDP); time at the start of each scan is **(a)** 0 s, **(b)** 64 s, and **(c)** 106.7 s, and Fe-doped solution was introduced at $t = 42.7$ s (steps in image **(c)** are immobile and *inset* shows $6 \times 6\,\mu m^2$ image of morphology of elementary steps on macrostep terraces); **(d)** $9 \times 9\,\mu m^2$ image collected at point c in Fig. 22.35 (elementary steps are immobile and macrosteps are distorted but moving slowly); **(e)** $13 \times 13\,\mu m^2$ image collected at point d in Fig. 22.35 (elementary steps are still immobile and macrosteps are straight and moving rapidly); **(f)** $12.5 \times 12.5\,\mu m^2$ image collected at point e in Fig. 22.35 (elementary steps are now moving but more slowly than macrosteps)

Thomson critical diameter, the steps cannot advance. As the supersaturation is increased and the critical diameter becomes smaller, the steps begin to squeeze through the *fence* of impurities and the step speed rises rapidly. The dotted curve in Fig. 22.35 illustrates the prediction of such a model. No growth occurs below a percolation threshold given by σ^*, while for $\sigma > \sigma^*$, the step speed rises rapidly to its unimpeded value. While the model correctly predicts the rapid rise in step speed for $\sigma > \sigma^*$, no growth is predicted below σ^*, in contrast to what has been observed experimentally. Land et al. [22.106] considered that recovery of the KDP surface occurred through the movement of macrosteps (Fig. 22.36), which were mobile even when elementary steps remained pinned. This led to a slow increase in growth rate below σ^*, even for supersaturations in the region $\sigma_d < \sigma < \sigma^*$. At supersaturation near σ^*, elementary steps begin to straighten out and become mobile, but their speed is still considerably less than that of the macrosteps. Finally, at supersaturations in excess of σ^*, the surface once again consists of a combination of rapidly moving elementary steps and step bunches that resemble the step trains for the undoped solution. The elementary steps are now moving with a velocity that is close to that of the macrosteps.

De Yoreo et al. [22.54] point out that, although all the defects observed in KDP are also present in DKDP, there are some aspects of crystal quality which are peculiar to the deuterated crystals. DKDP crystals grown at rates of 1–15 mm/d have always been found to exhibit vicinal sectorality in their x-ray topograph. This inhomogeneity has a moderate effect on the optical uniformity.

22.4.2 Supersaturation

According to the dependence of growth-step velocity of KDP prism faces on the supersaturation, three regions can be defined, corresponding to the dead zone, the transition region of high $dV/d\sigma$, and the linear part of $V(\sigma)$ with high supersaturation and growth rate.

When crystals grow in the dead zone, large variation occurs in steepness, resulting from different incorporation of impurities into separate vicinal slopes. This variation produces large misorientation between small crystals formed during regeneration on melted faces and the z-cut of the seed. This misorientation makes the process of joining difficult, which leads to the formation of cracks and subindividual crystals [22.107–110]. As a result, the crystallographic shape fails to regenerate. Even when regeneration occurred successfully, subindividual crystals and cracks often appeared if growth was performed within the dead zone.

The transition region of high $dV/d\sigma$ is a region of surface instability that develops when elementary steps pass through the fence of impurity blockers with the accompanying macrostep bunching [22.1]. Solution inclusions form most likely in this region because of higher variations of supersaturation on the growing surface. Another reason for increasing instability is the formation of new hillocks, which are frequently observed during the transition. Besides, the change of leading hillock on the growing face in the region of low supersaturation and in the vicinity of the dead zone is inevitably accompanied by the formation of growth bands. This is due to the appearance of new growth steps of various orientations which lead to the formation of inhomogeneity.

When crystal grows at high supersaturation, corresponding to the linear part of $V(\sigma)$, no visible inclusions are formed on the vicinal slopes, despite the fact that microscopic investigation of the surface clearly shows the structure of bent macrosteps on both prismatic and pyramidal faces [22.111]. The formation of large, bent macrosteps is not a sufficient condition for inclusion formation. There is also no evidence that such bent macrosteps cause microsolution inclusions or dislocation since etching did not reveal any dislocation pits.

It can be concluded that, in order to minimize defect and inhomogeneity formation, crystals should be grown in the region of the dead zone (traditional growth) or in the region of the linear part of the dependencies $V(\sigma)$ (rapid growth).

22.4.3 Filtration

KDP crystals often contain hairlike inclusions that run roughly perpendicular to the growth direction (Fig. 22.37) [22.112]. They have the appearance of fine fibres a few centimeters in length, extending through the

Fig. 22.37 Hairlike inclusions in rapidly grown KDP crystals (after [22.7], with permission of ASM Int.)

crystal in the direction of growth and deflecting away from the z-axis; they occur in both dipyramidal and prismatic growth sectors. Most of these inclusions, which lie at angles to the z-axis that vary between 20 and 90°, are not continuous but rather consist of long chains containing from dozens to hundreds of liquid inclusions.

Many models of liquid inclusion formation have been proposed [22.113, 114]. A most likely source of hairlike inclusions is particle incorporation. During crystal growth, steps often close around particles, but do not cover them. Researchers at LLNL have observed this process in KDP during the passage of hundreds of layers. Figure 22.38 shows holes caused by this process in a number of crystals, including KDP. Apparently, once these holes are formed, they are difficult to fill in and result in the generation of hollow channels running at a high angle to the growth front. Given that the KDP crystals were grown by a temperature drop method, any channels which formed during growth would have contained solutions that became increasingly supersaturated as the temperature was decreased and thus might be expected to condense into chains of inclusions.

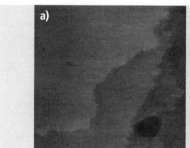 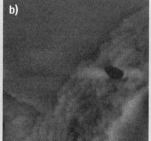 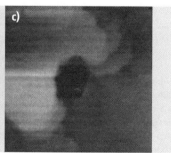

Fig. 22.38a–c Appearance of holes in crystals following inclusion of particle: (a) Canavalin, $6 \times 6\,\mu m^2$; (b) Satellite Tobacco Mosaic virus, $12 \times 12\,\mu m^2$; (c) KDP, $3 \times 3\,\mu m^2$ (after [22.7], with permission of ASM Int.)

Filtration of solution before crystallization to remove extraneous solid particles and thereby improve the crystal quality and laser damage threshold is a standard procedure for experimental or commercial crystal growth. However, no matter how well the solution is filtered before crystal growth, a large number of particles appear very soon after the filtration is stopped. This phenomenon is especially pronounced during the growth of large bulk crystals in equipment containing moving and rotating parts that generate particulate contamination and spoil previously purified growth solutions.

The requirement of high damage threshold in NIF crystals stimulated the design of a continuous filtration system. The system removes growth solution from the tank, heats it to prevent spontaneous nucleation, and filters the solution through 0.02 μm-pore filters. The solution is then cooled before its injection back into the growth tank. This system filters the solution not only before the growth (prefiltration) but also during the entire growth process in a continuous mode.

Comparison of crystals grown with and without continuous filtration showed substantial difference in structure and number of defects. When crystals grow from only prefiltered solutions, the pyramidal face of the crystal generally has a large number of dislocation pits uniformly distributed over the whole crystal (Fig. 22.39). This high density of dislocations, which shows that the distribution of growth sources on the face is not stable, is more or less typical for all KDP and DKDP crystals grown without continuous filtration.

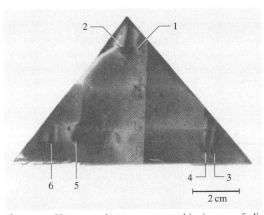

Fig. 22.39 X-ray synchrotron topographic image of dipyramid face of a KDP crystal grown without continuous filtration. *Black dots* are dislocation outcrops on the face. (1–6) Vicinal hillocks tips (after [22.78], with permission of Elsevier)

Initial dislocation hillocks formed after seed regeneration are located approximately in the center of each face, but the location of vicinal hillocks far from the central part of the face does not correspond to the distribution bunches created on the seed during regeneration. These dislocations were formed during the process of growth. When crystals grow with continuous filtration, positions of the leading hillocks do not change, and the crystal surface preserves its simple structure during the entire growth, with one growth hillock approximately in the middle of each face, corresponding to the locations formed during regeneration.

These results indicate that, under continuous filtration, the formation of new dislocations is significantly reduced. Large amounts of dislocations lead to the possibility of formation of vicinal hillocks that can compete in step generation under changing growth parameters, such as temperature, supersaturation or hydrodynamic conditions. The change of leading dislocation hillocks on the growing crystal face always results in the formation of structural defects. The shift of dislocation hillocks to the asymmetrical position close to the edges leads to a higher probability of formation of liquid inclusions and new dislocations on them. This is why crystals grown with continuous filtration have much higher optical homogeneity compared with crystals grown under usual conditions.

22.4.4 Hydrodynamic Effects

Obtaining the optimal hydrodynamic conditions in a crystallizer is a principal factor ensuring high quality of crystals.

The surface structure of each face of KDP crystal consists of growth hillocks, intervicinal valleys, and additional valleys produced by the *bending* of macrosteps originating from a single growth hillock (Fig. 22.40). By ex situ measurement on the surface topography of rapidly grown KDP crystals using a precision coordinate measuring machine (CMM) [22.106], these valleys are observed to be precursors of morphological instability and the formation of solution inclusions on the crystal surface. Results from numerical simulations of the hydrodynamics [22.115] and mass transfer [22.116] for the conditions used in the rapid growth process show that the surface supersaturation field generated on these crystals is inhomogeneous due to the spatially and temporally varying boundary layer thickness on the rotating crystal surface. These simulations clearly indicate that the process of step bending can produce the features which are observed in experiment. The simulations have

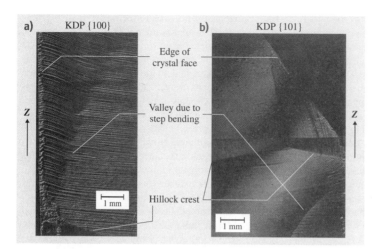

Fig. 22.40a,b Photographs of the typical surface structure resulting from step bending on the (**a**) {100} and (**b**) {101} face (after [22.106], with permission of Elsevier)

been used to explore the dependence of step bending on parameters such as crystal growth rate, rotation conditions, growth hillock location, and (for prismatic faces) impurity level [22.111]. It has been shown that the step bending and resulting valley formation is intensified by increasing crystal growth rate, by decreasing rotational speed, and (for the {100} face) by increasing the impurity level. It has also been shown that the resulting surface structure depends on the location and number of hillocks on the surface.

The crystallization process can be regarded as two stages: supply of crystallizing matter from solution to the crystallization surface through the diffusion boundary layer by molecular and convective diffusion, and the formation of a crystal. The former is rate-limiting if the supersaturation at the crystallization surface is lower compared with the bulk supersaturation, in which case the growth is said to proceed in the diffusion regime. The second stage is rate-limiting if the bulk supersaturation and crystal surface supersaturation are similar, in which case the growth regime is said to be kinetic. Crystals are generally considered to be in a mixed diffusion–kinetic regime. However, surface and bulk supersaturation should be made as close as possible to the kinetic regime so that the diffusion processes of supply of crystallizing matter do not limit the crystallization rate.

The traditional recommendation to approach the kinetic regime is to decrease the thickness of the diffusion layer by increasing the flow velocity or changing the flow geometry [22.36]. However, it can also be achieved by decreasing σ. Thus, to stay close to the kinetic regime, the growth rate should be lowered by increasing the rotation rate of platform. The dependence of the growth rate R_z on the rotation rate of the platform can be determined in a rapid growth crystallizer [22.37]. At very slow rotation of 0–25 rpm, inclusions always form on some crystal faces. Between 25 and 40 rpm, crystals grow with and without inclusions. With faster rotation, formation of solution inclusions is extremely rare and one can be considered to be close to the kinetic regime.

The hydrodynamic conditions most likely affect the solution inclusion process indirectly through changes of the dislocation structure. Low flow velocity at slow rotation results in a greater gradient of the thickness of the diffusion layer along the faces and of the supersaturation at the crystallization surface. The growth hillocks move to the positions of favorable hydrodynamic conditions, i.e., close to the crystal edges and apexes. Morphological instability develops on the large vicinal slope, which covers the whole face. Formation of macrosteps and increasing slope of the vicinals leads to large deviation of the face from its singular orientation and initiates step generation from edges. This phenomenon produces thin surface layers that cover the deviated vicinals with the formation of liquid inclusions. Shift of the hydrodynamic conditions towards the kinetic regime decreases the gradient of the supersaturation along the faces, which makes location of the growth sources close to the edges less probable. In practice, the approach to the kinetic regime is achieved in large crystallizers by increasing the rotation rate of the platforms, as well as by continuous filtration. The combination of these conditions almost completely eliminates solution inclusions without decreasing the growth rate.

22.5 Investigations on Crystal Quality

The optical parameters used to evaluate KDP-type crystals for the ICF project mainly include transparency, optical homogeneity, and resistance to damage by laser radiation.

22.5.1 Spectroscopic Studies

Transparent Spectrum

Transmission in the KDP crystal spectra is from 250 to 1700 nm. For DKDP crystals, the deuterated crystals have higher transmission efficiency (> 89%) at 1064 nm and lower optical absorption efficiency ($< 3\%\,\text{cm}^{-1}$) than those of the undeuterated crystals (Fig. 22.41). The infrared (IR) absorption edges of the deuterated crystals are obviously red-shifted by 0.4–0.5 µm to about 2.0 µm in comparison with those of the undeuterated crystals (about 1.5 µm) so that DKDP crystals can be applied over a wider wavelength region. The infrared cutoff ratio of the hydrogen mode against the deuterium mode is 1/1.33, which nearly coincides with the ratio of the square root of the reduced mass of O−H against O−D, which is known to be 1/1.3743 [22.117].

Raman Spectra

The Raman spectra of KDP crystals exhibit a red-shift, similar to that observed in the transmission spectra [22.117]. The Raman peak 916 cm^{-1} of the $H_2PO_4^-$ group in KDP crystal is red-shifted by 35 cm^{-1} to the 881 cm^{-1} Raman peak of the $D_2PO_4^-$ group in DKDP

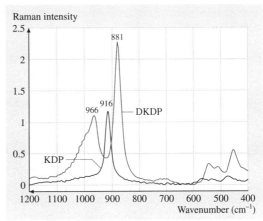

Fig. 22.42 Raman spectra of KDP/DKDP crystals (after [22.117], with permission of Elsevier)

crystal (Fig. 22.42). This red-shift is due to the isotope substitution effect. In the structure of the tetragonal phase, the PO$_4$ tetrahedra are coupled by potassium atoms, and four of these oxygen atoms are transformed to produce strong lines at 966 cm^{-1} in the DKDP crystal. These lines are not detected for the KDP crystal. This result confirms that the spectrum of Raman scattering in the deuterated crystal can be split into two parts related to internal and external vibrations.

Effect of Impurities on the Transmission Spectrum

Certain impurities have a discernible effect on the transmission of KDP crystals. For example, strong absorption in the wavelength region of 200–300 nm is a signature of metal impurities.

Many experiments have been done to investigate the dependence of near-ultraviolet (UV) absorption on the presence of impurities. *Garces* et al. [22.118] indicated that near-ultraviolet absorption is strongly related to the Fe^{3+} impurity in the growth solution. Near-ultraviolet optical absorption spectra taken from two KDP crystals are shown in Fig. 22.43. These data clearly indicate the presence of an intense 270 nm optical absorption band, along with additional absorption in the 200–230 nm region. They found a direct correlation between the Fe^{3+} content and the 270 nm band intensity. *Wang* et al. [22.73] also found that doping of Sn^{4+} lowers transmission near the ultraviolet region, as shown in Fig. 22.44.

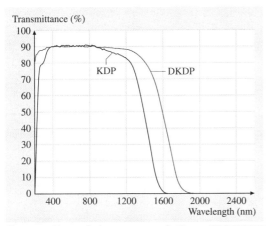

Fig. 22.41 Transmission spectra of Z-cut KDP/DKDP crystals (after [22.117], with permission of Elsevier)

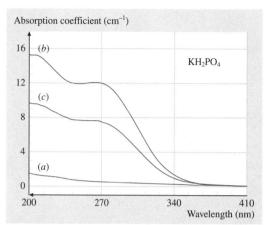

Fig. 22.43 Optical absorption of KDP crystals at room temperature (after [22.118], with permission of Elsevier)

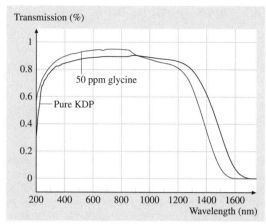

Fig. 22.45 Transmission curves of 50 ppm glycine doped KDP (after [22.119])

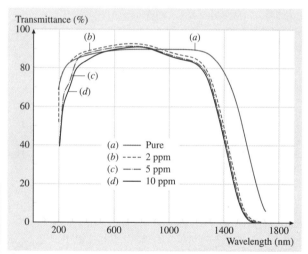

Fig. 22.44 Transmission spectra of KDP crystals grown from solution doped with different concentrations of Sn^{4+} ion by traditional temperature reduction method (after [22.73], with permission of Elsevier)

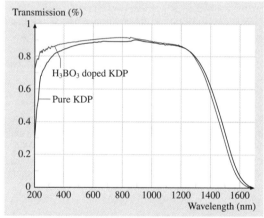

Fig. 22.46 Transmission curves of 10 ppm boric acid doped KDP (after [22.119])

The effects of some anion and amino-acid impurities have been studied and reported by *Sun* and others [22.98, 119], showing that pyrophosphoric acid shortens the transparency range while the addition of a certain concentration of EDTA can extend the range of transparency. As shown in Figs. 22.45 and 22.46, other impurities such as boric acid, glycine, and sulfate can affect the transmittance of KDP crystal.

Generally, metal impurities such as Fe^{3+} and Cr^{3+} are mainly responsible for the extra absorption in the near-ultraviolet band through an electron transition mechanism, while infrared and near-IR band shifts can be attributed to anion group and amino-acid impurities, as they can affect vibration in the IR band.

22.5.2 Homogeneity

Efficient operation of electrooptic devices such as Pockels cells and frequency converters requires crystals with a high degree of optical perfection. For a crystal to perform adequately when used in a laser system it must maintain beam polarization, introduce minimal distortion of the spatial phase profile, and provide uniform phase matching for frequency conversion. All growth defects have the potential to degrade the performance of KDP and DKDP crystals either by altering the index of refraction or by generating inhomogeneities in the unit normal to the c-plane of the crystal. This latter phenomenon has been referred to as z-axis wander [22.94, 95].

Variations in the refractive index are caused by two primary effects: impurities and strain. Impurities, even when uniformly distributed, generate compositional variations which can alter the diagonal components in the projection onto the plane of polarization of the refractive index tensor. These control the velocity and therefore the phase front of a transmitted beam. However, unless the impurities are of sufficient concentration to alter the crystal symmetry, the induced anomalous birefringence would be insignificant.

In contrast, strain due to externally applied stresses, inhomogeneities in impurity concentrations, foreign inclusions or structural defects such as dislocations, twin boundaries or low-angle grain boundaries can both alter the diagonal components and introduce off-diagonal components, the latter being the source of anomalous birefringence. From the theoretical relationships between optic index and applied strain, measured spatial variations in transmitted phase and depolarization of an incident beam can be related to the internal strain field of a crystal.

The primary effect of z-axis wander is to generate spatial nonuniformities in the critical phase-matching angle for frequency conversion. The same sources which are responsible for anomalous birefringence can be expected to lead to z-axis wander.

Regarding the techniques employed, x-ray topography [22.94, 95], scatterometry [22.120], optical absorption [22.121], and secondary-ion mass spectrometry (SIMS) have been used to correlate optical distortions with defects in KDP crystals:

- X-ray topography and scatterometry reveal the presence of structural defects and inclusions.
- X-ray topography and optical absorption probe the large-scale impurity variations from growth bands, crystal sectorality, and vicinal sectorality.
- Secondary-ion mass spectrometry, and to some extent scatterometry, probes the small-scale distribution of impurity ions.

Effect of Dislocations

In conventionally grown crystals, high dislocation density near the seed cap results in a large degree of optical distortion. Because the majority of these dislocations emerge at high angles to the {101} faces, as the crystal grows along the z-direction they pass to the {100} faces. The resulting crystal has low dislocation density only far from the seed, as illustrated in Fig. 22.47a. Consequently, useful material can only be taken from areas that are remote from the seed cap.

In contrast, dislocations in rapidly grown crystals of both KDP and DKDP are a minor source of optical distortion. This is because regeneration of the point seed leads to discrete bunches of dislocations, as shown schematically in Fig. 22.47b.

The research of *Smolsky* et al. [22.122] indicates that a temperature change of 0.1 K is sufficiently high to form large secondary dislocation sources. The distribution of growth steps over the faces of a crystal during growth is not constant. They change not only because of the variations of the external growth conditions but also because of some internal factors. The steps generated

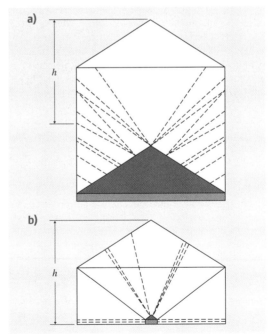

Fig. 22.47a,b Schematic of dislocation structure of (a) conventional and (b) rapidly grown KDP crystals (after [22.7], with permission of ASM Int.)

of strain but rather local variations in impurity levels [22.7].

Crystals grown at higher rates exhibit strong ultraviolet absorption in the {100} sectors, but the contrast between the two sectors is greatly reduced, as is the strain at the boundary. In fact, the level of optical distortion due to the presence of sector boundaries is now at the level of background distortion due to other factors. Even in crystals with rather uniform transmitted wavefront (TWF), significant inhomogeneities in phase-matching angle have been observed [22.123].

Perhaps the most troublesome aspect of crystal sectorality is the strong tendency of iron to incorporate into the {100} sectors, combined with its extremely high absorption coefficient in KDP at the third harmonic. In fact, of all the impurities that have been investigated in NIF development, iron appears to be the only impurity that generates significant optical absorption at 351 nm, which lies in the tail of an absorption band that peaks near 270 nm (Fig. 22.43).

Vicinal Sectorality

Within the {101} sectors of KDP crystals, the boundaries between the three sectors of each vicinal hillock as well as the boundaries between neighboring hillocks subdivide the crystal into regions of contrasting lattice parameters. This vicinal sectorality has a distinct dependence on growth rate. High growth rate cause high depolarization losses [22.7]. Impurities are the source of vicinal sectorality in KDP. The UV absorption and impurity content increase sharply at the boundary.

Deuterated KDP Crystals

De Yoreo et al. [22.7] point out that all the defects observed in KDP are also present in DKDP. Besides, there are some aspects of crystal quality which are peculiar to the deuterated crystals. These peculiarities arise from the fact that, in practice, all DKDP crystals contain significant levels of hydrogen, unless deuterated salt is utilized. Consequently, all deuterated crystals contain hydrogen as a substitutional impurity species at levels from $\approx 1\%$ to more than 10%. Indeed, in NIF development work, DKDP crystals grown at rates from less than 1 to 15 mm/day have always been found to exhibit vicinal sectorality in their x-ray topograph. This inhomogeneity has a moderate effect on optical uniformity. Figure 22.49a shows the TWF profile from a 92% DKDP crystal grown at 9 mm/day that exhibits strong vicinal sectorality.

The fact that such sectorality is due to variations in hydrogen content is demonstrated in Fig. 22.49b which

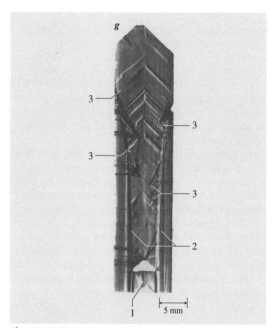

Fig. 22.48 Projection x-ray topograph of the X-cut of a KDP crystal. MoK_{α_1} radiation, reflection vector g[020]: (1) seed; (2) sectorial boundaries between the prismatic and bipyramidal faces; (3) striation due to a lower activity of the leading growth hillocks at the face edge and the formation of new dislocation sources of steps (after [22.7], with permission of ASM Int.)

by new dislocations have new spreading directions on the face and density in accordance with the dislocation source position and activity. Therefore some stresses are formed between the old and new layers which, in turn, give rise to striation (Fig. 22.48).

De Yoreo's results [22.94] show a distinct dependence on position in the boule. Crystals from near the seed (grown first) show considerably more strain than those from near the end of the boule (grown last). This dependence may result from the distribution of dislocation as shown in Fig. 22.48.

Differences Across Sector Boundaries

There is evidence to show that the level of impurities changes sharply at {101}/{100} boundaries, but remains fairly uniform within each sector, when the crystal is grown at 5 mm/day. In contrast, depolarization data shows that the strain is concentrated near the boundary, where the impurity level changes sharply, supporting the expectation that impurities per se are not the source

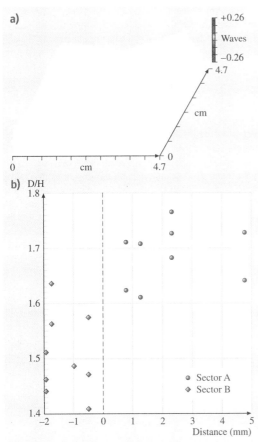

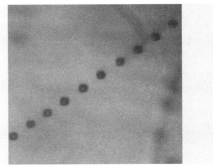

Fig. 22.50 Inclusions along Py–Pr sector (after [22.124])

Fig. 22.49a,b Strong vicinal sectorality of 92% DKDP crystal grown at 9 mm/day (a) TWF profile; (b) profile of D/H ratio across vicinal sector boundary as determined from secondary-ion mass spectrometry (SIMS) (after [22.7], with permission of ASM Int.)

shows a SIMS profile of the D/H ratio collected across a vicinal sector boundary. Results demonstrate that the hydrogen content changes by 15% across the boundary. Other results have shown that the magnitude of this change depends inversely on growth temperature. The difference is optically undetectable for growth temperatures greater than about 60 °C.

A kind of inclusions also is found in the area of pyramidal–pyramidal sector boundary, the formation of which is related to the absorption of impurities. During the growth of pyramidal sector, the impurities are rejected from the sector and aggregate in the pyramidal–pyramidal sector boundary, causing lattice mismatch.

This mismatch breaks the D-bond and creates a bunch of inclusions (Fig. 22.50) [22.124]. Similar results can be seen in the reports of *Joshi* and others [22.76, 105, 125].

In total, there are four possible sources of strain in crystals of $K(D_xH_{1-x})_2PO_4$. Each of them can act independently to produce strain in KDP–DKDP crystals [22.94]. First among these is dislocations, which produce strain fields that vary as $1/r$, where r is the distance from the dislocation. The second source is impurities, which gives rise to strain due to lattice mismatch. These first two factors are commonly found in the case of KDP crystals. The third one is the mixing of hydrogen and deuterium on the hydrogen sublattice, which generates strain due to the difference in structural parameters of KH_2PO_4 and KD_2PO_4. Both the cell parameters and orientation of the PO_4 tetrahedron are different for the two end members of this solid-solution series [22.126, 127], consequently the mixed crystals will be strained. The fourth source of strain is that the tetragonal–monoclinic phase transition in this solid-solution series occurs at increasingly lower temperature as the deuterium level increases.

22.5.3 Laser Damage Threshold

The mechanisms of laser-induced damage are still not well understood. Many papers are devoted to it and several models have been proposed, such as lattice defects, interstitial inclusions and vacancies, multiphoton ionization, and two-photon absorption, etc. [22.128, 129]. Impurities in the raw material and growth conditions have been proved to have a direct relation with the bulk laser-induced damage threshold (LDT) of the crystal, including inorganic impurities, such as Fe^{3+}, anion impurities, and some organic materials [22.7, 119].

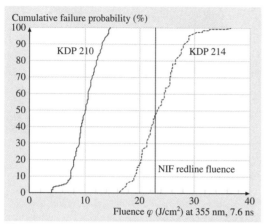

Fig. 22.51 Effect of continuous filtration of the 3ω damage performance of rapidly grown KDP. Samples were grown under nominally identical conditions with ultrahigh-purity salts. Sample 210 was prefiltered only, which resulted in high numbers of inclusions in the crystal, while sample 214 was continuously filtered during growth. Sample 214 represents the best damage performance of a rapid growth crystal to date (after [22.130])

Also, filtration before growth and continuous filtration during the growth process help to increase the LDT; investigations show that continuous filtration results in an increase in damage performance of $\approx 2\times$ over the entire cumulative failure distribution (CFD) range [22.130] (Fig. 22.51).

One should be aware that LDT is related to the laser conditions, such as wavelength, frequency, laser irradiation direction, and polarization, which makes it more complex [22.128, 133].

Effect of Postgrowth Treatment

Previous investigations of the effects of thermal annealing on KDP and KD*P damage performance [22.132–137] have indicated that it is possible to increase the 1ω damage threshold ($R/1$) by approximately $1.5\times$ compared with the level without conditioning (Table 22.3). The 3ω thresholds, however, were not substantially affected by the annealing process.

LLNL testing showed that postgrowth thermal annealing of KDP crystals at $160\,°C$ appears to increase the damage performance of rapidly grown samples by $2.7\times$ at 1ω. The benefits of thermal annealing at 3ω are not as well defined as the dramatic increases seen at 1ω. On the other hand, thermal annealing appeared to suppress the low-fluence tail of the CFD at 3ω, thus leading to less expected damage at 1ω. When large-area beams were used to study the feasibility of online conditioning for NIF triplers, it was found that substantial conditioning was achievable in 8–12 shots of equal fluence interval. Analysis of scatter density versus fluence showed that the number of damage sites evolved exponentially. Furthermore, damage sites were stable against increases in fluence. *Fujioka* et al. [22.91] reported similar result. The full-width at half-maximum (FWHM) of x-ray diffraction of KDP crystals was detected to decrease obviously after conditioning at $165\,°C$, as shown

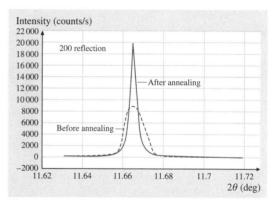

Fig. 22.52 Rocking curve of KDP crystal sample (200) double-crystal x-ray diffraction at same position before and after annealing at $165\,°C$ (after [22.131])

Table 22.3 Compilation of thermal annealing on KDP damage performance [22.132]

Sample	Sector	Growth temperature (°C)	Growth (d)	Unannealed (J/cm²)	Annealed (J/cm²)	Annealed with SHG (J/cm²)
F6-1	Pyramid	48	40	32	33	26
F6-2	Pyramid	63	30	17	23	18
F6-3	Pyramid	72	14	17	23	23
F6-4	Prism	71	20	23	40	30
F6-5	Prism	60	33	21	37	36
F6-6	Prism	25	48	17	40	28

in Fig. 22.52, which suggested that internal stress was released partly and the structural perfection of crystal was improved [22.131]. The result that the perfection of rapidly grown crystal improved more than those grown using conventional methods is consistent with the improvement of optical property by annealing [22.91].

DKDP is not amenable to thermal conditioning, because the crystals tend to fog or fall apart as a result of either decomposition or phase transformation at the temperatures used to anneal KDP. Consequently, laser conditioning is the only option [22.139].

Simulation of the Mechanism of LID

Understanding the susceptibility of KDP crystals to laser-induced damage (LID) at high laser fluence an order of magnitude below the expected intrinsic breakdown limits is a long-standing issue [22.7]. Recently, *Carr* et al. [22.128] employed a novel experimental approach in order to understand the mechanisms of laser-induced damage of KDP crystal. Two notable sharp steps in the damage threshold centered at 2.55 eV (487 nm) and 3.90 eV (318 nm) are clearly demonstrated in their experimental results (Fig. 22.53). Recent simulation study of KDP crystal seems to be targeted at the wavelength dependence of laser-induced damage to support the defect-assisted multistep photon mechanism proposed by *Carr* et al. [22.128].

The effect of neutral and charged H-interstitial and H-vacancy on laser damage was investigated by *Liu* et al. [22.140]. They reported that the bandgap of the neutral H-interstitial and positively charged H-vacancy are greatly reduced to 2.6 and 2.5 eV, respectively. This result is well consistent with the first sharp step at 2.55 eV and suggests that these two types of defects may be responsible for lowering the damage threshold in KDP crystal [22.128]. *Wang* et al. [22.141] explained that −2 charged O-interstitials are responsible for laser-induced local collapse.

In *Liang*'s work [22.138], an ab initio study of $[SO_4]^{2-}$ in KDP is presented [22.142]. $[SO_4]^{2-}$ is such a common impurity ion in KDP raw materials that point defects of $[PO_4]^{3-}$ replaced by $[SO_4]^{2-}$ are easily created during crystal growth [22.143]. The same simulation model as that adopted by *Liu* et al. [22.140] and *Wang* et al. [22.141] was used, which made it

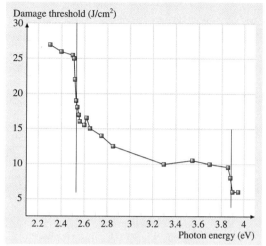

Fig. 22.53 Wavelength-dependent damage threshold (after 22.128)

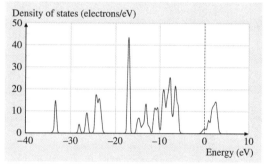

Fig. 22.54 Total density states of KDP ($[SO_4]^{2-}$) (after 22.138)

possible to compare between different works. In the density of states (DOS) of KDP containing $[SO_4]^{2-}$ (Fig. 22.54), it is shown that the bandgap has reduced to 3.90 eV (318 nm), which can induce laser absorption at 318 nm [22.138]. The results of this study are consistent with the experimental work of *Carr* et al. [22.128]. So, it can be speculated that sulfate substitution accounts to some extent for the second sharp step.

References

22.1 L.N. Rashkovich: *KDP Family Single Crystals* (Adam-Hilger, New York 1991)
22.2 Ferroelectrics **71** (1987)
22.3 Ferroelectrics **72** (1987)
22.4 J.F. Nye: *Physical Properties of Crystals* (Oxford Univ. Press, New York 1985), Chap. 13

22.5 D. Eimerl: Electro-optic, linear, and non-linear optical properties of KDP and its isomorphs, Ferroelectrics **72**, 95 (1987)

22.6 J.D. Lindl: Review of development of the indirect-drive approach to inertial confinement fusion and the target physics basis for ignition and gain, Phys. Plasmas **2**, 3933 (1995)

22.7 J.J. De Yoreo, A. Burnham, P.K. Whitman: Developing KDP and DKDP crystals for the world's most powerful laser, Int. Mater. Rev. **13**, 113–152 (2002)

22.8 N.P. Zaitseva, J.J. De Yoreo, M.R. Dehaven, R.L. Vital, K.E. Montgomery, M. Richardson, L.J. Atherton: Rapid growth of large-scale (40–55 cm) KH_2PO_4 crystals, J. Cryst. Growth **180**, 255–262 (1997)

22.9 V.I. Bespalov, V.I. Bredikhin, V.P. Ershov, V.I. Katsman, N.V. Kiseleva, S.P. Kuznetsov: Optical properties of KDP and DKDP crystals grown at high rates, Sov. J. Quantum Electron. **12**, 1527–1528 (1982)

22.10 G.M. Loiacono, J.J. Zola, G. Kostecky: Growth of KH_2PO_4 crystals at constant temperature and supersaturation, J. Cryst. Growth **62**, 545–556 (1983)

22.11 J.F. Cooper, M.F. Singlenton, J. Zandelevich: Rapid growth of potassium dihydrogen phosphate crystals, Proc. Int. Conf. Laser 1984 (1984) pp. 567–572

22.12 L.N. Rashkovich: High-speed growth of large crystals for nonlinear optics from solution, Vestn. Akad. Nauk SSSR **9**, 15–19 (1984)

22.13 S.A. de Vries, P. Goedtkindt, S.L. Bennett, W.J. Huisman, M.J. Zwanenburg, D.-M. Smilgies, J.J. De Yoreo, W.J.P. van Enckevort, P. Bennema, E. Vlieg: Surface atomic structure of KDP crystals in aqueous solution: An explanation of the growth shape, Phys. Rev. Lett. **80**, 2229 (1998)

22.14 P. Hartman: In: *Crystal Growth, An Introduction* (North-Holland, Amsterdam 1973) p. 367

22.15 P. Bennema: In: *Handbook of Crystal Growth*, Vol. 1a, ed. by T.J. Hurle (Elsevier Science Publishers, Amsterdam 1993) p. 477

22.16 P. Hartman: The morphology of zircon and potassium dihydrogen phosphate in relation to the crystal structure, Acta Crystallogr. **9**, 721–727 (1956)

22.17 B. Dam, P. Bennema, W.J.P. Van Enckevort: The mechanism of tapering on KDP-type crystals, J. Cryst. Growth **74**, 118–128 (1986)

22.18 W.J.P. Van Enckevort, R. Janssen-van Rosmalen, W.H. van der Linden: Evidence for spiral growth on the pyramidal faces of KDP and ADP single crystals, J. Cryst. Growth **49**, 502–514 (1980)

22.19 J.J. De Yoreo, T.A. Land, B. Dair: Growth morphology of vicinal hillocks on the {101} face of KDP: From step-flow to layer-by-layer growth, Phys. Rev. Lett. **73**, 838–841 (1994)

22.20 S.A. de Vries, P. Goedtkindt, W.J. Huisman, M.J. Zwanenburg, R. Feidenhans'l, S.L. Bennett, D.-M. Smilgies, A. Stierle, J.J. De Yoreo, W.J.P. van Enckevort: X-ray diffraction studies of potassium dihydrogen phosphate (KDP) crystal surfaces, J. Cryst. Growth **205**, 202–214 (1999)

22.21 I.K. Robinson: Crystal truncation rods and surface roughness, Phys. Rev. B **33**, 3830–3836 (1986)

22.22 B. Dam, W.J.P. van Enckevort: In situ observation of surface phenomena on {100} and {101} potassium dihydrogen phosphate crystals, J. Cryst. Growth **69**, 306–316 (1984)

22.23 B. Dam, E. Polman, W.J.P. van Enckevort: *Industrial Crystallization*, Vol. 84, ed. by S.J. Jancic, E.J. de Jong (Elsevier Science Publishers, Amsterdam 1984) p. 97

22.24 N. Cabrera, D.A. Vermilyea: The growth of crystals from solution. In: *Growth and Perfection of Crystals*, ed. by R.H. Doremus, B.W. Roberts, D. Turnbull (Wiley, New York 1958) pp. 393–410

22.25 T.A. Land, T.L. Martin, S. Potapenko, G.T. Palmore, J.J. De Yoreo: Recovery of surfaces from impurity poisoning during crystal growth, Nature **399**, 442–445 (1999)

22.26 W.J.P. van Enckevort, A.C.J.F. van den Berg: Impurity blocking of crystal growth: A Monte Carlo study, J. Cryst. Growth **183**, 441–455 (1998)

22.27 M.K. Cerreta, K.A. Berglund: The structure of aqueous solutions of some dihydrogen orthophosphates by laser Raman spectroscopy, J. Cryst. Growth **84**, 577–588 (1987)

22.28 L.G. Wu, X.H. Rui, G.J. Teng, Z.S. Qi, C. Yong, Z.Y. Qi: Lattice vibration modes and growth mechanism of KDP single crystals studied by Raman spectroscopy, Chin. J. Light Scatt. **01**, 28–34 (2002)

22.29 X. Yu, X. Yue, H. Gao, H. Chen: Quantitative studies of solute boundary layers around crystals by holographic phase-contrast interferometric microphotography, J. Cryst. Growth **106**, 690–694 (1990)

22.30 X. Yu, J. Yu, Y. Wang, Z. Cheng, B. Yu, S. Zhang, D. Sun, G. Jiang: Microprobe of structure of crystal/liquid interface boundary layers, Sci. China Ser. E **3**, 43–51 (2001)

22.31 M.F. Reedijk, J. Arsic, F.F.A. Hollander, S.A. de Vries, E. Vlieg: Liquid order at the interface of KDP crystals with water: Evidence for icelike layers, Phys. Rev. Lett. **90**, 66103 (2003)

22.32 N.P. Zaitseva, L.N. Rashkovich, S.V. Bogatyreva: Stability of KH_2PO_4 and $K(H,D)_2PO_4$ solutions at fast crystal growth rates, J. Cryst. Growth **148**, 276–282 (1995)

22.33 D. Kaminski, N. Radenovic, M.A. Deij, W.J.P. van Enckevort, E. Vlieg: pH-dependent liquid order at the solid-solution interface of KH_2PO_4 crystals, Phys. Rev. B **72**, 245404 (2005)

22.34 J.W. Mullin: *Crystallization* (Butterworths, London 1993)

22.35 D. Kashchiev, D. Verdoes, G.M. van Rosmalen: Induction time and metastability limit in new phase formation, J. Cryst. Growth **110**, 373–380 (1991)

22.36 A.S. Myerson, A.F. Izmailov, H.-S. Na: Thermodynamic studies of levitated microdroplets of highly supersaturated electrolyte solutions, J. Cryst. Growth **166**, 981–988 (1996)

22.37 M. Bohenek, A.S. Myerson, W.M. Sun: Thermodynamics, cluster formation and crystal growth in highly supersaturated solutions of KDP, ADP and TGS, J. Cryst. Growth **179**, 213–225 (1997)

22.38 M.H. Jiang, C.S. Fang, X.L. Yu, M. Wang, T.H. Zheng, Z.S. Gao: Polymorphism and metastable growth of DKDP, J. Cryst. Growth **53**, 283–291 (1981)

22.39 M. Shanmugham, F.D. Gnanam, P. Ramasamy: Non-steady state nucleation process in KDP solutions in the presence of XO_4 impurities, J. Mater. Sci. **19**, 2837–2844 (1984)

22.40 O. Shimomura, M. Suzuki: The increase of temperature range in the region of supersaturation of KDP solution by addition of impurity, J. Cryst. Growth **98**, 850–852 (1989)

22.41 S. Nagalingam, S. Vasudevan, R. Dhanasekaran, P. Ramasamy: Effect of impurities on the nucleation of ADP from aqueous solution, Cryst. Res. Technol. **16**, 647–650 (1981)

22.42 N.P. Zaitseva, L. Carman: Rapid growth of KDP-type crystals, Prog. Cryst. Growth Charact. Mater. **43**, 1–118 (2001)

22.43 A.B. Ahza: Kinetics of milk fat crystallization in a continous crystallizer, Ph.D. Thesis (University of Wisconsin-Madison 1995)

22.44 A.D. Randolph, M.A. Larson: *Thery of Particulate Processes. Analysis and Techniques of continuous crystallization*, 2nd edn. (Academic, San Diego 1988)

22.45 R. Boistele: Fundamentals of nucleation and crystal growth. In: *Crystallization and Polymorphism of Fats and Fatty Acids*, Surfactant, Vol. 31, ed. by N. Garti, K. Sato (Marcel Dekker, New York 1988) pp. 189–226

22.46 J. Nývlt, O. Söhnel, M. Matichová, M. Bruol: *The Kinetics of Industrial Crystallization* (Academia, Prague 1985)

22.47 E.G. Denk Jr., G.D. Botsaris: Fundamental studies in secondary nucleation from solution, J. Cryst. Growth **13/14**, 493–499 (1972)

22.48 J. Synowiec: A method calculation of the limiting supersaturation of inorganic salt solutions, Krist. Tech. **8**, 701–708 (1973)

22.49 O. Söhnel: Metastable regions of aqueous solutions of inorganic soluble salts, Krist. Tech. **11**, 141–148 (1976)

22.50 A. Mersmann, K. Bartosch: How to predict the metastable zone width, J. Cryst. Growth **183**, 240–250 (1998)

22.51 T.A. Land, J.J. De Yoreo, T.L. Martin, G.T. Palmore: A comparison of growth hillock structure and step dynamics on KDP {100} and {101} surfaces using force microscopy, Crystallogr. Rep. **44**, 704–716 (1999)

22.52 I.L. Smolsky, N.P. Zaitseva: *Growth of Crystals*, Vol. 19, ed. by E.I. Givargizov, S.A. Grinberg (Plenum Publ. Corp., New York 1995) p. 173

22.53 A.A. Chernov, N.P. Zaitseva, L.N. Rashkovich: Secondary nucleation induced by the cracking of a growing crystal: KH_2PO_4 (KDP) and $K(H,D)_2PO_4$ (DKDP), J. Cryst. Growth **102**, 793–800 (1990)

22.54 H. Hilscher: Microscopic investigations of morphological structures on the pyramidal faces of KDP and DKDP single crystals, Cryst. Res. Technol. **20**, 1351–1361 (1985)

22.55 N.P. Zaitseva, L. Carman, I.L. Smolsky, R. Torres, M. Yan: The effect of impurities and supersaturation on the rapid growth of KDP crystals, J. Cryst. Growth **204**, 512–524 (1999)

22.56 C. Belouet, M. Monnier, J.C. Verplanke: Autoradiography as a tool for studying iron segregation and related defects in KH_2PO_4 single crystals, J. Cryst. Growth **29**, 109–120 (1975)

22.57 A.I. Malkin, A.A. Chernov, I.V. Alexeev: Growth of dipyramidal face of dislocation-free ADP crystals; free energy of steps, J. Cryst. Growth **97**, 765–769 (1989)

22.58 N.P. Zaitseva, L. Carman, I.L. Smolsky: Habit control during rapid growth of KDP and DKDP crystals, J. Cryst. Growth **24**, 363–373 (2002)

22.59 H.V. Alexandru, S. Antohe: Prismatic faces of KDP crystal, kinetic and mechanism of growth from solutions, J. Cryst. Growth. **258**, 149–157 (2003)

22.60 A.A. Chernov: Stability of faceted shapes, J. Cryst. Growth **24/25**, 11–31 (1974)

22.61 H.J. Scheel, P. Niedermann: Growth mechanisms of $YBa_2Cu_3O_{7-x}$ platelet crystals from STM/SEM investigations, J. Cryst. Growth **94**, 281–284 (1989)

22.62 N.P. Zaitseva, I.L. Smolsky, L. Carman: Growth phenomena in the surface layer and step generation from the crystal edges, J. Cryst. Growth **222**, 249–262 (2001)

22.63 E.P. Lokshin: Growth and properties of KDP-group crystals, Crystallogr. Rep. **41**, 1061–1069 (1996)

22.64 L.N. Rashkovich, G.T. Moldazhanova: Growth kinetics and morphology of potassium dihydrogen phosphate crystal faces in solutions of varying acidity, J. Cryst. Growth **151**, 145–152 (1995)

22.65 J.W. Mullin, A. Amatavivadhana: Growth kinetics of ammonium and potassium dihydrogen phosphate crystals, J. Appl. Chem. **17**, 151–156 (1967)

22.66 L.N. Rashkovich, N.V. Kronsky: Influence of Fe^{3+} and Al^{3+} ions on the kinetics of steps on the {100} faces of KDP, J. Cryst. Growth **182**, 434–441 (1997)

22.67 T.A. Eremina, V.A. Kuznetsov, N.N. Eremin, T.M. Okhrimenko, N.G. Furmanova, E.P. Efremova, M. Rak: On the mechanism of impurity influence on growth kinetics and surface morphology of KDP crystals-II: Experimental study of influence of bivalent and trivalent impurity ions on growth kinetics and surface morphology of KDP crystals, J. Cryst. Growth **273**, 586–593 (2005)

22.68 J.J. De Yoreo, T.A. Land, L.N. Rashkovich, T.A. Onischenko, J.D. Lee, O.V. Monovskii, N.P. Zaitseva: The effect of dislocation cores on growth hillock vicinality and normal growth rates of KDP {101} surfaces, J. Cryst. Growth **182**, 442–460 (1997)

22.69 P.G. Vekilov, Y.G. Kuznetsov, A.A. Chernov: The effect of temperature on step motion; (101) ADP face, J. Cryst. Growth **121**, 44–52 (1992)

22.70 M. Nakatsuka, K. Fujioka, T. Kanabe, H. Fujita: Rapid growth over 50 mm/d of water-soluble KDP crystal, J. Cryst. Growth **171**, 531–537 (1997)

22.71 P.G. Vekilov, Y.G. Kuznetsov: Growth kinetics irregularities due to changed dislocation source activity; (101) ADP face, J. Cryst. Growth **119**, 248–260 (1992)

22.72 G.M. Loiacono, J.J. Zola, G. Kostecky: The taper effect in KH_2PO_4 type crystals, J. Cryst. Growth **58**, 495–499 (1982)

22.73 B. Wang, C.-S. Fang, S.-L. Wang, X. Sun, Q.-T. Gu, Y.-P. Li, X.-G. Xu, J.-Q. Zhang, B. Liu, X.-M. Mou: The effects of Sn^{4+} ion on the growth habit and optical properties of KDP crystal, J. Cryst. Growth **297**, 352–355 (2006)

22.74 Z.S. Gao, Y.P. Li, C. Wang, Z.K. Lu: Effect of metaphosphate on the growth in KDP crystal, J. Synth. Cryst. **23**(3/4), 52–55 (1994)

22.75 Y.-J. Fu, Z.-S. Gao, J.-M. Liu, Y.-P. Li, H. Zeng, M.-H. Jiang: The effects of anionic impurities on the growth habit and optical properties of KDP, J. Cryst. Growth **198/199**, 682–686 (1999)

22.76 S. Wang, Z.S. Gao, Y. Fu, X. Sun, J. Zhang, H. Zeng, Y. Li: Scattering centers caused by adding metaphosphate into KDP crystals, J. Cryst. Growth **223**, 415–419 (2001)

22.77 V.A. Kuznetsov, T.M. Okhrimenko, M. Rak: Growth promoting effect of organic impurities on growth kinetics of ADP and KDP crystals, J. Cryst. Growth **193**, 164–173 (1998)

22.78 N.P. Zaitseva, J. Atherton, R. Rozsa, L. Carman, I.L. Smolsky, M. Runkel, R. Ryon, L. James: Design and benefits of continuous filtration in rapid growth of large KDP and DKDP crystals, J. Cryst. Growth **197**, 911–920 (1999)

22.79 I. Owsczarek, B. Wojciechowski: Nucleation and growth behaviour of KDP from degassed and undegassed aqueous solutions, J. Cryst. Growth **84**, 329–331 (1987)

22.80 H.J. Scheel, T. Fukuda (Eds.): *Crystal Growth Technology* (Wiley, New York 2003) p. 446

22.81 I. Kolodyazhngl: *Abstracts of ICCG XI, P201A.21* (The Hague, 1995) pp. 18–23

22.82 M.S. Yan, D. Wu, J.B. Zeng, X. Zhang, Y. Guan, L. Wang: Growth of large cross section KDP-type crystals, J. Synth. Cryst. **115**, 1–4 (1986)

22.83 T. Sasaki, A. Yokotani: Growth of large KDP crystals for laser fusion experiments, J. Cryst. Growth **99**, 820–826 (1990)

22.84 Z.K. Lu, Z.S. Gao, Y.P. Li, C. Wang: Growth of large KDP crystals by solution circulating method, J. Synth. Cryst. **25**, 19–22 (1996)

22.85 V.M. Loginer: *Abstracts of ICCG XI, P203B.22* (The Hague, June 1995) pp. 18–25

22.86 V.I. Bespalov, V.I. Bredikhin, V.P. Ershov, V.I. Katsman, L.A. Lavrov: KDP and DKDP crystals for nonlinear optics grown at high rate, J. Cryst. Growth **82**, 776–778 (1987)

22.87 P.F. Bordui, G.M. Loiacono: In-line bulk supersaturation measurement by electrical conductometry in KDP crystal growth from aqueous solution, J. Cryst. Growth **67**, 168–172 (1984)

22.88 S.L. Wang, Y.J. Fu, W.C. Zhang, X. Sun, Z.S. Gao: In-line bulk concentration measurement by method of conductivity in industrial KDP crystal growth from aqueous solution, Cryst. Res. Technol. **35**, 1027–1034 (2000)

22.89 K. Montgomery, N.P. Zaitseva, J.J. De Yoreo, R. Vital: Device for isolation of seed crystals during processing of solutions, LLNL Docket No. IL-9643, DOE case No. S-82,943, US Patent 5904772

22.90 S.L. Wang, Z.S. Gao, Y.J. Fu, A.D. Duan, X. Sun, C.S. Fang, X.Q. Wang: Study on rapid growth of highly-deuterated DKDP crystals, Cryst. Res. Technol. **38**, 941–945 (2003)

22.91 K. Fujioka, S. Matsuo, T. Kanabe, H. Fujita, M. Nakatsuka: Optical properties of rapidly grown KDP crystal improved by thermal conditioning, J. Cryst. Growth **181**, 265–271 (1997)

22.92 M.J. Runkel, W.H. Williams, J.J. De Yoreo: Predicting bulk damage in NIF triple harmonic generators, Proc. SPIE **3578**, 322–335 (1998)

22.93 B. Kahr, S.-H. Jang, J.A. Subramony, M.P. Kelley, L. Bastin: Dyeing salt crystals for optical applications, Adv. Mater. **8**, 941–944 (1996)

22.94 J.J. De Yoreo, B.W. Woods: A study of residual stress and the stress-optic effect in mixed crystals of $K(D_xH_{1-x})_2PO_4$, J. Appl. Phys. **73**, 7780–7789 (1993)

22.95 J.J. De Yoreo, Z.U. Rek, N.P. Zaitseva, B.W. Woods: Sources of optical distortion in rapidly grown crystals of KH_2PO_4, J. Cryst. Growth **166**, 291–297 (1996)

22.96 R.A. Hawley-Fedder, H.F. Robey III, T.A. Biesiada, M.R. DeHaven, R. Floyd, A.K. Burnham: Rapid growth of very large KDP and KD*P crystals in support of the National Ignition Facility, Proc. SPIE **4102**, 152–161 (2001)

22.97 S. Wang, Z.S. Gao, Y. Fu, J. Zhang, X. Sun, Y. Li, H. Zeng, G. Huang, J. Yang, Y. Zhuang, Z. Xue: Effects of metaphosphate doping on growth and properties of KDP crystals, Acta Opt. Sin. **22**, 753–757 (2002)

22.98 J. Zhang, S. Wang, C. Fang, X. Sun, Q. Gu, Y. Li, B. Wang, B. Liu, X. Mu: Growth habit and transparency of sulphate doped KDP crystal, Mater. Lett. **61**, 2703–2706 (2007)

22.99 S. Hirota, H. Miki, K. Fukui, K. Maeda: Coloring and habit modification of dyed KDP crystals as functions

22.99 of supersaturation and dye concentration, J. Cryst. Growth **235**, 541–546 (2002)
22.100 W.J.P. van Enckevort: Surface microtopography of aqueous solution grown crystals, Prog. Cryst. Growth Charact. Mater. **9**, 1–50 (1984)
22.101 I. Sunagawa: In: *Materials Science of the Earth's Interior*, ed. by I. Sunagawa (Terra Scientific, Tokyo 1984) pp. 63–105
22.102 J.A. Subramony, S.-H. Jang, B. Kahr: Dyeing KDP, Ferroelectrics **191**, 293–300 (1997)
22.103 O.A. Gliko, N.P. Zaitseva, L.N. Rashkovich: Morphology and dynamics of crystal surfaces in complex molecular systems. In: *Materials Research Society Symposium Proceedings*, Vol. 620, ed. by J.J. De Yoreo, W. Casey, A. Malkin, E. Vlieg, M. Ward (Materials Research Society, Warrendale 2001)
22.104 L.N. Rashkovich, A.A. Mkrtchan, A.A. Chernov: Kristallografiya **30**, 380–387 (1985)
22.105 I.L. Smolsky, N.P. Zaitseva, E.B. Rudneva, S.V. Bogatyreva: Formation of "hair" inclusions in rapidly grown potassium dihydrogen phosphate crystals, J. Cryst. Growth **166**, 228–233 (1996)
22.106 H.F. Robey, S.Y. Potapenko, K.D. Summerhays: "Bending" of steps on rapidly grown KH_2PO_4 crystals due to an inhomogeneous surface supersaturation field, J. Cryst. Growth **213**, 340–354 (2000)
22.107 T.A. Land, J.J. De Yoreo: The evolution of growth modes and activity of growth sources on canavalin investigated by in situ atomic force microscopy, J. Cryst. Growth **208**, 623–637 (2000)
22.108 H.G. Van Bueren: *Imperfections in Crystals* (Interscience, New York 1960)
22.109 P. Feng, J.-K. Liang, G.-B. Su, Q.-L.. Zhao, Y.-P.. He, Y.-S. Hunag: Growth of KDP crystals by splicing parallel-seed in aqueous solution and its mechanism, Cryst. Res. Technol. **25**, 1385–1391 (1990)
22.110 P. Feng, J.-K. Liang, G.-B. Su, Y.-P. He, B.-R. Huang, Y.-S. Huang, Q.-L. Zhao: The growth of KDP crystals and its mechanism by splicing misoriented in aqueous solution, Cryst. Res. Technol. **26**, 289–295 (1991)
22.111 H.F. Robey, S.Y. Potapenko: Ex situ microscopic observation of the lateral instability of macrosteps on the surfaces of rapidly grown KH_2PO_4 crystals, J. Cryst. Growth **213**, 355–367 (2000)
22.112 I.L. Smolski, J.J. De Yoreo, N.P. Zaitseva, J.D. Lee, T.A. Land, E.B. Rudneva: Oriented liquid inclusions in KDP crystals, J. Cryst. Growth **169**, 741–746 (1996)
22.113 R. Brooks, A.T. Horton, J.L. Torgesen: Occlusion of mother liquor in solution-grown crystals, J. Cryst. Growth **2**, 279–283 (1968)
22.114 M.S. Joshi, B.K. Paul: Effect of supersaturation and fluid shear on the habit and homogeneity of potassium dihydrogen phosphate crystals, J. Cryst. Growth **22**, 321–327 (1974)
22.115 H.F. Robey, D. Maynes: Numerical simulation of the hydrodynamics and mass transfer in the large scale, rapid growth of KDP crystals. Part 1: Computation of the transient, three-dimensional flow field, J. Cryst. Growth **222**, 263–278 (2001)
22.116 H.F. Robey: Numerical simulation of the hydrodynamics and mass transfer in the large scale, rapid growth of KDP crystals. Part 2: Computation of the mass transfer, J. Cryst. Growth **259**, 388–403 (2003)
22.117 G. Li, L. Xue, G. Su, X. Zhuang, Z. Li, Y. He: Study on the growth and characterization of KDP-type crystals, J. Cryst. Growth **274**, 555–562 (2005)
22.118 N.Y. Garces, K.T. Stevens, L.E. Halliburton, M. Yan, N.P. Zaitseva, J.J. De Yoreo: Optical absorption and electron paramagnetic resonance of Fe ions in KDP crystals, J. Cryst. Growth **225**, 435–439 (2001)
22.119 X. Sun, X.-F. Cheng, Z.-P. Wang, Q. Gu, S. Wang, Y. Li, B. Wang, X. Xu, C. Fang: Effect of impurities on optical qualities of KDP crystal, Laser Part. Beams **16**, 830–834 (2004)
22.120 B. Woods, M. Runkel, M. Yan, M. Staggs, N.P. Zaitseva, M. Kozlowski, J.J. De Yoreo: Investigation of laser damage in KDP using light scattering techniques, Proc. SPIE **2966**, 20–31 (1997)
22.121 M. Yan: Chemistry and materials science, Progress Report UCID–20622–95 12–16 LLNL CA 1996
22.122 I.L. Smolsky, A.E. Voloshin, N.P. Zaitseva, E.B. Rudneva, H. Klapper: X-ray topography study of striation formation in layer growth of crystals from solutions, Philos. Trans. R. Soc. Lond. A **357**, 2631–2649 (1999)
22.123 J. Auerbach, P.J. Wegner: Modeling of frequency doubling and tripling with measured crystal spatial refractive-index nonuniformities, Appl. Opt. **40**, 1404–1411 (2001)
22.124 X. Sun, X. Xu, Z. Wang, Y. Fu, S. Wang, H. Zeng, Y. Li, X. Yu, Z. Gao: Inclusion in DKDP crystals, Chin. Sci. Bull. **46**(20), 1757–1760 (2001)
22.125 M.S. Joshi, A.V. Antony: Oriented inclusions in single crystals of potassium dihydrogen phosphate, Krist. Tech. **14**, 527–530 (1979)
22.126 R.J. Nelmes, Z. Tun, W.F. Kuhs: A compilation of accurate structural parameters for KDP and DKDP, Ferroelectrics **71**, 125–141 (1987)
22.127 Z. Tun, R.J. Nelmes, W.F. Kuhs, R.F.D. Stansfield: A high resolution neutron diffraction study of the effects of deuteration on the crystal structure of KH_2PO_4, J. Phys. C **21**, 245–258 (1988)
22.128 C.W. Carr, H.B. Radousky, S.G. Demos: Wavelength dependence of laser-induced damage: Determining the damage initiation mechanisms, Phys. Rev. Lett. **91**, 127402 (2003)
22.129 K. Wang, C. Fang, J. Zhang, X. Sun, S. Wang, Q. Gu, X. Zhao, B. Wang: Laser-induced damage mechanisms and improvement of optical qualities of bulk potassium dihydrogen phosphate crystals, J. Cryst. Growth **287**, 478–482 (2006)

22.130 M. Runkel, R. Jennings, J.J. De Yoreo, W. Sell, D. Milam, N. Zaitseva, L. Carmen, W. Williams: An overview of recent KDP damage experiments and implications for NIF tripler performance, Proc. SPIE **3492**, 374–385 (1999)

22.131 S. Wang, L. Li, X. Hu, Z. Gao, Y. Fu, X. Sun, Y. Li, H. Zeng: The effect of thermal conditioning on microstructure of KDP crystals, J. Funct. Mater. **34**, 331–333 (2003)

22.132 M. Runkel, M.S. Maricle, R.A. Torres, J. Auerbach, R. Floyd, R. Hawley-Fedder, A. Burnham: Effect of thermal annealing and second harmonic generation on bulk damage performance of rapid-growth KDP type-I doublers at 1064 nm, Proc. SPIE **4347**, 389–399 (2001)

22.133 H. Yoshida, T. Jitsuno, H. Fujita, M. Nakatsuka, M. Yoshimura, T. Sasaki, K. Yoshida: Investigation of bulk laser damage in KDP crystal as a function of laser irradiation direction, polarization, and wavelength, Appl. Phys. B **70**, 195–201 (2000)

22.134 F. Rainer, L.J. Atherton, J.J. De Yoreo: Laser damage to production- and research grade KDP crystals, Proc. SPIE **1848**, 46–58 (1992)

22.135 F. Rainer, F. De Marco, M. Staggs, M. Kozlowski, L. Atherton, L. Sheehan: A historical perspective on fifteen years of laser damage thresholds at LLNL, Proc. SPIE **2114**, 9–23 (1994)

22.136 J. Swain, S. Stokowski, D. Milam, F. Rainer: Improving the bulk laser damage resistance of potassium dihydrogen phosphate by pulsed laser irradiation, Appl. Phys. Lett. **40**(4), 350–352 (1982)

22.137 J. Swain, S. Stokowski, D. Milam, G. Kennedy: The effect of baking and pulsed laser irradiation on the bulk laser threshold of potassium dihydrogen phosphate crystals, Appl. Phys. Lett. **41**(1), 12–16 (1982)

22.138 L. Liang, Z. Xian, S. Xun, S. Xueqin: Sulfate may play an important role in the wavelength dependence of laser induced damage, Opt. Exp. **14**, 12196–12198 (2006)

22.139 L.J. Atherton, F. Rainer, J.J. De Yoreo, I. Thomas, N. Zaitseva, F. De Marco: Thermal and laser conditioning of production and rapid-growth KDP and DKDP crystals, Proc. SPIE **2114**, 36–45 (1994)

22.140 C.S. Liu, N. Kioussis, S.G. Demos, H.R. Radousky: Electron- and hole-assisted reactions of H defects in hydrogen-bonded KDP, Phys. Rev. Lett. **91**, 15505 (2003)

22.141 K. Wang, C. Fang, J. Zhang, C. Liu, R. Boughton, S. Wang, X. Zhao: First-principles study of interstitial oxygen in potassium dihydrogen phosphate crystals, Phys. Rev. B **72**, 184105 (2005)

22.142 M.C. Payne, M.P. Teter, D.C. Allen, T. Arias, J. Joannopoulos: Iterative minimization techniques for ab initio total-energy calculations: Molecular dynamics and conjugate gradients, Rev. Mod. Phys. **64**, 1045–1097 (1992)

22.143 J.-Q. Zhang, S.-L. Wang, C.-S. Fang, X. Sun, Q. Gu, Y. Li, K. Wang, B. Wang, Y. Li, B. Liu: Effects of sulphate doping on the growth habit of KDP crystal, J. Funct. Mater. **36**, 1505–1508 (2005)